VOLUME NINETY NINE

VITAMINS AND HORMONES

Bone Morphogenic Protein

VOLUME NINETY NINE

VITAMINS AND HORMONES

Bone Morphogenic Protein

Series Editor

GERALD LITWACK, PhD
Toluca Lake, North Hollywood, California

ELSEVIER

AMSTERDAM • BOSTON • HEIDELBERG • LONDON
NEW YORK • OXFORD • PARIS • SAN DIEGO
SAN FRANCISCO • SINGAPORE • SYDNEY • TOKYO
Academic Press is an imprint of Elsevier

Cover photo credit:
Mulloy, B., Rider, C.C.
The Bone Morphogenetic Proteins and Their Antagonists
Vitamins and Hormones (2015) **99**, pp. 63–90

Academic Press is an imprint of Elsevier
225 Wyman Street, Waltham, MA 02451, USA
525 B Street, Suite 1800, San Diego, CA 92101-4495, USA
125 London Wall, London, EC2Y 5AS, UK
The Boulevard, Langford Lane, Kidlington, Oxford OX5 1GB, UK

First edition 2015

ISBN: 978-0-12-802442-3
ISSN: 0083-6729

For information on all Academic Press publications
visit our website at store.elsevier.com

CONTENTS

CONTRIBUTORS

Paul F. Austin
Department of Surgery, Division of Urology, Washington University School of Medicine, St. Louis Children's Hospital, St. Louis, Missouri, USA

Ana Claudia Oliveira Carreira
NUCEL-NETCEM (Cell and Molecular Therapy Center), Internal Medicine Department, School of Medicine, University of São Paulo, São Paulo, Brazil

Suvro Chatterjee
Vascular Biology Lab, AU-KBC Research Centre, MIT Campus, and Department of Biotechnology, Anna University, Chennai, India

Jian Q. Feng
Department of Biomedical Sciences, Texas A&M Baylor College of Dentistry, Dallas, Texas, USA

Renato Astorino Filho
NUCEL-NETCEM (Cell and Molecular Therapy Center), Internal Medicine Department, School of Medicine, University of São Paulo, São Paulo, Brazil

José Mauro Granjeiro
Bioengineering Division, National Institute of Metrology, Quality, and Technology, Duque de Caxias, and Department of Dental Materials, Dental School, Fluminense Federal University, Niteroi, Brazil

Judith B. Grinspan
Children's Hospital of Philadelphia, and Perelman School of Medicine, University of Pennsylvania, Philadelphia, Pennsylvania, USA

Qiusha Guo
Department of Surgery, Division of Urology, Washington University School of Medicine, St. Louis Children's Hospital, St. Louis, Missouri, USA

Robert J. Hinton
Department of Biomedical Sciences, Texas A&M Baylor College of Dentistry, Dallas, Texas, USA

Eijiro Jimi
Division of Molecular Signaling and Biochemistry, Department of Health Promotion, Center for Oral Biological Research, Kyushu Dental University, Kitakyushu, Fukuoka, Japan

Junjun Jing
State Key Laboratory of Oral Diseases, West China Hospital of Stomatology, Sichuan University, Chengdu, China

Piotr Kraj
Department of Biological Sciences, Old Dominion University, Norfolk, Virginia, USA

Michal Kuczma
Cancer Center, Georgia Regents University, Augusta, Georgia, USA

Isabel La Rosa
Laboratory of Animal Biotechnology, Agriculture Faculty, University of Buenos Aires (UBA), Buenos Aires, Argentina

Scott R. Manson
Department of Surgery, Division of Urology, Washington University School of Medicine, St. Louis Children's Hospital, St. Louis, Missouri, USA

Katelynn H. Moore
Department of Surgery, Division of Urology, Washington University School of Medicine, St. Louis Children's Hospital, St. Louis, Missouri, USA

Thomas D. Mueller
Department Plant Physiology and Biophysics, Julius-von-Sachs Institute of the University Wuerzburg, Wuerzburg, Germany

Barbara Mulloy
Centre for Biomedical Sciences, School of Biological Sciences, Royal Holloway, University of London, Egham, Surrey, United Kingdom

Saranya Rajendran
Vascular Biology Lab, AU-KBC Research Centre, Anna University, MIT Campus, Chennai, India

Chris C. Rider
Centre for Biomedical Sciences, School of Biological Sciences, Royal Holloway, University of London, Egham, Surrey, United Kingdom

Mariana Correa Rossi
Department of Chemistry and Biochemistry, Biosciences Institute, UNESP, Universidade Estadual Paulista, Botucatu, Brazil

Jamila H. Siamwala
Department of Orthopaedic Surgery, University of California, San Diego, California, USA

Mari Cleide Sogayar
NUCEL-NETCEM (Cell and Molecular Therapy Center), Internal Medicine Department, School of Medicine, University of São Paulo, and Chemistry Institute, Biochemistry Department, São Paulo, Brazil

Willian Fernando Zambuzzi
Department of Chemistry and Biochemistry, Biosciences Institute, UNESP, Universidade Estadual Paulista, Botucatu, Brazil

PREFACE

Bone morphogenic (or morphogenetic) proteins (BMPs) represent a subfamily in the transforming growth factor beta superfamily. About 20 BMPs are already known. First discovered in connection with their activities on bone, they play key roles in bone formation and skeletal development and in the differentiation of cartilage and chondrocytes. BMPs are considered as potential treatments for bone healing and for the loss of bone during space flight, especially in flights of extended duration. Additionally, it is now recognized that BMPs are involved in the development of several tissues including limb buds, kidney, heart, eye, and skin. In kidney disease, for example, BMP levels fall, opening the possibility that BMP treatment might offer a beneficial therapeutic effect.

The basic information on the interaction of BMPs triggering activation of their receptors as well as the antagonists of this interaction is known. Features of BMP signaling also are now being studied. The signaling processes involve nuclear factor kappa B in some cases and also are known to affect the process of myelination.

The chapters in this volume are arranged by first considering the basic information in the mechanism of BMP action. Accordingly, the first chapters deal with BMP–receptor interaction. T.D. Mueller describes the "Mechanisms of BMP–Receptor Interaction and Activation." In addition, B. Mulloy and C.C. Rider report on "The Bone Morphogenetic Proteins and Their Antagonists." Both of these chapters demonstrate the use of three-dimensional crystal structures.

S.R. Manson, P.F. Austin, Q. Guo, and K.H. Moore report on "BMP-7 Signaling and Its Critical Roles in Kidney Development, the Responses to Renal Injury, and Chronic Kidney Disease." E. Jimi reports on "The Role of BMP Signaling and NF-κB Signaling on Osteoblastic Differentiation, Cancer Development, and Vascular Diseases—Is the Activation of NF-κB a Friend or Foe of BMP Function?" Additionally, in this vein, M. Kuczma and P. Kraj review "Bone Morphogenic Protein Signaling Regulates Development and Activation of $CD4^+$ T Cells." J. Grinspan writes on "Bone Morphogenetic Proteins: Inhibitors of Myelination in Development and Disease." Continuing on the topic of development, I. La Rosa describes "Bone Morphogenetic Proteins in Preimplantation Embryos."

The following chapters refer to the effects on bone and cartilage. J.H. Siamwala, S. Rajendran, and S. Chatterjee introduce "Strategies of Manipulating BMP Signaling in Microgravity to Prevent Bone Loss." J. Jing, R.J. Hinton, and J.Q. Feng review "Bmpr1a Signaling in Cartilage Development and Endochondral Bone Formation." The final chapter covers "Bone Morphogenetic Proteins: Promising Molecules for Bone Healing, Bioengineering, and Regenerative Medicine" authored by A.C.O. Carreira, W.F. Zambuzzi, M.C. Rossi, R.A. Filho, M.C. Sogayar, and J.M. Granjeiro.

The illustration on the cover is Figure 1 of Chapter 2 by B. Mulloy and C.C. Rider entitled "The Bone Morphogenetic Proteins and Their Antagonists." It presents the crystal structure of the BMP antagonist noggin complexed with the BMP-7 dimer (*magenta*) at the top. Noggin dimer is a ribbon in *red* below while at the very bottom is the heparin-binding site in *yellow*.

Final processing of this volume was facilitated by Helene Kabes (Oxford, UK) and Vignesh Tamilselvvan (Chennai, India).

Gerald Litwack
North Hollywood, CA
June 17, 2015

CHAPTER ONE

Mechanisms of BMP–Receptor Interaction and Activation

Thomas D. Mueller[1]
Department Plant Physiology and Biophysics, Julius-von-Sachs Institute of the University Wuerzburg, Wuerzburg, Germany
[1]Corresponding author: e-mail address: mueller@biozentrum.uni-wuerzburg.de

Contents

Abstract

Bone morphogenetic proteins (BMPs), together with the eponymous transforming growth factor (TGF) β and the Activins form the TGFβ superfamily of ligands. This protein family comprises more than 30 structurally highly related proteins, which determine formation, maintenance, and regeneration of tissues and organs. Their importance for the development of multicellular organisms is evident from their existence in all vertebrates as well as nonvertebrate animals. From their highly specific functions *in vivo* either a strict relation between a particular ligand and its cognate cellular receptor and/or a stringent regulation to define a distinct temperospatial expression pattern for the various ligands and receptor is expected. However, only a limited number of receptors are found to serve a large number of ligands thus implicating highly promiscuous ligand–receptor interactions instead. Since in tissues a multitude of ligands are often found, which signal via a highly overlapping set of receptors, this raises the question how such promiscuous interactions between different ligands and their receptors can generate concerted and highly specific cellular signals required during embryonic development and tissue homeostasis.

Vitamins and Hormones, Volume 99
ISSN 0083-6729
http://dx.doi.org/10.1016/bs.vh.2015.06.003

1. EVOLUTIONARY EXPANSION AND DIVERSIFICATION OF THE TRANSFORMING GROWTH FACTOR β SUPERFAMILY

Multicellular organisms require continuous intercellular communication not only during their development but also for homeostasis and survival. Processes such as cell differentiation, proliferation, migration or apoptosis depend on endocrine, paracrine or possibly autocrine stimuli, which at their heart are often, but not exclusively exerted by protein–protein interactions at the cell surface involving a secreted (sometimes also membrane-associated) growth factor, and a transmembrane receptor. During evolution, nature has "recycled" successful examples of above combinations thereby forming larger protein families, in which further homologous growth factors plus their respective receptors were formed possibly by gene duplication and acquired additional functionalities necessary to cope with the increasing complexity of the evolving organisms. The transforming growth factor β (TGFβ) superfamily comprising TGFβs, Activins, and bone morphogenetic proteins (BMPs) as well as growth and differentiation factors (GDFs) presents a prime example of such a protein family with a few growth factors in simple organisms like worms (five TGFβ ligands, for review: Savage-Dunn, 2005) and a large number of ligands in mammals (>30 TGFβ factors in human, for review: Feng & Derynck, 2005; Hinck, 2012; Mueller & Nickel, 2012; Fig. 1A). An evolutionary expansion in the TGFβ superfamily can be also noted from the observation that homologs of BMPs—in contrast to *senso strictu* TGFβs and Activins—are already found in worms, whereas homologs of Activins appear for the first time in flies and *senso strictu* TGFβs emerge in fish and amphibian (Newfeld, Wisotzkey, & Kumar, 1999). This suggests that BMPs are likely the founding members of this growth factor family, which then diverged into Activins and TGFβ. Thus, TGFβs seem to be the evolutionary youngest members despite serving as eponym of the whole superfamily. The later emergence of Activins and TGFβs is also consistent with their encoded functionalities. Activins modulate the reproductive axis (Bilezikjian, Blount, Donaldson, & Vale, 2006) and exert regulatory roles in inflammation and immunity (for reviews: Aleman-Muench & Soldevila, 2012; Hedger, Winnall, Phillips, & de Kretser, 2011), and TGFβs being implicated in the control of immunity (for review: Yoshimura & Muto, 2011) and wound healing (for review: Leask & Abraham, 2004), functions that are not or differently implemented in

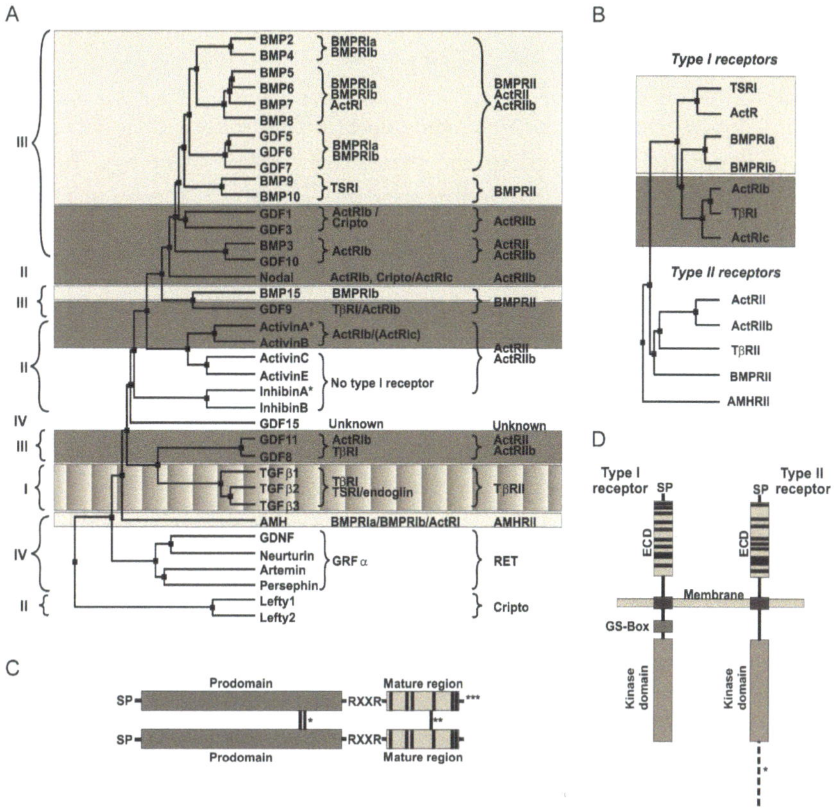

Figure 1 (A) Phylogenetic analysis of the TGFβ ligand superfamily. The TGFβs can be classified into four subgroups indicated on the left: (I) *sensu stricto* TGFβs, (II) Activin/Inhibins, (III) BMPs/GDFs, and (IV) others. Type I and type II receptor recruitment is indicated, the activation of either the SMAD1/5/8 or SMAD2/3 pathway is marked by light or dark gray-shaded boxes, respectively. (B) Phylogenetic analysis of the TGFβ receptors showing the classification into type I and type II receptors. Light and dark gray boxes indicate the activation of either SMAD1/5/8 or SMAD2/3. (C) TGFβ proteins are expressed as pre-proproteins containing a signal peptide (SP), a prodomain, which in TGFβs is covalently dimerized by disulfide bonds (marked by asterisks), a proteolytic processing site (RXXR) and a mature region containing the characteristic cystine-knot motif comprising six conserved cysteine residues (marked by bars). Some TGFβs lack a seventh cysteine residue (marked by two asterisks) involved in covalent dimer formation. (D) Architecture of the TGFβ receptors comprising a signal peptide (SP), an extracellular ligand-binding domain (ECD), a single-span transmembrane element, and an intracellular kinase domain. Type I receptors differ by an additional membrane-proximal glycine/serine-rich motif (GS-box). Furthermore, BMPRII has a unique C-terminal domain (marked by an asterisks), which recruits additional signaling proteins.

simpler organisms such as worms or insects. But not only TGFβs and Activin additionally appeared later in evolution, but also the number of BMP homologs expanded dramatically.

In *Caenorhabditis elegans*, four of the five TGFβ members, dbl1, daf7, tig2, and tig3, could be mapped to the mammalian BMP orthologs, BMP5, GDF8/11, BMP8, and BMP2 (for review: Gumienny & Savage-Dunn, 2013); however, the functional similarities seem limited. For instance, dbl1 and daf7, which are involved in the regulation of body size in the so-called Dauer larval development pathway, possibly exert a similar growth-limiting function as found for GDF8/11 in vertebrates. Despite their limited homology with BMP8 and BMP2, no functions have yet been described for the *C. elegans* orthologs tig-2 and tig-3, but both members might be involved in patterning. Unc129, whose mature region exhibits limited sequence homology to mammalian BMP8 and GDF6, seems to be involved in axon guidance and signals via a non-TGFβ related non-canonical signaling pathway (Gumienny & Savage-Dunn, 2013). In flies, seven TGFβ members have been identified of which the ligands dpp, gbb, and screw can be mapped to the mammalian BMP2/4 and BMP5/6/7 (Newfeld et al., 1999), myoglianin likely presents an ortholog of GDF8/11 (Lo & Frasch, 1999), and dActivinβ, Dawdle and Maverick are fly Activin-like ligands (Kutty et al., 1998; Nguyen, Parker, & Arora, 2000; Parker, Ellis, Nguyen, & Arora, 2006; Serpe & O'Connor, 2006). Possibly due to the evolutionary smaller distance, the fly BMP orthologs dpp, gbb, and screw exert *in vivo* function more closely related to their vertebrate/mammalian counterparts. Dpp, the fly ortholog of BMP2 and BMP4, is essential for correct dorsoventral patterning in fly (Irish & Gelbart, 1987), a function it shares with BMP2/swirl in fish (Kishimoto, Lee, Zon, Hammerschmidt, & Schulte-Merker, 1997) and BMP4 in mouse (Winnier, Blessing, Labosky, & Hogan, 1995). Drosophila gbb is involved in the development of the fly's intestinal tract or the eyes similarly as found for BMP6/7 in vertebrates (Helder et al., 1995; Luo et al., 1995; Perr, Ye, & Gitelman, 1999; Wharton et al., 1999). On the contrary, the functions encoded by dActivinβ and the further distant Activin-like members Dawdle and Maverick seem to be more limited to neuronal morphogenesis compared to their vertebrate homologs (Kutty et al., 1998; Nguyen et al., 2000; Ting et al., 2007; Zhu et al., 2008).

With the emergence of vertebrates, the number of TGFβ members not only doubled as evident from the 14 and 19 TGFβ ligands in fish (two Activin orthologs; Thisse, Wright, & Thisse, 2000) of *Danio rerio* are not listed in Massague (2000) and amphibian (*Xenopus laevis*), but their encoded functions are now more closely resembling those from mammalian orthologs. For instance, BMP4 exerts a mesoderm-inducing activity in early gastrulation in fish and amphibian identical with its patterning function in mammals (Fainsod, Steinbeisser, & De Robertis, 1994; Koster et al., 1991; Neave, Holder, & Patient, 1997; Nikaido, Tada, Saji, & Ueno, 1997; Schmidt, Suzuki, Ueno, & Kimelman, 1995; Winnier et al., 1995). Besides its role in pattern formation, BMP2 seems to be similarly involved in the organogenesis of the heart as well as limb formation in all vertebrates (Beck, Christen, Barker, & Slack, 2006; Crotwell, Sommervold, & Mabee, 2004; Kishimoto et al., 1997; Sparrow, Kotecha, Towers, & Mohun, 1998; Zhang & Bradley, 1996). On the contrary, BMP7, which acts in the morphogenesis of the eye, limbs, and kidney (Dudley, Lyons, & Robertson, 1995; Godin, Takaesu, Robertson, & Dudley, 1998), is an example that functions of an ortholog might have still diversified during evolution. The regulatory elements driving the tissue-specific expression of BMP7 differ between mammals, amphibian, and fish such that in frog and fish BMP7 is only expressed in the developing eye and limbs thereby possibly indicating that the additional functionality in developing the kidney was acquired during adaptation from aquatic to terrestrial life (Adams, Karolak, Robertson, & Oxburgh, 2007). In mammals, additional TGFβ members emerged, which have no direct orthologs in amphibian, fish, or fly. For instance, anti-Mullerian hormone (AMH), GDF9, or BMP15 play important roles in sex determination, spermatogenesis and ovarian follicle development (Dong et al., 1996; Dube et al., 1998; Josso et al., 1993; Laitinen et al., 1998; Nicholls, Harrison, Gilchrist, Farnworth, & Stanton, 2009), ligands such as glial-derived neurotrophic factor (GDNF) and the related Artemin, Neurturin, and Persephin act as neurotrophic factors in the homeostasis of various populations of neurons (Baloh et al., 1998; Kotzbauer et al., 1996; Lin, Doherty, Lile, Bektesh, & Collins, 1993; Milbrandt et al., 1998; for review: Sariola & Saarma, 2003). These additional TGFβ members were likely formed by gene duplication during evolution to implement these new functionalities. In addition, the now increased number of ligands also allowed specifying the function encoded by a particular member more precisely thereby enabling a more stringent regulation.

2. PHYLOGENETIC ANALYSIS REVEALS FOUR FUNCTIONAL SUBFAMILIES FOR TGFβ LIGANDS

Amino acid sequence analyses suggest that the more than 30 mammalian TGFβ ligands can be grouped into four subfamilies (Fig. 1A). Within these subgroups, the associated growth factors not only exert similar functionalities but also share similarities with respect to receptor activation and regulatory mechanisms. With only the three members TGFβ1/2/3, the *sensu stricto* TGFβs form the smallest subgroup within the TGFβ superfamily (Fig. 1A). Despite serving as eponym for the growth factor superfamily, the *sensu stricto* TGFβs present the evolutionary youngest ligands possibly explaining why receptor binding and activation as well as modulation of signaling for the members of this subgroup differs from all other TGFβ subgroups (for review: Hinck, 2012). TGFβ1, -2, and -3 are pleiotropic growth factors that control proliferation, differentiation of many different cell types, but do not seem to directly act as morphogens although they are expressed early during development (Pelton, Johnson, Perkett, Gold, & Moses, 1991). Targeted mutation in mice nevertheless showed that deletion of either TGFβ2 or -3 results in various developmental defects leading to perinatal lethality. Whereas genetic ablation of TGFβ2 yields defects in the heart, lung, and eye, and also includes malformation of the limbs and craniofacial defects (Sanford et al., 1997), disruption of TGFβ3 results in nonoverlapping defects in the palatogenesis and delayed pulmonary development (Kaartinen et al., 1995). In contrast, TGFβ1 seems mainly involved in regulation of immunity, as TGFβ1 null mice suffer from a diffuse inflammation due to infiltration of lymphocytes and macrophages into many organs (Shull et al., 1992). The *sensu stricto* TGFβs do not only exert functions during embryonic development but are also essential for homeostasis in the adult organism. TGFβ1 has been implicated in immunity by inducing the expression of FoxP3 in lymphocytes leading to the formation of so-called regulatory T-cells (iTregs) (Schramm et al., 2004), which are important to suppress immune responses to self-antigens. Through the induction of extracellular matrix synthesis, TGFβs contribute to fibrosis making strategies to neutralize TGFβ signaling an interesting therapy approach (Roberts et al., 1986; for review: Akhurst & Hata, 2012). However, their dual role in the development and progression of cancer is probably the functionality most investigated. Under physiological conditions, TGFβs inhibit cell proliferation of many cell types through arresting the cell cycle in the G1 phase thereby acting as tumor suppressors (Smeland et al., 1987). On the other hand, TGFβs are strong inducers of the epithelial-to-mesenchymal transition, which is important for TGFβs to promote wound healing. In tumors, however, this enables cancer cells to spread and metastasize (for review: Nawshad, Lagamba, Polad, & Hay, 2005). It is noteworthy that despite the three TGFβ isoforms utilize the same receptor isoform-specific functions have been described (e.g., Merwin, Newman, Beall, Tucker, & Madri, 1991; Roberts et al., 1990; for review: Laverty, Wakefield, Occleston, O'Kane, & Ferguson, 2009; Letterio & Roberts, 1996). Although this could be possibly achieved through a temperospatial separation, it has been reported that often the expression of at least two isoforms overlaps, leaving open how isoform-specific signals are generated.

Activin/Inhibin-like ligands constitute a second subgroup within the TGFβ superfamily (Fig. 1A), which is clearly more diverse with respect to functionality and amino acid sequence homology compared to the *sensu stricto* TGFβs. Members of this subgroup have a higher sequence similarity with ligands of the BMP/GDF subfamily than with the TGFβs sometimes rendering a clear affiliation of a particular ligand to either the Activin/Inhibin or the BMP/GDF subgroup difficult. Besides the Activins (homodimeric ActivinβA, ActivinβB, ActivinβC, ActivinβE, and heterodimers thereof) and Inhibins (InhibinA and InhibinB), the subgroup additionally comprises the TGFβ factors Nodal and Lefty1 and Lefty2 (Fig. 1A). Activins are multifunctional factors being involved in patterning during early embryonic development (e.g., mesoderm induction) and it is considered as the first signal establishing patterning during amphibian development (for review: Ariizumi & Asashima, 1995). Bioactive Activins are either a homodimer of two identical or a heterodimer of two different so-called Inhibinβ (InhibinβA or InhibinβB) subunits and were isolated for the first time from gonadal fluids shown to regulate the expression of the follicle-stimulating hormone (FSH) in the gonads (Ling et al., 1986a, 1986b). Later, additional regulatory functions in processes such as folliculogenesis (for review: Peng & Mukai, 2000), spermatogenesis (for review: Barakat, Itman, Mendis, & Loveland, 2012), and also sex determination (e.g., Wu et al., 2013) were described. Their pleiotropic nature is further visible from their involvement in the organogenesis of various organs (e.g., kidney, pancreas, the heart, or the eye) (for review: Asashima, Ariizumi, & Malacinski, 2000) or their functions in inflammation and immunity (for review: Phillips, de Kretser, & Hedger, 2009). And even though Activins are no direct osteogenic factors as different BMPs, they—by various mechanisms—regulate the differentiation and proliferation of bone cells, skeletal development, and bone turnover with the two isoforms ActivinβA and ActivinβB seemingly exerting differing functions (for review: Lotinun, Pearsall, Horne, & Baron, 2012; Nicks, Perrien, Akel, Suva, & Gaddy, 2009). Due to their role in bone homeostasis, Activin traps such as soluble Activin receptor-Fc fusion proteins have become interesting alternatives for future therapies of bone-loss diseases such as osteoporosis (Fajardo et al., 2010; Lotinun et al., 2010; Pearsall et al., 2008). Two further Inhibinβ subunits, InhibinβC and InhibinβE, have been discovered in liver (Fang, Wang, Smiley, & Bonadio, 1997). They can either form homo- or heterodimers together with the InhibinβA and βB subunits (Fig. 1A). As these isoforms are impaired in binding to Activin receptors (specifically the type I receptor ActRIb/Alk4, Fig. 1B) (Muenster, Harrison, Donaldson, Vale, & Fischer, 2005), they have no signaling activity and were thus termed nonactive Activins. By competing with the bioactive Activins for binding to receptors and modulatory proteins such as Follistatin, they can regulate the signaling of those Activins (Gold et al., 2009). Inhibins have been initially identified to inhibit FSH secretion from the pituitary (de Kretser & Robertson, 1989; Vale et al., 1988). The two known isoforms InhibinA and InhibinB are heterodimers consisting of a common α subunit linked via a disulfide bond to one InhibinβA or InhibinβB subunit (Fig. 1A). Similar as for the Inhibinβ subunits C and E, the presence of the α subunit abrogates the binding to the Activin type I receptors thereby rendering the Inhibins inactive with respect to Activin signals (Martens et al., 1997). The unaffected binding of Inhibins to Activin type II receptors enables a simple competition mechanism by which inactive Inhibins displace Activins from their receptors and thus effectively act as Activin antagonists (for review: Cook, Thompson, Jardetzky, & Woodruff, 2004). Although the mechanism, by which ActivinC/E (as well as the heterodimeric forms) and InhibinA/B antagonize bioactive Activins, seems highly similar there is a difference in that the inhibition efficiency of the Inhibins is dependent on the presence of the TGFβ coreceptor betaglycan thereby explaining the tissue-specific action of Inhibins (Lewis et al., 2000). Whether Inhibins act exclusively as Activin antagonist or whether they can initiate signaling cascades themselves is currently unknown. The three remaining members of the Activin/Inhibin subgroup, Nodal, Lefty1, and Lefty2 form a separate functional group involved in the formation of left/right (L/R) asymmetry (for review: Hamada, Meno, Watanabe, & Saijoh, 2002; Mercola, 2003; Fig. 1A). L/R asymmetry is likely induced first by a directional flow of Nodal in the lateral plate mesoderm (Okada et al., 1999). The asymmetric distribution of Nodal then leads to a spatially defined expression of Lefty1 at the midline barrier and Lefty2 in the left side of the lateral plate mesoderm. Lefty1 and 2 are unusual TGFβ ligands in that they not only lack the conserved cysteine residue involved in formation of a disulfide-linked dimer but also have an extended C-terminus distinct from classical TGFβ ligands (Kosaki et al., 1999; Sakuma et al., 2002;

Fig. 1A and C). Despite their asymmetric distribution, both Lefty isoforms seem to function as Nodal antagonists possibly in a negative feedback loop; however, whether Lefty1 and 2 suppress Nodal signaling by competing with Nodal for binding to Activin receptors or by direct interaction and neutralization of Nodal is still a matter of debate (Bisgrove, Essner, & Yost, 1999; Chen & Shen, 2004; Sakuma et al., 2002). Nodal has additional roles during development such as mesoendoderm induction (Feldman et al., 1998; Gritsman et al., 1999), neural patterning (Sampath et al., 1998) and was also shown to maintain pluripotency in human pluripotent stem cells (Vallier, Reynolds, & Pedersen, 2004; for review: Pauklin & Vallier, 2015).

With about 20 members, the BMPs together with the GDFs form the largest subgroup within the TGFβ superfamily (Fig. 1A). On the basis of amino acid sequence similarities and functional homology, the members of these ligands can be further subdivided into the BMP2/4, the BMP5/6/7/8, the GDF5/6/7, the GDF8/11, the GDF9/BMP15, the GDF2/BMP10, the GDF1/3, and the GDF10/BMP3 subgroups (Fig. 1A). Despite the name BMP implies a bone growth inducing or supporting property, studies showed that only a subgroup of BMPs (as well as GDFs) exhibit significant osteogenic activity, which varies in potency as well as to which cell type is susceptible (Cheng, Jiang, et al., 2003; Li et al., 2003). Of the 20 BMP/GDF members, only BMP2 (InductOs, also known as Diotermin alfa or Infuse) and BMP7 (OP-1 Putty and OP-1 Implant) are currently therapeutically used to treat bone defects such as nonunion fractures or spinal fusions (e.g., Argintar, Edwards, & Delahay, 2011). In fact, the biological functions of the 20 different BMPs and GDFs of this subgroup is highly diverse as evident from gene inactivation studies in animals (for review: Chang, Brown, & Matzuk, 2002; Zhao, 2003) and even led to the suggestion to change their name in to body morphogenetic proteins (Reddi, 2005). For instance, BMP2 despite being used as osteogenic factor is essential for the development of the heart (Zhang & Bradley, 1996), *in vivo* BMP4 induces mesoderm during early gastrulation (Winnier et al., 1995), but it is also critically involved in limb formation (Bandyopadhyay et al., 2006; Benazet et al., 2009) as well as the organogenesis of different organs where it might cooperate with other BMPs (Danesh, Villasenor, Chong, Soukup, & Cleaver, 2009; Goldman, Donley, & Christian, 2009; Gordon, Patel, Mishina, & Manley, 2010). BMP7 even though commercially used as osteogenic factor also exerts functions in the development of the kidney (reviewed in Archdeacon & Detwiler, 2008), the eye (Jena, Martin-Seisdedos, McCue, & Croce, 1997; Solursh, Langille, Wood, &

Sampath, 1996, for review: Lang, 2004), or the heart (Kim, Robertson, & Solloway, 2001). Genetic ablation of other BMPs and GDFs usually has less severe, nonlethal effects often displayed by defects in bone formation or organogenesis (for review: Chang et al., 2002). *Bmp3* null mice for instance exhibit increased bone density highlighting its role as negative regulator of bone growth (Daluiski et al., 2001). Removal of BMP5 leads to skeletal abnormalities genetic ablation of BMP6 manifests in delayed ossification as well as defects in the heart development as also seen for BMP7 (Kim et al., 2001; King, Marker, Seung, & Kingsley, 1994; Kingsley et al., 1992; Solloway et al., 1998). Inactivation of GDF5 leads to skeletal malformation (brachypodism); however, only specific limbs are affected and the effect seems mediated through GDF5s role in cartilage and joint formation (Storm et al., 1994). Deletion of either *Gdf6* or 7 in mice yields subtler effects in an altered tendon formation suggesting functional compensation (Mikic, Entwistle, Rossmeier, & Bierwert, 2008; Mikic, Rossmeier, & Bierwert, 2009). Other BMP/GDF members such as GDF8 (also known as myostatin) have non-bone-related functions. A loss of GDF8 results in a massive increase in skeletal muscle (McPherron, Lawler, & Lee, 1997). Consistently naturally occurring, inactivating mutations in *Gdf8* found in cattle and human confirm that GDF8 act as a negative regulator of muscle growth (McPherron & Lee, 1997; Schuelke et al., 2004). It is noteworthy that GDF11 despite sharing 93% sequence identity on amino acid level with GDF8 in the mature region (see Fig. 1C)—only 10 residues differ—seems to exert completely different functions *in vivo*, as deletion of *Gdf11* in mice manifests in an altered number of vertebral segments (McPherron, Lawler, & Lee, 1999). However, since recombinant GDF11 can indeed act as a negative regulator of muscle growth (Gamer, Cox, Small, & Rosen, 2001), this clearly shows that the sometimes significantly different functionalities of homologous BMPs and GDFs is likely due to temperospatial expression differences. Also, GDF9 and BMP15 exhibit no osteogenic activities and form a small functional subgroup involved in the regulation of follicullogenesis (Dong et al., 1996; Laitinen et al., 1998). In contrast to GDF8/11, however, GDF9 and BMP15 share much lower sequence similarity although both factors exert very similar functions and seem to cooperate or even synergize *in vivo* (McNatty et al., 2005a; Mottershead et al., 2013; Peng et al., 2013).

The "fourth" subgroup in the TGFβ superfamily is not formed by a selection of functionally related factors, but rather accommodates all TGFβ ligands not matching one of the other three subgroups (Fig. 1A). Thus, this subgroup comprises the most heterogeneous group of TGFβ ligands with respect to the encoded functions as well as to the underlying signaling mechanisms. For instance, due to its endocrine, long-range action the AMH resembles more a hormone rather than a (TGFβ) growth factor (Josso, 1990). In addition, its limited sphere of action—it is responsible of the regression of the Mullerian ducts in the fetal male genital tract (for review: Josso et al., 1993; Josso, Racine, di Clemente, Rey, & Xavier, 1998)—together with fact that it utilizes a unique receptor for signaling, the AMH type II receptor AMHR-II (Baarends et al., 1994), make AMH an outsider in the TGFβ superfamily. Similarly to AMH, the TGFβ ligand GDF15, also known as MIC1 (macrophage inhibitory cytokine 1), PDF (prostate-derived factor), NAG-1 (non-steroidal anti-inflammatory drug-activated gene 1), or placental TGFβ, is also an isolated ligand member within the TGFβ superfamily. GDF15 indeed shares rather low sequence similarity with other TGFβ ligands and was initially discovered for its macrophage inhibiting activity induced by proinflammatory cytokines such as TNFα (Bootcov et al., 1997; Fairlie et al., 1999). Besides in macrophages, GDF15 mRNA transcripts were only detected in the placenta in larger quantities (Fairlie et al., 1999). The *in vivo* functions of GDF15 remain rather obscure, its expression under pathophysiological conditions suggests however a possible involvement in the onset and progression of cardiovascular diseases and cancer (Ago & Sadoshima, 2006; Baek, Kim, Nixon, Wilson, & Eling, 2001; Brown et al., 2002, 2003). Consistent with its use as a prognostic marker in various cancers (Brown et al., 2012; Shnaper et al., 2009), other studies showed that elevated GDF15 serum levels seem to correlate with an increased risk of all-cause mortality (Daniels, Clopton, Laughlin, Maisel, & Barrett-Connor, 2011; Wiklund et al., 2010). In addition to these functions, GDF15 also exhibits neurotrophic activities similar to members of the small GDNF subfamily (see below). GDF15 is expressed in different areas of the brain and maintains survival of mainly dopaminergic neurons under different stress conditions (Strelau, Schober, Sullivan, Schilling, & Unsicker, 2003; Strelau et al., 2000). Despite its involvement in cancer and its potential usefulness in therapies of neurodegenerative diseases, the receptors for GDF15 as well as the underlying signaling cascades still remain unclear. Within the fourth TGFβ subgroup, the only coherent subpopulation comprises the factors GDNF, Artemin, Persephin, and Neurturin, which also present functionally and with respect to sequence homology the most distant TGFβ subgroup (Baloh et al., 1998; Kotzbauer et al., 1996; Lin et al., 1993; Milbrandt et al., 1998; Fig. 1A). Ligand members of the GDNF subfamily exert neurotrophic activities on dopaminergic neurons regulating the neurite outgrowth, cell size as well as dopamine uptake. *In vitro*, these factors promote survival of motoneurons and peripheral neurons and have thus found interest in future applications for the therapy of, e.g., Parkinson's disease or amyotrophic lateral sclerosis (ALS) (for review: Mickiewicz & Kordower, 2011; Tovar, Ramirez-Jarquin, Lazo-Gomez, & Tapia, 2014). The distant relationship between the GDNF subfamily and all other members of the TGFβ superfamily is also evident from the fact that GDNF ligands bind and signal via an own set of receptors comprising glycosyl-phosphatidylinositol (GPI)-anchored membrane proteins and the receptor tyrosine kinase and not through the classical TGFβ serine/threonine kinase receptors (for recent review: Wang, 2013; Fig. 1A and B).

3. EXPRESSION AS PROTEASE-ACTIVATED PROPROTEINS AND A CYSTINE-KNOT MOTIF IN THE C-TERMINAL MATURE REGION AS KEY FEATURES OF TGFβ LIGAND MEMBERS

That all TGFβ members derive from one common ancestor is best recognized from two key features. First, *in vivo* all ligands are produced as large dimeric proproteins comprising a signal peptide for secretion, followed by a prodomain and an about 100–150 aa large C-terminal mature region harboring a so-called cystine-knot consensus motif (Fig. 1C). With the exception of the very distant four GDNF members, whose prodomain size ranges between no residues at all (Persephin) to 76 aa (Neurturin), the prodomains of TGFβ ligands (≥200 residues) are usually significantly larger than the C-terminal mature region responsible for the ligand's activity. It is noteworthy that amino acid sequence similarity is much lower within the prodomain compared to the cystine-knot containing mature region (see Supplementary of Shi et al., 2011), indicating that the prodomain's functional significance possibly differs among the various TGFβ factors. For activation after synthesis, one or several members of subtilisin-like proprotein convertases process the proproteins at a conserved dibasic motif RXXR (Dubois et al., 2001; Dubois, Laprise, Blanchette, Gentry, & Leduc, 1995; Fig. 1C). Members of this family comprising nine proconvertases have also been shown to process other proproteins such as pro-βNGF and others (Bresnahan et al., 1990; Kiefer & Saling, 2002; Wise et al., 1990). There is an ongoing debate to where the processing of TGFβ ligands occurs (for review: Constam, 2014). Although Furin one member of this proconvertase family was first proposed to process its substrate in the trans-Golgi network (Molloy,

Thomas, VanSlyke, Stenberg, & Thomas, 1994), other members exist either as soluble or as plasma membrane-bound forms. Furthermore, additional processes such as proteolytic shedding or alternative splicing result in proconvertases being present in the intra- as well as extracellular lumen or being even localized to the cell surface (for review: Seidah et al., 2008). Thus, processing of TGFβs may occur at any of these different compartments and the site may be even subject to regulatory processes. Some BMPs even require sequential processing at different sites by different proconvertases leading to differently sized TGFβ ligands with altered signaling capacities (Akiyama, Marques, & Wharton, 2012; Cui et al., 2001; Kunnapuu et al., 2014; Nelsen & Christian, 2009).

In spite of our current knowledge, about proconvertase processing and the underlying mechanism the putative role(s) of the prodomain remain(s) unclear. Sufficient data for the analysis of the structure/function relationships of the prodomain are so far only available for TGFβ1 (Fig. 2A). In the *sensu stricto* TGFβs, the prodomain confers latency (Wakefield et al., 1989), i.e., the prodomain remains tightly associated with the C-terminal mature region even after processing by the proconvertases thereby keeping the TGFβs in an inactive state. For activation, the mature region must free from this proprotein complex (Grainger, Wakefield, Bethell, Farndale, & Metcalfe, 1995). *In vivo*, this is achieved in a rather complex process at the cell surface involving interactions between the prodomain (also termed LAP or latency associated peptide) and components of the extracellular matrix such as α_V integrins and LTBPs (latent TGFβ-binding proteins), which then strip off the prodomain by mechanical forces (for review: Annes, Munger, & Rifkin, 2003; Keski-Oja, Koli, & von Melchner, 2004; see Fig. 2A). The structure analysis of the TGFβ1 proprotein complex now for the first time provides insights into how the prodomain "inactivates" TGFβ1 and how the prodomain removal is realized mechanistically (Shi et al., 2011). The prodomain-mature region assembly resembles a "straitjacket," in which the straitjacket-like prodomain wraps around the mature region covering all receptor epitopes thereby efficiently impeding the interaction of TGFβ1 with its receptors (Fig. 2A). However, this interaction and the architecture of assembly are possibly unique to the *sensu stricto* TGFβs as at least many if not most other TGFβ ligands (e.g., BMPs, GDFs, and Activins) are not latent (e.g., Constam, 2014; Sengle, Ono, Sasaki, & Sakai, 2011). As mentioned above, sequence similarity is much lower between the prodomains of TGFβ superfamily members compared to the C-terminal activity-harboring mature region. Noteworthy, one

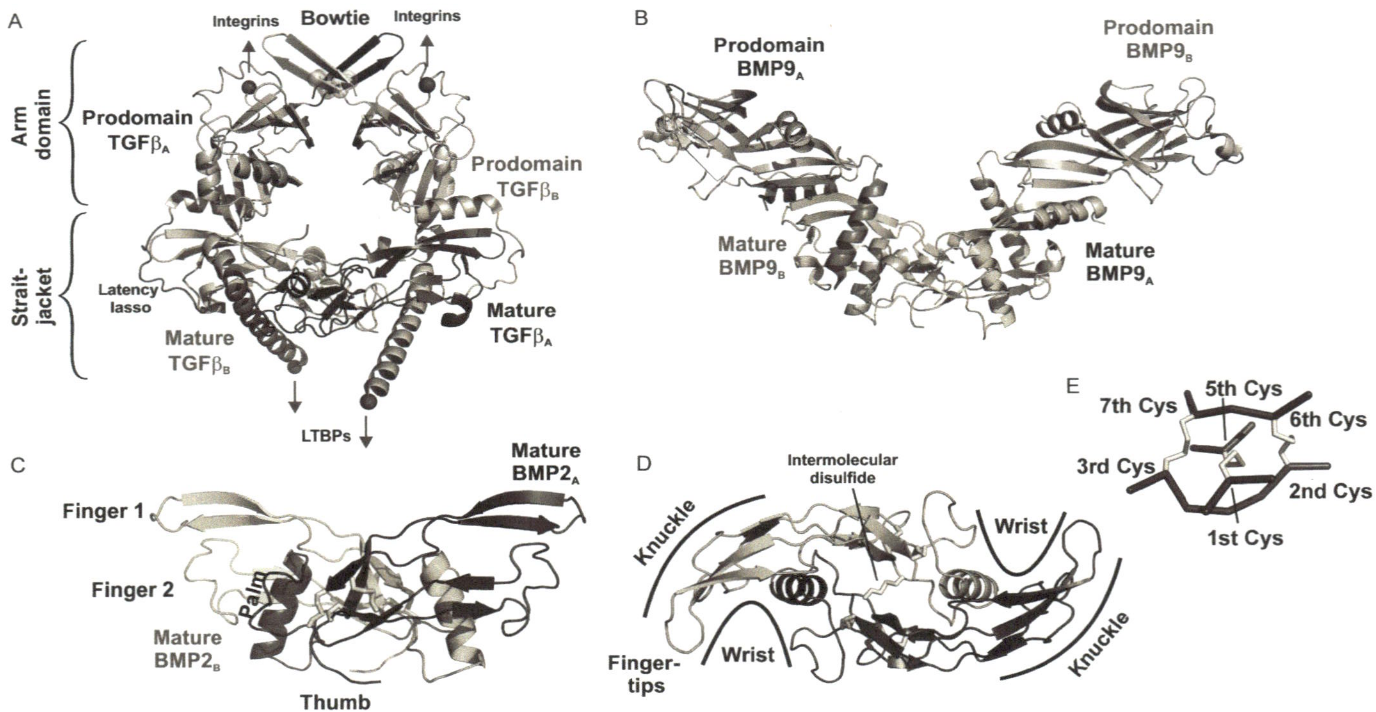

Figure 2 See legend on opposite page.

key element in the straitjacket assembly the so-called bowtie, an element involved in dimerization of the prodomain differs significantly between the *sensu stricto* TGFβs and the other ligand members (Fig. 2A). Most striking is the loss of four cysteine residues in the bowtie element of most non-*sensu stricto* TGFβs (Fig. 1C), which lead to the formation of a covalent, disulfide-linked prodomain in case of TGFβs (Zou & Sun, 2004; Fig. 2A). Very recent data by Timothy Springer and colleagues show the proprotein assembly of BMP9 (PDB entries 4YCG and 4YCI), which likely match the prodomain of most non-*sensu stricto* TGFβs better (Fig. 2B; Mi et al., 2015). Although the prodomain also attaches to and covers both receptor-binding epitopes of BMP9 as seen in pro-TGFβ1, the assembly markedly differs with each prodomain moiety operating as a monomer and being oriented differently on top of type II receptor interface. Thus, while the prodomain of pro-TGFβ1 somewhat resembles the interaction of Noggin with BMPs (see below), the prodomain of BMP9 is more alike a larger type II receptor with additional elements covering the type I receptor epitope. Consistently, the prodomain of BMP2 similarly lacking the cysteine residues in the bowtie, also is monomeric (Hillger, Herr, Rudolph, & Schwarz, 2005). As binding of a covalent dimer (the prodomain), to another dimer (the mature region) will likely experience affinity enhancement due to avidity effects, one might

Figure 2 (A) Structure of the proprotein complex of TGFβ1. The mature region is marked in green and dark gray. The prodomain can be separated into the arm domain and the straitjacket. Whereas the latter wraps around the mature region thereby blocking binding to the TGFβ type I and type II receptors, the former domain is involved in dimerization of two prodomains. The prodomain dimer is then stabilized by two disulfide bonds formed in the bowtie segment. Activation of the TGFβ mature region is likely to occur via tensile forces using the RGD motifs in the arm domain to bind to integrins and two cysteine residues at the N-terminus to connect to LTBP proteins. (B) Structure of the proprotein complex of BMP9 showing that the architecture vastly differs from TGFβ1. Although type I and type II receptor binding is abrogated by the prodomains, the elements conferring this blockage are different. In contrast to the prodomains of TGFβ1 the prodomains of BMP9 do not dimerize. (C) The butterfly-shaped architecture of the mature region of BMP2 resembles two hands assembled palm-to-palm, the β-sheets represent the fingers, the α-helix and the dimer interface form the palm. (D) The convex side of the fingers present the type II receptor binding sites (termed knuckle) of BMP/GDFs and Activins, the TGFβs bind their type II receptors via the fingertips. The type I receptor binding in BMPs/GDFs and Activins occurs in the wrist, which is formed by the concave side of the finger and the palm of the other monomer. (E) The cystine-knot is built from three disulfides of which two disulfide bonds between the second and fifth and the third and seventh form a ring, which is then penetrated by the third disulfide bond between the first and the fourth cysteine residue. (See the color plate.)

assume that the "non-latency" of BMP2 is due to a less tight interaction between the prodomain and the mature region. However, the prodomain of some latent BMPs (e.g., GDF8, GDF11, BMP10) similarly lack these cysteine residues and hence form only monomers. In addition, *in vitro* interaction analyses using prodomains of latent and non-latent BMPs revealed no affinity difference suggesting that implementation of latency is more complex (Sengle et al., 2011). But if latency is not a general feature of the prodomain of all TGFβ ligands, what might be its putative function? The group of Elizabeth Schwarz has recently shown that the prodomain is required for the biogenesis of BMP2 *in vivo*, recombinant expression of BMP2 in mammalian cells without its propart resulted in full retention of the protein in the endoplasmic reticulum (ER) (Kuhfahl & Schwarz, 2014). The assumption that the prodomain might be therefore required for folding seems unlikely as various BMPs have been recombinantly produced by refolding solely the mature region showing that the latter part is fully sufficient to correctly produce the cystine-knot fold (Bessa et al., 2009; Honda, Andou, Mannen, & Sugimoto, 2000; von Einem, Schwarz, & Rudolph, 2010). In fact, solubility enhancement might be the most crucial function of the prodomain, since the mature region is known to be very poorly soluble under physiological conditions as shown by *in vitro* studies, but also evident from the harsh extraction methods used by Hari Reddi, John Wozney, and colleagues for the isolation and subsequent cloning of BMPs (Hillger et al., 2005; Luyten et al., 1989; Sampath & Reddi, 1981) Wang et al., 1988. Thus, the prodomain might be necessary for biogenesis to avoid aggregation of the TGFβ ligand in the ER and to facilitate the secretion into the extracellular lumen. Here, the proprotein complex might function as storage and its enhanced solubility properties might also allow for "long-range" activities when processing/activation occurs at different sites (see also Akiyama et al., 2012). In addition, the specific protein/ECM-binding properties of the prodomain might also exert a targeting function to transport TGFβs to their final site of action (e.g., Sengle et al., 2008). The second key feature of TGFβ ligands is the presence of a cystine-knot comprising of six cysteine residues of which four are present in the motif C_2-X-G-X-C_3 and C_6-X-C_7 and form an eight-membered ring (Fig. 2C–E, see also Fig. 1C). The first and fifth cysteine residue engages in a third disulfide bond, which then penetrates the ring thereby tying the knot (Fig. 2E). Cystine-knot motifs are not only found in other growth factors, e.g., vascular-endothelial growth factor (VEGF), the members of the platelet-derived growth factors (PDGFs), the nerve-growth factor (NGF, BDNF, and NT1-3), or the glycoprotein hormone (hGC, FSH, TSH, and LH) families (for review: Sun & Davies, 1995) but also in some BMP- and Wnt-modulator proteins, e.g., Noggin and members of the DAN family (Groppe et al., 2002; Nolan et al., 2013; Veverka et al., 2009; Weidauer et al., 2009), as well as in "miniproteins" (also termed knottins) that act as toxins in animals or in plant defense (Craik, 2009; Craik, Daly, & Waine, 2001; Gruber, Cemazar, Anderson, & Craik, 2007). Despite the fact that the cystine-knot results in a general framework comprising three loops emanating from the central knot, with the first and the third loop running into the same and the second loop into the opposite direction, the overall architecture can vary significantly forming either homo- or heterodimeric assemblies, e.g., the above-mentioned growth factors, or functioning as monomers (e.g., the Wnt modulators Sclerostin and Wise belonging to the DAN modulator family as well as the knottins). Comparing the structures of NGF and TGFβ2 shows that the dimer architecture differs also among the growth factors. Whereas in the butterfly-like architecture of the TGFβ2 dimer the "wings" (represented by the two monomer subunits) are fully spread, in NGF these wings are closed (for review: Sun & Davies, 1995). These large structural differences in the dimer assembly stem first, from the different ring size of the cystine-knot; second, from the possible presence and location of a seventh cysteine, which is present in all TGFβ ligands except for GDF3, GDF9, and BMP15 and is involved in an intermolecular disulfide bond to stabilize the dimer assembly; and third from the length and composition of the three loops emanating from the central cystine-knot. Noteworthy, all TGFβ ligands with the exception of the very distant members Lefty1 and 2, for which no structure data is available, are characterized by a stop codon following one amino acid residue past the last cysteine residue. As the C-termini of the monomer subunits are directly facing each other and are almost completely buried inside the butterfly-shaped dimer assembly additional residues at the C-terminus would likely distort the dimer orientation also explaining, why C-terminal tags were reported to (at least partially) inactivate TGFβ ligands or impair their secretion (Mottershead et al., 2008; Pulkki et al., 2011; Swencki-Underwood et al., 2008). Although the listing of the TGFβs in Fig. 1A might suggest that the majority of the ligands are homodimeric, structure data for the mature factors, and analysis of the residues contributing to the dimer interface provide barely any reason why heterodimers cannot exist. In fact, the heterodimeric members of the Activin/Inhibin subgroup clearly refute this simplification, but evidence for heterodimers of the *sensu stricto* TGFβs or the BMPs/GDFs *in vivo* is sparse with few reports of BMP heterodimers existing in fish and fly (Kunnapuu et al., 2014; O'Connor, Umulis, Othmer, & Blair, 2006; Schmid et al., 2000; Shimmi, Umulis, Othmer, & O'Connor, 2005). Only one report about the isolation of osteogenic protein from bovine bone from the group of Kuber Sampath provides evidence for BMP heterodimers also in mammals (Sampath et al., 1990). Due to the presence of the intermolecular disulfide bond in most TGFβs any potential heterodimer formation, however, demands that both ligands are coexpressed in the same cell. The successful production of various heterodimeric BMPs *in vitro* by simple coexpression in eucaryotic systems shows that this principally works (Hazama, Aono, Ueno, & Fujisawa, 1995; Israel et al., 1996). Upon application to cells or organisms, these heterodimeric factors then exhibited elevated activities or encoded for unique functions not observed with their homomeric isoforms (Aono et al., 1995; Buijs et al., 2012; Butler & Dodd, 2003; Kusumoto et al., 1997; Nishimatsu & Thomsen, 1998). Although the knowledge to whether and how frequent BMP/GDF heterodimers occur naturally, the (at least partially) overlapping expression in various tissues (Dudley et al., 1995; Pizette & Niswander, 1999; Settle et al., 2003), suggest that BMP/GDF heterodimers are not unlikely to exist in many mammals (Israel et al., 1996). For BMP15 and GDF9, which lack the intermolecular disulfide, heterodimer formation has been proposed in the extracellular lumen even after secretion resulting in a noncovalent BMP15:GDF9 heterodimer with unique functions, an effect that was termed synergism (McNatty et al., 2005a, 2005b, see also Mottershead et al., 2013; Peng et al., 2013).

4. TGFβ RECEPTOR ACTIVATION AND ITS DOWNSTREAM SIGNALING CASCADE

Members of the TGFβ superfamily transmit their signal via binding and oligomerizing two different subgroups of transmembrane serine/threonine kinase receptors, termed type I and type II (Cheifetz, Like, & Massague, 1986; see also Fig. 1D). Despite that the two receptor subgroups show minor structural differences in the extracellular ligand-binding domain, the major difference (and used for classification) is in a membrane-proximal glycine/serine-rich amino acid stretch (termed GS-box) present only in type I receptors (Saitoh et al., 1996; Wieser, Wrana, & Massague, 1995; Fig. 1D). Due to the nature of TGFβ ligands as dimers, it is assumed that the ligand assembles two receptors of each subgroup into a hetero-tetrameric receptor complex in a stepwise assembly procedure. Besides the formation of this so-called BMP-induced signaling complex or BISC, a second receptor activation mechanism termed PFC exists in which BMPs bind to a preformed, (signaling-wise) silent type I-type II heteromeric receptor complex (Gilboa et al., 2000). The latter is then activated by an unknown allosteric mechanism, however in both cases, the constitutively active type II receptor kinase transphosphorylates the type I receptor kinase in the GS-box resulting in the activation of the type I receptor kinase to trigger downstream signaling cascades (Fig. 3). It is not known whether both activation mechanisms lead to different transphosphorylation patterns, however, they seem to initiate

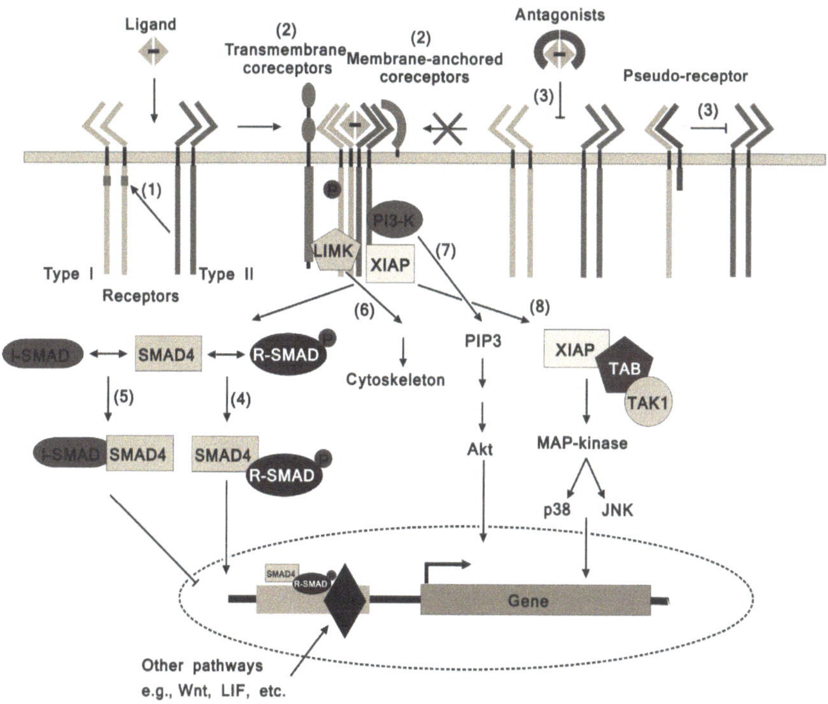

Figure 3 Schematic representation of the TGFβ signaling cascade. Activation occurs by binding of a TGFβ ligand to TGFβ type I and type II receptors leading to transphosphorylation of the type I receptor by the type II receptor kinase (1). Activation can be modulated by coreceptors (2), or inhibited by modulator proteins/secreted antagonists and negatively regulating pseudo-receptors (3). Activation can be transferred to R-SMAD proteins, which then hetero-oligomerize with SMAD4 (4), translocate into the nucleus and regulate gene transcription. The SMAD pathway can be inhibited by inhibitory SMADs (5) that interfere with R-SMAD/Co-SMAD complex formation. Also SMAD-independent signaling cascades can be triggered by TGFβ ligands, such as the LIMK kinase (6), the phosphatidylinositol-3-kinase (7), or the MAP kinase pathway (8). (See the color plate.)

different signaling cascades with BISC triggering the MAP kinase p38 pathway and PFC activating the SMAD signaling pathway (Nohe et al., 2002; Fig. 3). It is interesting to note that the activation of either SMAD or non-SMAD (MAPK) pathways coincide with the localization of BMP receptors in distinct membrane domains. BMP receptors present in nonlipid raft regions lead to activation of the SMAD signaling cascade, whereas BMP signaling leading to alkaline phosphatase expression, a marker of the MAPK p38 pathway, is initiated from cholesterol-enriched membrane micro domains also known as lipid rafts (Hartung et al., 2006). SMAD proteins are transcription factors (Graff, Bansal, & Melton, 1996; Liu et al., 1996; Savage et al., 1996), which hetero-oligomerize upon phosphorylation by the activated type I receptor kinase and subsequently translocate into the nucleus, where they act as transcriptional coactivators or corepressors to regulate the transcription of TGFβ/BMP-dependent genes. Three different types of SMAD proteins exist, one class of SMAD proteins associate with the TGFβ type I receptors and are thus termed receptor-associated or R-SMAD (Fig. 3). The unique common mediator (Co-SMAD) SMAD4 presents a second class of SMAD factors and together with the phosphorylated R-SMADs forms a hetero-trimeric complex, which then migrates into the nucleus (Chacko et al., 2001; Correia, Chacko, Lam, & Lin, 2001). The third class of SMAD proteins comprising SMAD6 and 7 are the so-called inhibitory or I-SMADs (Fig. 3), which were initially discovered for the capability to impede the phosphorylation of the R-SMADs and were therefore considered to negatively regulate SMAD activation and signaling (Huse et al., 2001; Itoh et al., 2001). Meanwhile the role of SMAD6 in BMP receptor activation seems more complex (Xu et al., 2013). Rik Derynck and coworkers have shown that SMAD6 comes associated with BMP type I receptors and gets methylated at a distinct arginine residues by the methyltransferase PRMT1, which itself is associated with the BMP type II receptor. Upon arginine methylation, SMAD6 dissociates and the type I receptor is derepressed leading to subsequent activation of the downstream R-SMAD signaling cascade. The upstream methylation required now explains for the first time the rather slow R-SMAD phosphorylation kinetics clearly different from other kinase receptors; in addition, the requirement of the methyltransferase PRMT1 ahead of the downstream SMAD cascade potentially introduces an additional regulatory step. The receptor-associated SMADs (R-SMADs) can be classified into two subgroups, the SMAD proteins 1, 5, and 8 are substrates of type I receptors usually engaging in BMP/GDF signaling, whereas the SMAD factors 2 and 3 are activated by type

I receptors binding to the *sensu stricto* TGFβs, Activins, Nodal as well as some GDFs (e.g., GDF8, GDF11, BMP9). This high specificity of a particular type I receptor for one or the other SMAD subgroup derives from the interaction of a cytoplasmic loop segment termed L45 loop in the type I receptor, which is in close proximity to its GS-box, and a corresponding segment termed L3 loop in the SMAD protein, ensuring that only the cognate SMAD can bind to a given type I receptor (Feng & Derynck, 1997; Lo, Chen, Shi, Pavletich, & Massague, 1998). The limitation to two major R-SMAD subgroups transducing the signals of all TGFβ ligands contrasts the functional diversity of this growth factor superfamily. Furthermore, all R-SMADs recognize the DNA sequence CAGA; however, binding occurs only with low affinity thus requiring additional DNA motifs—usually targeted by co-(transcription) factors—to gain sufficient binding affinity for transcriptional regulation. Thus, the presence (or absence) of these cofactors and/or the coupling of the TGFβ SMAD pathway with other signaling pathways enables the transcriptional regulation of a cell-specific set of target genes also specific for a particular TGFβ member (for review: Massague & Wotton, 2000). This kind of cross-talk between the TGFβ/SMAD pathway and other signaling cascade has been reported for instance for the Wnt/β-catenin, the leukemia inhibitory factor (LIF) or the tumor necrosis factor (TNF) pathways (for review: (Guo & Wang, 2009). Besides, the p38 MAP kinase cascade further non-SMAD/SMAD-independent pathways can be triggered by TGFβ ligands (Fig. 3). Through association of the phosphoinositide 3-kinase p85 regulatory subunit with the TGFβ type I receptor TβRI, TGFβs can activate the PI3K/Akt pathway in a ligand-dependent manner (Yi, Shin, & Arteaga, 2005). A similar linkage to the PI3 kinase pathway was also described for BMP2 in cardiomyocytes (Ghosh-Choudhury, Abboud, Chandrasekar, & Ghosh Choudhury, 2003). BMP receptors can form complexes with the MAP kinase-kinase-kinase TAK1 (TGFβ-activated kinase 1) and its activator TAB1 via the ubiquitin ligase XIAP, which is also known from its inhibitory activity on apoptotic caspases, thereby not only directly connecting BMP signaling with the MAP kinase pathway but also showing a cross-talk between TGFβ and BMP signaling (Yamaguchi et al., 1999). The *sensu stricto* TGFβs can activate Erk via Ras and MEK possibly through a tyrosine phosphorylation at the TGFβ type II receptor TβRII, which then recruits the adaptor proteins Grb2 and Shc linking the type II receptor to MAPK activation (Galliher & Schiemann, 2007; Lee et al., 2007). The activation of the various signaling cascades and the cross-talk between the signaling pathways of

TGFβs and different other growth factors show that the TGFβ signaling cascade does not act or induce biological actions in an isolated manner, but is rather entangled in a highly interwoven signaling network (see also Poorgholi Belverdi, Krause, Guzman, & Knaus, 2012).

5. TOO FEW RECEPTORS FOR TOO MANY LIGANDS LEAD TO PROMISCUITY

One "highlight" of the TGFβ superfamily is the likewise small number of receptors—seven type I and five type II receptors exist in humans—serving a very large number (≥25) of ligands (see also Fig. 1A and B). This numeral discrepancy usually requires a particular receptor of either subtype binding to more than one TGFβ ligand. Even when considering the combinatorial diversity resulting from the ligand-induced assembly of TGFβ receptors into a hetero-tetrameric complex with this small number of receptors it is impossible to provide each ligand with a unique combination being consistent with the simplified assumption—one factor—one receptor—one function. Only very few TGFβ receptors seem to be restricted in ligand binding. The TGFβ type II receptor TβRII exclusively interacts with the three TGFβ isoforms TGFβ1, 2, and 3 but no other TGFβ ligand. Only the AMH type II receptor AMHR-II seems strictly confined to bind only AMH (Josso, di Clemente, & Gouedard, 2001). In contrast, it has been shown that many BMP and GDF ligands bind to more than one receptor of each subtype and thus ligand–receptor interaction for instance in the BMP/GDF subgroup is highly promiscuous with almost all ligands seemingly binding to all receptors available to this subgroup (Fig. 1A). The ligand–receptor promiscuity is particularly evident from the BMP/GDF type II receptor interaction. *In vitro* binding analyses showed that the ligands of the BMP2/4, BMP5/6/7, and the GDF5/6/7 group bind similarly to the three type II receptors ActRII, ActRIIb, and BMPRII with usually less than 10-fold difference in affinity (Heinecke et al., 2009). Studies on ligands from the GDF1/3, the BMP3/GDF10 or the GDF8/11 groups employing chemical cross-linking, biophysical *in vitro* interaction analyses or functional assays suggest that these ligands are more selective in binding to a type II receptor interacting specifically with ActRIIb (Allendorph, Isaacs, Kawakami, Izpisua Belmonte, & Choe, 2007; Cheng, Olale, Bennett, Brivanlou, & Schier, 2003; Sako et al., 2010). On the other hand, the TGFβ ligands BMP15 and GDF9 as well as BMP9 seem to have a strong preference for BMPRII (Brown et al., 2005; Moore, Otsuka, & Shimasaki, 2003; Vitt,

Mazerbourg, Klein, & Hsueh, 2002), but, *in vitro* interaction analysis for BMP15 and GDF9 revealed that the preference of BMPRII over the Activin type II receptors does not exceed a factor of five (D.G. Mottershead and T.D. Mueller, unpublished). Among the type II receptors, ActRII and ActRIIb are most widely used and can function as type II receptors for ligands of the Activin/Inhibin as well as the BMP/GDF subgroups. This dual specificity has an important impact as usage of the Activin type II receptors with Activin/Inhibin ligands leads to activation of the SMAD2/3 pathway, whereas engaging with most members of the BMP/GDF subgroup results in SMAD1/5/8 signaling. Thus, opposing functions between Activins and BMP/GDF ligands might be also due to a direct competition for the Activin type II receptor (Piek et al., 1999). It also shows that on type II receptor level one receptor can participate in both SMAD signaling cascades by engaging in different ligand–receptor complexes.

However, not only type II receptors are shared among many TGFβ ligands, but also the type I receptors responsible for diverting the signaling toward either to the SMAD1/5/8 or to the SMAD2/3 cascade are often used in a highly overlapping manner. But despite jointly usage of type I receptors among different TGFβ ligands, the SMAD lineage is usually restricted within a TGFβ subgroup (Fig. 1A). The only exception so far known are the three TGFβ isoforms TGFβ1, 2, and 3, which can signal via the SMAD2/3 pathway using the type I receptor TβRI (also known as Alk5) but can also alternatively activate the SMAD1/5/8 signaling cascade via employing the type I receptor TSRI (also known as Alk1) (Kimchi, Wang, Weinberg, Cheifetz, & Massague, 1988; Oh et al., 2000). Noteworthy, the TGFβ type I receptor TβRI, which emerged late during evolution, can also bind GDF8 and GDF9 showing that this type I receptor is not limited to just the *sensu stricto* TGFβs (Kaivo-Oja et al., 2005; Rebbapragada, Benchabane, Wrana, Celeste, & Attisano, 2003). For TGFβ ligands activating the SMAD2/3 pathway, the Activin type I receptor ActRIb (Alk4) is the most promiscuous receptor interacting with BMP3, GDF1/3, Nodal, the various Activin isoforms as well as GDF8 and GDF11 (Cheng, Jiang, et al., 2003; Daluiski et al., 2001; Rebbapragada et al., 2003; Reissmann et al., 2001; Willis, Zimmerman, Li, & Mathews, 1996). For some of the ligands, however, the coreceptor Cripto is required to allow binding to ActRIb (Cheng, Jiang, et al., 2003; Reissmann et al., 2001; see Fig. 1A). The most promiscuous sharing of type I receptors is among the members of the BMP2/4, the BMP5/6/7 and the GDF5/6/7 group, which can all bind to the type I receptors BMPRIa (Alk3) and

BMPRIb (Alk6) albeit with varying affinities (Heinecke et al., 2009). The third type I receptor available to this group of ligands, the Activin type I receptor (Alk2) seems only essential for signaling by members of the BMP5/6/7 group, but unpublished data from the group of Walter Sebald showed that in cells lacking BMPRIa/Ib, also BMP2 can signal via ActRI (S. Harth, Ph.D. thesis; Harth, 2010). It is interesting to note that despite the requirement of ActRI for BMP6/7 signaling, *in vitro* interaction analyses showed that the affinity of BMP6 and BMP7 to ActRI is much lower than to BMPRIa and BMPRIb raising the question whether and how ActRI can effectively engage in BMP6/7 ligand–receptor complexes if all three receptor are present at the cell surface (Heinecke et al., 2009; Saremba et al., 2008). Further studies revealed that ActRI binding to BMP6/7 requires the presence of a carbohydrate at a N-glycosylation site in the type I receptor-binding epitope, which is conserved in BMP2/4 and BMP5/6/7, but not in members of the GDF5/6/7 group (Saremba et al., 2008). Enzymatic removal of the carbohydrate moiety in BMP6 fully abrogated binding to ActRI, but did not alter the interaction of BMP6 with BMPRIa and BMPRIb. Deglycosylated BMP6 had thus the same receptor-binding properties as BMP6 derived from *E. coli*, which was biologically inactive when tested for its capacity to induce expression of alkaline phosphatase, a target of the BMP-induced p38 MAP kinase pathway. This observation not only confirms that ActRI, despite its very low affinity for BMP6 *in vitro* is the type I receptor required for BMP6 signaling. It also indicates that despite the higher binding affinities of BMP6 for BMPRIa or BMPRIb, the latter type I receptors seem neither capable to substitute for ActRI with respect to signaling nor do they effectively compete off ActRI in BMP6 ligand–receptor complexes. This suggests that binding of a BMP ligand to a particular receptor does not necessarily correlate with its capability to activate or to transduce a signal via this receptor. This scenario also provides a new perspective at the issue of ligand–receptor promiscuity in the TGFβ superfamily. Although *in vitro* analyses show that many of the above BMP/GDF ligands interact similarly with a limited set of receptors indicating a highly promiscuous usage of receptors of either subtype, these quantitative data might, however, not properly reflect the *in vivo* condition as temperospatial expression differences for the ligands and the receptors might result in the availability of only a limited set of receptors thereby breaking the "*in vitro*" promiscuity. Furthermore, even receptors jointly used by different ligands might result in distinct signals as the assembly of an seemingly identical receptor complexes might not lead to their activation as seen for BMP6.

6. MOLECULAR MECHANISMS TO ENSURE LIGAND–RECEPTOR PROMISCUITY AND SPECIFICITY: THE CONCEPT OF MULTIPLE HOT SPOTS OF BINDING

From a structural biologist' point of view, the above-described promiscuity raises the question how interaction epitopes are formed on atomic level to allow binding of several distinct interaction partners on the one hand, but on the other ensure specificity to select only the cognate type I and type II receptors for a particular ligand. Structure analyses of various TGFβ ligands showed that the activity-bearing mature region of all ligands adopts a highly similar butterfly-shaped dimer structure. The architecture resembles two left hands depicting the two monomer subunits assembled palm-to-palm (see also Fig. 2C). The two-stranded β-sheets represent the fingers, the α-helix together with the dimer interface forms the palm and the N-terminal segment ahead of the first knot-forming cysteine residue is described as thumb. The four receptor-binding sites were identified first from mutagenesis studies (Gray et al., 2000; Harrison et al., 2004; Kirsch, Nickel, & Sebald, 2000; see also Fig. 2D). Activins and BMPs/GDFs bind type I receptors in the so-called wrist epitope made from the front side of the fingers of one monomer and the palm of the other ligand monomer (see Fig. 2C and D). Their interaction with type II receptors occurs in the knuckle epitopes, which are formed by the backsides of the fingers of each monomer (see Figs. 2C and D, and 4A). Various structures of ligand–receptor complexes of Activin and BMP ligands confirmed that promiscuous binding of different receptors does not imply the use of alternative epitopes (Allendorph, Vale, & Choe, 2006; Greenwald et al., 2003, 2004; Keller, Nickel, Zhang, Sebald, & Mueller, 2004; Kotzsch, Nickel, Sebald, & Mueller, 2009; Thompson, Woodruff, & Jardetzky, 2003; Townson et al., 2012; Weber et al., 2007). The *sensu stricto* TGFβs (TGFβ1, 2, and 3) are, however, an exception in that the type II receptor TβRII is bound via the finger tips (Hart et al., 2002; Fig. 4B). Differences in the length and conformation of the TGFβs finger-tip loops compared with those in BMPs ensure the strict specificity of TβRII allowing binding of TβRII only to TGFβs but not to BMPs. Structural differences in the extracellular ligand-binding domain of TβRII compared to those of ActRII,

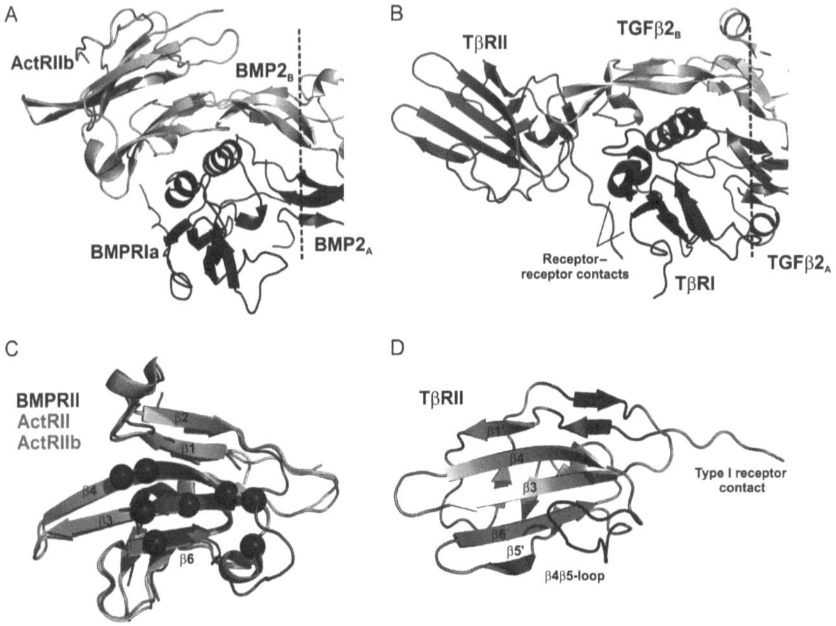

Figure 4 Type II receptor specificity between BMPs/GDFs and TGFβs is implemented via different receptor binding sites. (A) In BMPs/GDFs and Activins/Inhibins, the type II receptor (marked in light blue) binds to the ligand's knuckle epitope and seems not to share any contacts with the type I receptor (marked in green). (B) In contrast in TGFβs, the type II receptor binds to the fingertips, the type I receptor moves toward the type II receptor engaging in receptor–receptor contacts likely explaining cooperativity and high receptor selectivity. (C) The core structure as represented by the two- and three-stranded β-sheet involved in the ligand contact (indicated by red spheres) is almost identical for the three type II receptors BMPRII, ActRII, and ActRII. Only the loops connecting the strands differ in length and conformation allowing implementing ligand specificity where needed. (D) Binding of the type II receptor TβRII to BMPs/GDFs and Activins is impeded by the different length and conformation of the β4β5-loop (marked in blue), which would protrude into the BMP/GDF ligand interface. Binding to TGFβs is achieved by interaction of the β-strands β1 and β2 (marked in red) with the TGFβ fingertips. (See the color plate.)

ActRIIb, and BMPRII, e.g., different length of several loops, different arrangement of the β-strands forming the central β-sheet, additionally contribute to the high specificity between TGFβs and TβRII (Fig. 4C and D). On the contrary, binding of the Activin and BMP type II receptors to TGFβs is impeded due to several amino acid exchanges in the finger region of the TGFβs. For instance, in the equivalent of the knuckle epitope, all TGFβ isoforms harbor a conserved glutamate residue, the same position in BMP2 holds an alanine, which when converted to an aspartate yields an BMP2 antagonist (Kirsch et al., 2000). Besides the specific TGFβ–type II receptor interaction also type I receptor binding differs between TGFβs on one and Activins, BMPs, and GDFs on the other side resulting in a rather high specificity of the TGFβs for TβRI. Structure analysis of the ternary TGFβ ligand–receptor complex revealed that the type I receptor TβRI takes a different location and orientation when compared with the type I receptors of the BMP2 ternary ligand–receptor assembly or the GDF5:BMPRIb complex (Allendorph et al., 2006; Groppe et al., 2008; Kotzsch, Nickel, Sebald, et al., 2009; Weber et al., 2007; Fig. 4A). The extracellular domain of TβRI is relocated toward the ligand's fingertips and engages there in direct contacts with the type II receptor TβRII (Fig. 4B). As a consequence, binding of TβRI to the TGFβs is highly cooperative and requires prior presence of the type II receptor TβRII (Groppe et al., 2008; Zuniga et al., 2005). In BMP ligand–receptor complexes, such receptor–receptor contacts are absent explaining the noncooperative binding of both receptor subtypes to BMPs (Weber et al., 2007). These differences in ligand–receptor interaction of the *sensu stricto* TGFβs again confirm their deviating nature within this superfamily. Interestingly, TβRI has been described to also serve as alternative type I receptor for GDF8 and GDF9 (Mazerbourg et al., 2004; Rebbapragada et al., 2003), which bind their type II receptors ActRIIb and BMPRII via the classical knuckle epitope. Hence, analogous contacts between type I and type II receptor ectodomains essential in TGFβ ligand–receptor complex formation will be impossible for these members of the Activin and BMP subgroup raising the question whether alternative docking modes exist for TβRI to engage in complexes with non-TGFβs.

In contrast to the high specificity of TβRII for the *sensu stricto* TGFβs, type II receptor usage is highly promiscuous among Activins, BMPs, and GDFs. This indicates that either the receptor-binding epitopes of these ligands interacting with either ActRII, ActRIIb, or BMPRII have special characteristics allowing for this low specificity or suggests that the protein–protein interfaces are highly conserved (Fig. 5A and B). Of the 24 residues in BMP2 being involved in the ligand–type II receptor interface, mutagenesis demonstrated that only six amino acids are significantly influencing receptor recognition and binding (Kirsch et al., 2000). Of those six residues, however, only a single amino acid, e.g., the equivalent to Leu90 in BMP2, is invariant in all Activin and BMP type II receptor-binding ligands implying that a strong conservation of the residues participating in the ligand–type II receptor interface is not the basis of the above-mentioned

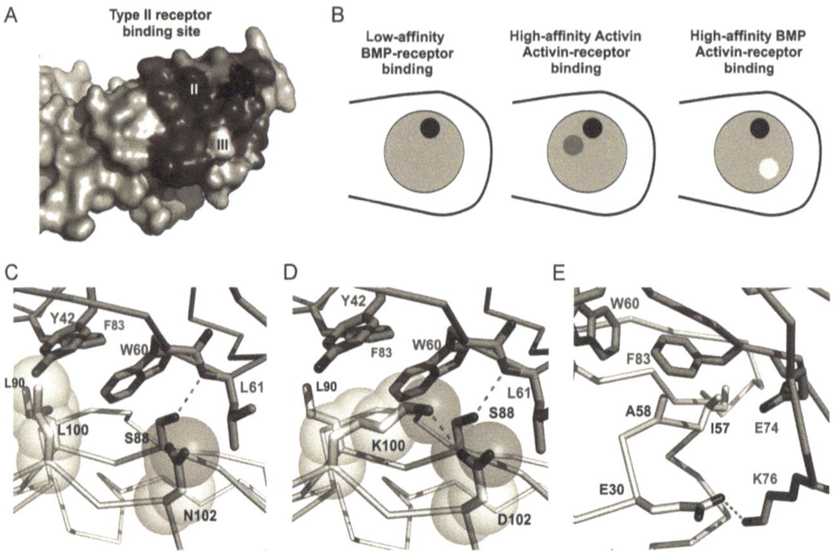

Figure 5 (A) The type II receptor interface in BMPs is modular. Different (combinations) of hot spots of binding (indicated with I, II, and III) allow modulating binding affinity and receptor specificity. In the low-affinity binding of BMP2 to its type receptors, only a single hot spot is utilized (B), mutating two residues that affect shielding (marked in cyan) activates a second potential hot spot of binding and ensures high-affinity binding of BMP2 to ActRIIb (B, middle panel). In BMP3 and BMP7, another hot spot of binding (indicated in green) is used to discriminate between the type II receptors ActRIIb and ActRII (B, right panel). (C) The hydrogen bond between Ser88 of BMP2 and the backbone amide of L61 in ActRIIb does not contribute significantly to the overall binding in wild-type BMP2 due to inefficient shielding of the polar bond from the access of solvent. (D) Mutating Leu100 and N102 to the equivalent amino acids as found in ActivinA activates this hot spot. (E) The charge–charge interaction between various BMPs (e.g., BMP3, BMP7) and ActRII and ActRIIb allows discriminating between the two highly similar Activin type II receptors. (See the color plate.)

promiscuity. Analysis of the structure/function relationship within this epitope shows that the Activin/BMP–type II receptor interface is dominated by hydrophobic interactions (Allendorph et al., 2006; Greenwald et al., 2003, 2004; Kirsch et al., 2000; Thompson et al., 2003; Townson et al., 2012; Weber et al., 2007). Since hydrophobic contacts are not as sensitive to smaller structural rearrangements as polar bonds, e.g., hydrogen bonds, which only function within a small angular band and distance range, such hydrophobic interactions might allow for a more variable interface ensuring promiscuity. In addition, the concept of "hot spot of binding" introduced by James Wells in the mid-1990s might explain how promiscuity is implemented in the Activin/BMP–type II receptor interaction

(Clackson & Wells, 1995; Fig. 5A). Here, a (single) residue (pair) usually in the center of the protein–protein interfaces of both binding partners dominates the interaction "delivering" more than 50% of the overall binding energy, the surrounding amino acids shield this interaction from the environment, which means the hydrophobic or polar bond in the center is buried from the access of water. As different types of amino acids might provide a similar efficient shielding effect, as long as their chemical nature and approximate size is preserved, surrounding residues could easily vary in sequence. However, a highly important difference in the promiscuous type II receptor usage among Activins and BMPs/GDFs is that Activin ligands usually bind their Activin type II receptors with affinities (only 1:1 interaction values shall be considered to exclude avidity) almost an order of magnitude higher than observed for BMPs (Heinecke et al., 2009). There is evidence that all TGFβ ligands, which activate the SMAD2/3 downstream cascade, bind their type II receptors, i.e., ActRII and ActRIIb, with high affinity in the nanomolar range, whereas binding to their type I receptors occurs with affinities an order of magnitude lower. In contrast, BMPs and GDFs signaling via the SMAD1/5/8 pathway, usually bind their type I receptors with higher affinities than their type II receptors (Heinecke et al., 2009). As a consequence, receptor activation employing a ligand-induced oligomerization scheme (i.e., BISC, see above) will take place via a reversed receptor-binding order, which might influence the downstream signaling cascade and result in an additional discrimination between the SMAD2/3 and the SMAD1/5/8 pathways. For instance, effective concentrations for half-maximal responses (i.e., EC_{50}) for TGFβs and Activins were reported to be in the picomolar range (Carcamo et al., 1994; Lach-Trifilieff et al., 2014; Ye et al., 2006), which is lower than the affinities of these ligands for their high-affinity type II receptors. On the contrary, EC_{50} values for BMP-induced expression of alkaline phosphatase, which is initiated via a BISC receptor activation mechanism of the p38 MAPK pathway, are resembling concentrations close to the affinity values of the high-affinity BMP–type I receptor interaction (e.g., see Keller et al., 2004). One possible explanation for this different efficiency might be that in the case of Activins and TGFβs, the high-affinity type II receptor remains in complex with the ligand after initial assembly, whereas the activated type I receptor could dissociate and be replaced by another (nonactivated) type I receptor thereby resulting in an signaling amplification due to a single TGFβ ligand activating multiple type I receptors. In contrast, in BMPs, the type I receptor presents the high-affinity receptor, which likely remains in the ligand–receptor complex after assembly and activation, thus one BMPs ligand dimer likely activates only a single receptor pair. Analysis of the molecular cause of these affinity differences in Activins and BMPs revealed that the type II receptor interface has a modular architecture with a variable number of hot spots of binding (Weber et al., 2007). Mutating two amino acid residues in the type II receptor epitope of BMP2, Leu100, and Asn102 to those present in Activins (i.e., Lys and Asp) resulted in a variant exhibiting almost the same high affinity for ActRIIb as ActivinA (Fig. 5A and B). A more detailed analysis then showed that the exchange of these two residues enhances the affinity through a more effective shielding of a central hydrogen bond (i.e., between Ser88 of BMP2 and a backbone amide in ActRIIb), which is highly conserved between many Activin and BMP/GDF ligands (Weber et al., 2007; Fig. 5C and D). Due to the differing shielding residues, this (additional) polar hot spot is silenced in BMP2 and the residual single hydrophobic hot spot of binding (see above) only allows for low affinity type II receptor binding. Sequence comparison of several TGFβ ligands, which are assumed to bind ActRIIb with high affinity, showed indeed residues with similar side chain size at these two shielding positions. As the mutations in BMP2 only affected binding affinity to ActRIIb, but neither to ActRII nor to BMPRII, it might point toward a concept in which a modular binding site allows to introduce specificity. Accordingly, mutation of two other amino acid positions in BMP2 specifically modulated the affinity to BMPRII (Weber et al., 2007). The high specificity of BMP3 for ActRIIb was found to be due to a charge-charge interaction involving a lysine residue of BMP3 (Lys30) and a glutamate residue in ActRIIb (Glu76) (Allendorph et al., 2007). This lysine residue is conserved in several TGFβ ligands, e.g., Nodal, GDF8/11, which all exhibit a high affinity/specificity for ActRIIb. In BMP7, which binds preferentially to ActRII (Heinecke et al., 2009; Weber et al., 2007), we just see the inverted implementation, instead of the lysine residue of BMP3, BMP7 has a glutamate residue which engages in a salt bridge with the lysine residue present in ActRII (Greenwald et al., 2003; Fig. 5E). Thus, the special architecture of the type II receptor-binding site in Activins and BMPs/GDFs making (combinatorial) use of a variable number of hotspots not only allows to vary binding affinities by more than 50-fold, it also implements a simple mechanism to create specificity/promiscuity toward a single or a set of type II receptors (independent from the binding affinity) by providing several interaction sites that like a jigsaw piece interacts with all fitting counter pieces (Fig. 5A and B). The high adaptability of this interface is also evident from the interaction of BMPs and GDFs (as well in part also for Activins) with a plethora of structurally highly variable set of antagonists and modulator proteins. Although only a relatively small number of examples of these modulator–TGFβ ligand complexes have been structurally characterized so far, e.g., the BMP7–Noggin complex, Follistatin and Follistatin-like proteins bound to either ActivinA or GDF8 and the first Chordin-like Von Willebrand type C domain of Crossveinless 2 bound to BMP2 (Cash et al., 2012; Cash, Rejon, McPherron, Bernard, & Thompson, 2009; Groppe et al., 2002; Harrington et al., 2006; Stamler et al., 2008; Thompson, Lerch, Cook, Woodruff, & Jardetzky, 2005; Zhang et al., 2008), the data clearly show that the BMP and Activin antagonists do neither share any structural similarity among each other nor with the receptors of the TGFβ superfamily (Fig. 6A–D). However, despite this structural dissimilarity, all modulator proteins bind to the same epitopes also recognized by the TGFβ type I and type II receptors strongly indicating that binding to a common epitope does not require structural mimicry. Functional data on the

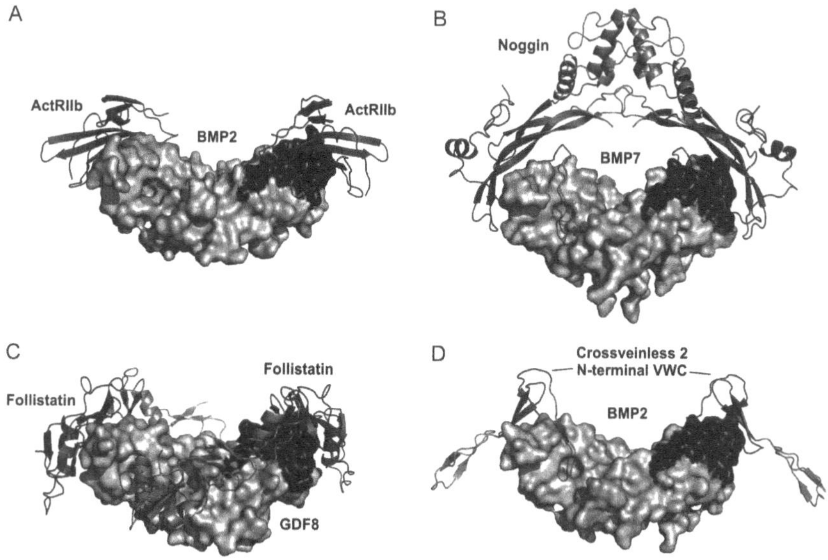

Figure 6 The concept of a modular hot spot of binding allows highly different structures to bind to the same interface area. The type II receptor-binding site in the knuckle epitope somewhat presents an interaction hub, with the almost identical epitope in the ligand being recognized by very different binding partner and despite the fact that the architecture and orientation of the interacting element differs significantly between ActRIIb (A), Noggin (B), Follistatin (C), and the Von Willebrand type C domain of the Chordin-modulator family member Crossveinless 2 (D). (See the color plate.)

BMP2–Crossveinless 2 interaction suggest that even though the Von Willebrand type C domain is structurally completely different from the BMP type II receptors, the same hydrophobic residue in BMP2, Leu100 in the type II receptor interface, serves also as hot spot for the recognition and binding of the modulator protein domain (Zhang et al., 2008).

7. MOLECULAR MECHANISMS TO ENSURE LIGAND–RECEPTOR PROMISCUITY AND SPECIFICITY: THE CONCEPT OF STRUCTURAL ADAPTABILITY

In addition to the concept of hot spot of binding, other mechanisms possibly encode for promiscuity or specificity in the ligand–receptor interaction of the TGFβ superfamily. One important observation in the structure analysis of various TGFβ ligand–receptor complexes was the presence of structural variability either in one of the component or along the reaction coordinate aka the complex formation. The latter principle might be used to restrict type I receptor usage of Activins and the *sensu stricto* TGFβs. Structure analyses of ActivinA and TGFβ2 bound to their type II receptors ActRII and TβRII surprisingly revealed ligand architectures strongly deviating from the butterfly-shaped closed conformation usually common for all TGFβ ligands (Greenwald et al., 2004; Hart et al., 2002; Thompson et al., 2003; Fig. 7A–C). Although the existence of the open conformation of TGFβ2 when bound to TβRII seemed puzzling at first—the C-termini of the two TβRII receptor ectodomains would not be similarly oriented toward the surface of the cell membrane due to a rotation of the two TGFβ monomers by 110°—the open conformation possibly just highlights the inherent flexibility of the TGFβ2 dimer (Hart et al., 2002; Fig. 7A). Evidence that the dimer architecture of the *sensu stricto* TGFβs might be indeed inherently flexible comes from NMR relaxation studies of TGFβ3 showing that the dimer interface is destabilized potentially allowing for such a partial unfolding (Bocharov et al., 2002; Huang, Schor, & Hinck, 2014). Furthermore, in the structure of the TGFβ1 proprotein complex, the orientation of the mature region monomer subunits is also skewed compared to the canonical conformation found in unbound TGFβ2 although here the two subunits are twisted into the opposite direction as observed in the TGFβ2:TβRII complex (Shi et al., 2011; Fig. 7A). However, contrary to what the structure of the binary TGFβ2–TβRII complex seemingly suggests, binding and anchoring of the flexible TGFβ ligand by two TβRII transmembrane receptors fixed into a two-dimensional plane on the cell surface, will likely lead to

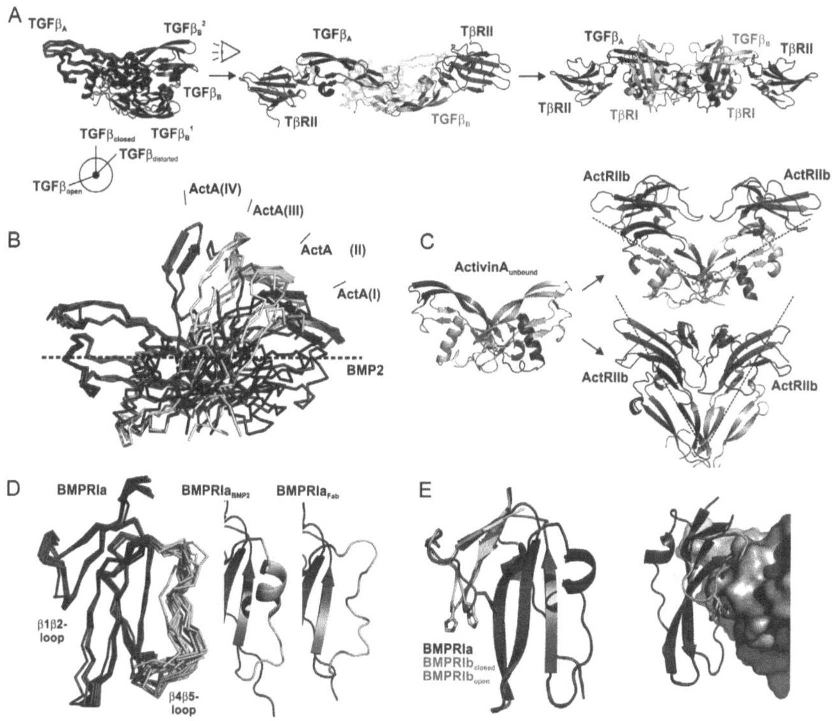

Figure 7 Structural variability can be utilized to generate specificity and promiscuity. (A) Data suggest that the butterfly-shaped architecture of the *sensu stricto* TGFβs is flexible and exists in an open and a closed conformation depending on its binding partner or environment. Structure analysis of TGFβ2 bound to its TβRII revealed an open conformation for TGFβ2 (middle panel), whereas in the ternary complex of TGFβ3 bound to TβRII and TβRI the ligand adopts the canonical closed conformation. The inherent flexibility likely contributes to type I receptor specificity and also possibly ensures a defined order of receptor binding. (B) ActivinA also exhibits a highly flexible architecture with the two monomers adopting different angles to each. Considering the ligand architecture as a butterfly, the wings are fully spread in BMPs (BMP2 shown in gray), but adopt other conformations with the wings being in a more closed form (blue: ActivinA when bound to Follistatin, PDB 2B0U; green: ActivinA in its unbound conformation, PDB 2ARV; yellow: ActivinA when bound to Follistatin Fs12, PDB 2ARP; red: ActivinA when bound to ActRIIb, PDB 1NYS). (C) Similar to TGFβ2 also for ActivinA different ligand conformations were found when bound to its type II receptor ActRIIb suggesting that also for ActivinA the type II/type I receptor-binding order might be due to the type I receptor interface being formed only when the ligand is in complex with its type II receptor on the cell surface. (D) In the BMP type I receptors, a dynamic and flexible loop (left panel, BMPRIa in its unbound conformation), enable adaptation of the type I receptor to its respective binding partner via adopting different structures (α-helical upon binding to BMPs or extended when bound to a neutralizing antibody fragment). Here, flexibility is used to generate promiscuity. (E) The β1β2-loop in the BMP type I receptors BMPRIa and BMPRIb also exist in different conformations. In BMPRIa, the loop folds into a locked conformer, which attenuates binding to GDF5 due to strong van der Waals contacts. In contrast in BMPRIb, the loop adopts two more open conformations, which both fold away from the GDF5 surface thereby leaving sufficient space for bulky residues in GDF5 and thus allowing high-affinity binding of BMPRIb to GDF5. (See the color plate.)

the formation and stabilization of the butterfly-shaped closed conformation, which can then recruit the type I receptors into the complex (Groppe et al., 2008). Together with the formation of a combined epitope derived from the TGFβ ligand and the TβRII receptor, this switching between an open and closed conformation thereby not only ensures specific binding of only the TβRI receptor but also absolutely restricts the order of receptor binding requiring binding of TGFβs to TβRII prior to their interaction with TβRI (Fig. 7A). A similarly flexible dimer architecture has also been described for ActivinA albeit the axis of rotation for the wing movement is differently placed (Fig. 7B). Several structures of ActivinA bound to ActRIIb were determined from different crystal forms revealing highly variable ActivinA dimer conformations (Greenwald et al., 2004; Thompson et al., 2003; Fig. 7C). When superimposed these ActivinA dimers differ in their inter-domain angles by about 45°—the axis of rotation protrudes the Cα atoms of the intermolecular disulfide bond—with the different dimers somewhat resembling the wing beat of a butterfly (Fig. 7B and C). Together with structures of unbound ActivinA and ActivinA bound to the Activin and BMP modulator Follistatin, these data suggest that the monomer subunits in the ActivinA dimer can rotate by more than 60° against each other with the dimer arrangement in which both wings are most spread resembling almost the conformation of BMP ligands (Harrington et al., 2006; Stamler et al., 2008; Thompson et al., 2005). As said for TGFβ2 above, the physiological significance of the flexible dimer architecture of ActivinA is not yet clear; however, it possibly provides an explanation how a strict receptor-binding order—ActivinA supposedly binds first to its type II receptors and then recruits ActRIb—is implemented and the flexible dimer architecture could also contribute to the low binding affinity observed between ActivinA and its type I receptor ActRIb. As the type I receptor epitope might not be fully formed or stabilized if ActivinA is not bound to a (surface-located) type II receptor, part of the binding energy from the type I receptor interaction would then be used to payoff the conformational rearrangement and dimer stabilization. As similar structure data for ActRIb-interacting ligands such as Nodal or GDF8/11 are not yet available, it is unclear whether such flexible dimer architecture is a general feature of all Activin-like ligands or whether it is specific for Activins or a limited set thereof only.

In contrast to the large conformational rearrangements seen in TGFβ2 and ActivinA, the mature region of BMP/GDF ligands seems rather rigid. Various structures from unbound ligands as well as BMPs/GDFs in complex with receptor ectodomains so far provided no hints for a similar conformational rearrangement of the ligand dimer (Allendorph et al., 2007; Brown et al., 2005 Greenwald et al., 2003; Griffith, Keck, Sampath, Rueger, & Carlson, 1996; Groppe et al., 2002; Keller et al., 2004; Kotzsch, Nickel, Seher, Sebald, & Muller, 2009; Nickel, Kotzsch, Sebald, & Mueller, 2005; Saremba et al., 2008; Scheufler, Sebald, & Hulsmeyer, 1999; Schreuder, Liesum, Pohl, Kruse, & Koyama, 2005; Townson et al., 2012). Except for the N-terminal segment in front of the first cysteine residue, only an eleven residue long loop ahead of helix $\alpha1$ in the wrist (type I receptor binding) epitope termed pre-helix loop shows signs for elevated conformational flexibility (Keller et al., 2004). Compared with the rigid BMP/GDF ligands, the extracellular domains of the BMP receptors seem to be more flexible. A comparison of the structures of the type II receptors ActRII, ActRIIb, and BMPRII in their bound and unbound states indicates that only the conformation of individual loop elements, e.g. the $\beta1\beta2$-, $\beta2\beta3$-, $\beta4\beta5$-, and the $\beta5\beta6$-loop, of the three finger toxin fold vary (Greenwald, Fischer, Vale, & Choe, 1999; Greenwald et al., 2003, 2004; Mace, Cutfield, & Cutfield, 2006; Thompson et al., 2003; Weber et al., 2007). The concave side of the three-stranded β-sheet comprising the strands $\beta3$, $\beta4$, and $\beta6$, however, which contributes most to the ligand–receptor interface, is structurally almost identical in all three type II receptors independent on whether the structures were determined for their bound or unbound conformation suggesting that this interface element is preformed. Nevertheless, as some residues present in the flexible loops were identified to confer ligand–type II receptor specificity, flexibility of these elements might thus be important to implement the required degree of adaptability and specificity (Yeh, Falcon, Garces, Lee, & Lee, 2012). The matter takes on a different complexion when it comes to the BMP type I receptors. NMR structure analysis of the type I receptor BMPRIa revealed that the majority of its BMP-binding epitope is highly flexible prior to ligand binding (Klages et al., 2008; Fig. 7D). Upon binding to BMP2 the $\beta4\beta5$-loop in BMPRIa, a 15mer loop with no secondary structure, folds into a highly defined 1.5-turn short α-helix via a disorder-to-order transition with some parts moving by more than 5 Å (Fig. 7D middle panel). Importantly, the loop not only contributes about 60% of the interface area, it also carries the hot spot of binding for the BMP2–BMPRIa interaction showing that contrary to common belief, the hot spot does not have to be located within a rigid environment (Keller et al., 2004). NMR relaxation data indicates that the loop is however not fully unstructured in its unbound state, but rather switches fast between two or more imprinted conformations, one of which is then selected by the binding partner upon interaction proposing a possible adaptation mechanism allowing to bind different binding partners (Klages et al., 2008). That indeed a different conformation can be adopted by this loop element was confirmed by structures analysis of BMPRIa in complex bound to a neutralizing Fab fragment that binds the BMPRIa ectodomain via the same epitope as BMP2 (Harth, Kotzsch, Hu, Sebald, & Mueller, 2010; Fig. 7D right panel). Interestingly, although the conformation of the interacting loop of BMPRIa greatly varies depending on whether it is bound to either BMP2 or the Fab antibody fragment, mutagenesis revealed that the same interface residues of BMPRIa contribute very similar to the binding energetics of either binding partner (Harth et al., 2010). While the loop in the center of the BMP-binding interface of the BMP type I receptor is flexible, possibly to ensure adaptability in the binding to rigid BMP/GDF ligands, NMR studies demonstrated that the binding interfaces of the TGFβ type I receptors TβRI and TSRI (Alk1) are highly defined and their epitopes are preformed (Mahlawat, Ilangovan, Biswas, Sun, & Hinck, 2012; Zuniga et al., 2011). One could speculate whether this difference is due to the fact that in case of BMPRI receptors the ligand is rigid therefore requiring a flexible receptor ectodomain to adjust to different ligand surfaces, whereas for TβRI and TSRI the binding partner TGFβ1, 2, and 3 itself is flexible and thus the ligand can adapt to the receptor. But TSRI interacts with very high affinity with the BMP/GDF ligands BMP9 and BMP10, which belong to the BMP/GDF family considered to be rigid nonflexible ligands. Crystal structures of BMP9 in its unbound conformation, in complex with its prodomain (PDB entries 4CYG and 4YCI; Mi et al., 2015) as well as in complex bound to ActRIIb and TSRI (Brown et al., 2005; Townson et al., 2012; Wei, Salmon, Upton, Morrell, & Li, 2014) indeed do not reveal any conformational differences for BMP9 in any of these states suggesting that not always adaptation of a "soft" to a rigid binding partner is required. One might even consider that the very affinity observed for the binding of TSRI to BMP9 might be due to the absence of larger flexible elements. Whereas the flexible β4β5-loop of BMP type I receptors seems to ensure adaptability and hence promiscuity, a second flexible loop element in the BMP type I receptors confers ligand specificity. Data from gene deletion studies in mice suggested that GDF5 signals rather specifically via the BMP receptor BMPRIb (Storm et al., 1994; Yi, Daluiski, Pederson, Rosen, & Lyons, 2000) and physical interaction of GDF5 and BMPRIb was confirmed by chemical cross-linking (Nishitoh et al., 1996). Interaction analysis, however, indicates that *in vitro* the selectivity for BMPRIb over BMPR1a by GDF5 is lower than what is assumed from above-mentioned studies and GDF5 binds to the BMPRIb ectodomain only with a 10-fold higher affinity than to BMPRIa (Heinecke et al., 2009; Nickel et al., 2005). A single amino acid in the center of the type I receptor epitope of GDF5, arginine 57, seems to be solely responsible for the BMPRIb preference as its mutation to alanine abolishes any selectivity for either BMPRIb or Ia (Nickel et al., 2005). That the type I receptor specificity as implemented in wild-type GDF5 is functionally relevant *in vivo* comes from the observation that a similar type I receptor selectivity-affecting mutation R57L (R438L when numbering takes the prodomain into account) was found in patients suffering from the skeletal malformation disease Symphalangism showing that GDF5 signaling via BMPRIb is very likely distinct from signaling via BMPRIa (Seemann et al., 2005). On a molecular level, the BMPRIb preference of GDF5 is due to the interaction of the Arg57 of GDF5 with the β1β2-loop of the BMP type I receptor. The crystal structure analysis of the GDF5:BMPRIb complex showed that the BMP type I receptor β1β2-loop can adopt different conformations suggesting that in BMPRIa this loop—due to its different amino acid sequence—folds into a locked conformation, which would result in a van der Waals overlap with Arg57 of GDF5 thereby attenuating BMPRIa binding (Fig. 7E). In contrast in BMPRIb the loop adopts an open or closed conformation both of which provide sufficient space for the large bulky arginine side chain present in GDF5 thereby allowing for high-affinity binding of this BMP type I receptor (Kotzsch, Nickel, Seher, et al., 2009). Interestingly, BMP15, which is also highly selective for BMPRIb, also carries an arginine residue at the equivalent position (Moore et al., 2003). The data listed above indicates that flexibility can work either way, it may be used to allow for promiscuous binding when allowing one binding partner to adapt to different surface geometries, but it may also be used to ensure that only one interaction partner can utilize a highly defined conformational rearrangement and thus this interaction will occur with high specificity.

8. CONSEQUENCES OF PROMISCUITY AND SPECIFICITY IN THE TGFβ SUPERFAMILY: CONCLUSIONS

Despite our knowledge how specificity and promiscuity is generated in protein-protein interactions in the TGFβ family at an atomic level, we need to be careful about the functional consequences. It is unclear, whether these promiscuous ligand–receptor interactions as observed by *in vitro* experiments will also occur to the same extent *in vivo* and whether such promiscuous interactions also inevitably lead to functional overlap. Strict temperospatial regulated expression might possibly prevent a signaling interference as assumed from the *in vitro* data. During limb and joint development or fracture healing, however, at least partially overlapping expression of BMP/GDF ligands and receptors, e.g. BMP2 and GDF5, has indeed been shown (Dathe et al., 2009; Kwong, Hoyland, Evans, & Freemont, 2009; Lorda-Diez, Montero, Garcia-Porrero, & Hurle, 2010). From *in vitro* interaction data one would hence expect that both ligands will lead to the formation of identical receptor assemblies on these cells; GDF5 can bind to BMPRIa and BMPRIb (albeit BMPRIa with 10-fold lower affinity than BMPRIb) and BMP2 can bind both BMP type I receptor with very similar affinity (Heinecke et al., 2009). Even though both BMP type I receptors possibly—by a so far unknown mechanism—encode for different functions (Kaps et al., 2004; Sahni et al., 2010; Zou, Wieser, Massague, & Niswander, 1997), if solely SMAD activation is considered, both receptors seem to deliver identical signals (Feng & Derynck, 1997; Fujii et al., 1999). BMP2 should nevertheless be capable to functionally replace GDF5 via forming BMP2:BMPRIb complexes, which should deliver the same signals as GDF5:BMPRIb unless the receptors can "sense" the nature of the ligand, transfer this information to the cytoplasm and subsequently integrate this knowledge into a modulation of the downstream signaling cascade. Similarly, GDF5 might mimic BMP2 through the formation of GDF5:BMPRIa assemblies albeit with slightly lower efficiency due to its lower affinity for BMPRIa. But despite the different *in vivo* functions (Storm et al., 1994), there is also evidence from *in vitro* data, e.g. from the use of the two marker cell lines C2C12 and ATDC5, that both ligands are functionally distinct. Both, the pre-chondrogenic ATDC5 cells as well as the myoblast cell line C2C12 express a highly similar set of BMP type I and type II receptors and have been described to lack the BMPRIb receptor at the cell surface (Akiyama, Shukunami, Nakamura, & Hiraki, 2000; Ebisawa et al., 1999; Namiki et al., 1997). Stimulation of ATDC5 cells with BMP2 or GDF5 leads to the expression of alkaline phosphatase, a marker gene of osteogenesis and a target of the p38 MAPK pathway (Gallea et al., 2001; Nakamura et al., 1999). This observation indicates that in ATDC5 cells, which lack BMPRIb, GDF5 and BMP2, both showing identical type II receptor binding *in vitro*, seemingly assemble an identical BMP type I-type II receptor complex. The lower affinity of GDF5 for BMPRIa nicely correlates with the requirement of a 10-fold higher GDF5 concentration to induce a similar ALP expression as BMP2 confirming that GDF5 here signals via BMPRIa (Nickel et al., 2005). Conformingly, when using the GDF5 variant R57A, which binds BMPRIa with BMP2-like affinity, GDF5 R57A-induced ALP expression in ATDC5 is indistinguishable from BMP2 (Nickel et al., 2005). While this observation implies that identical receptor assemblies encode for redundant signals independent of the ligand's nature, signaling of GDF5 and BMP2 in C2C12 cells provides a completely different picture. Here, BMP2 stimulation similarly leads to a dose-dependent expression of alkaline phosphatase; however, GDF5 fails to induce ALP expression even at elevated concentration (Seemann et al., 2005). Surprisingly, also employing the GDF5 variant R57A, which closely mimics BMP2 when it comes to receptor binding, did not lead to the expression of alkaline phosphatase in this cell line (U. Klammert et al., manuscript in preparation). That BMP2 and GDF5 likely assemble identical BMP receptor complexes also in C2C12 cells could be confirmed by the observation that GDF5 R57A can directly antagonize BMP2 activity (U. Klammert et al., manuscript in preparation). Thus, the same BMP receptor assembly that is activated by one particular TGFβ ligand, might be silent for another clearly indicating that identically composed BMP receptor complexes do not necessarily yield identical signals. This raises the question, how a shared receptor can "sense" the nature of the ligand and accordingly modulate the signaling in a ligand-specific manner. One possible mechanism might be structural differences in the orientation or location of the receptors in the different ligand–receptor complexes. A detailed comparison of structures of BMP2 and GDF5 ligand–receptor complexes revealed indeed small differences in the orientation of the type I receptor ectodomain (Kotzsch, Nickel, Seher, et al., 2009; Weber et al., 2007). In the ternary BMP9 ligand–receptor interaction, the type II receptor ActRIIb is slightly shifted compared to its location in complexes with BMP2 and the orientation of the type I receptor TSRI significantly deviates from that of BMPRIa in BMP2 ligand–receptor assemblies (Townson et al., 2012). Whether these differences in architecture are purely affecting ligand–receptor specificity or whether these disparities also qualitatively and quantitatively influence signaling remains to be determined. However, if these differences are transferred into the cytoplasm they might not only affect transactivation of the receptor kinases, but could also result in altered phosphorylation of substrates and thus qualitatively alter the downstream signaling cascade. While signaling by single-span transmembrane receptors is thought to be limited to a transfer of lateral movements in the cell membrane, i.e., ligand-induced oligomerization, studies from other growth factor families showed that signaling could be also initiated by vertical or rotational movements (He, Chow, Martick, & Garcia, 2001; Livnah et al., 1999; Ogawa, Qiu, Ogata, & Misono, 2004; Ottemann, Xiao, Shin, & Koshland, 1999) and orientation differences might alter signaling output in a ligand-specific manner (Livnah et al., 1998, see also Greiser, Stross, Heinrich, Behrmann, & Hermanns, 2002). A second mechanism to modulate signaling through shared receptors in a ligand-specific manner might derive from variations in the ligand–receptor binding kinetics. Since BMP receptors contain kinases, recognition and conversion of cytoplasmic substrates into their phosphorylated forms depends on enzymatic properties such as the Michaelis–Menten constant (a measure of substrate affinity) and the turnover number (determining the conversion rate). Differences in the receptor complex lifetime due to ligand-specific association and dissociation rates could yield different activation states of the type I receptor kinase or result in a nonactivated type I receptor kinase when the lifetime of the receptor assembly is too short to permit transphosphorylation by the type II receptor kinase. We have hints that not only affinities but also the binding kinetic are important determinants for BMP receptor activation. GDF5 variants that exhibit the same binding affinity for BMPRIa as wild-type GDF5 but differ in the association and dissociation rates for receptor complex formation were not able to activate the downstream signaling cascade (Kotzsch, Nickel, Seher, et al., 2009). A similar effect was reported for human growth hormone (hGH); mutations specifically leading to an increased dissociation rate of hGH from its capturing receptor thereby shortening complex lifetime resulted in inactivation of hGH (Pearce, Cunningham, Fuh, Teeri, & Wells, 1999). Therefore, formation of the same receptor complex with different kinetics might not activate signaling or restrict the downstream cascade to a subset of signals. A third mechanism involves the participation of additional components, such as membrane-bound or transmembrane coreceptors, in the ligand–receptor complex. The coreceptors betaglycan and endoglin can engage in quaternary complexes with TGFβ ligands and receptors and recruit additional components through their short cytoplasmic domains thereby modulating the signaling cascade depending on the presence or absence of the coreceptor (Blobe, Liu, Fang, How, & Lodish, 2001; Blobe, Schiemann, et al., 2001; Conley et al., 2004). But even without a cytoplasmic domain, coreceptors such as Cripto or the BMP-specific coreceptors of the repulsive guidance molecules (RGM) family can influence the signaling cascade in a ligand-specific manner. By directly recruiting the ligands to the cell membrane, they can act as affinity modulators and alter the local concentration of the ligands. If the coreceptor is required for ligand binding, the cell's responsiveness to a particular ligand would then depend on the presence of the coreceptor as is suggested for the interaction of Cripto with Nodal (Calvanese et al., 2014, 2010). Alternatively, the coreceptor might alter the receptor binding specificity/preference of a ligand and thereby alter the composition of the receptor assembly induced by a particular ligand as was described for the coreceptors of the RGM family. Here, the RGM anchored in the membrane not only act as affinity enhancer thereby sensitizing the cells (Babitt et al., 2005; Samad et al., 2005), they were also reported to alter the type II receptor binding properties of BMP ligands (Xia et al., 2007). As the type II receptor BMPRII differs from the Activin type II receptors ActRII and ActRIIb by containing an additional C-terminal domain, which was shown to associate with various signaling molecules (Hassel et al., 2004, 2006), a change in the receptor assembly composition will very likely also qualitatively alter the signaling outcome. This list of mechanisms, which is not intended to be exhaustive, presents a toolbox allowing to diversify signaling through a limited set of receptors and indicates that our understanding of how TGFβ ligands signal is still rather incomplete.

REFERENCES

Adams, D., Karolak, M., Robertson, E., & Oxburgh, L. (2007). Control of kidney, eye and limb expression of Bmp7 by an enhancer element highly conserved between species. *Developmental Biology*, *311*(2), 679–690. http://dx.doi.org/10.1016/j.ydbio.2007.08.036.

Ago, T., & Sadoshima, J. (2006). GDF15, a cardioprotective TGF-beta superfamily protein. *Circulation Research*, *98*(3), 294–297. http://dx.doi.org/10.1161/01.RES.0000207919.83894.9d.

Akhurst, R. J., & Hata, A. (2012). Targeting the TGFbeta signalling pathway in disease. *Nature Reviews. Drug Discovery*, *11*(10), 790–811. http://dx.doi.org/10.1038/nrd3810.

Akiyama, T., Marques, G., & Wharton, K. A. (2012). A large bioactive BMP ligand with distinct signaling properties is produced by alternative proconvertase processing. *Science Signaling*, *5*(218), ra28. http://dx.doi.org/10.1126/scisignal.2002549.

Akiyama, H., Shukunami, C., Nakamura, T., & Hiraki, Y. (2000). Differential expressions of BMP family genes during chondrogenic differentiation of mouse ATDC5 cells. *Cell Structure and Function*, *25*(3), 195–204.

Aleman-Muench, G. R., & Soldevila, G. (2012). When versatility matters: Activins/inhibins as key regulators of immunity. *Immunology and Cell Biology*, *90*(2), 137–148. http://dx.doi.org/10.1038/icb.2011.32.

Allendorph, G. P., Isaacs, M. J., Kawakami, Y., Izpisua Belmonte, J. C., & Choe, S. (2007). BMP-3 and BMP-6 structures illuminate the nature of binding specificity with receptors. *Biochemistry*, *46*(43), 12238–12247. http://dx.doi.org/10.1021/bi700907k.

Allendorph, G. P., Vale, W. W., & Choe, S. (2006). Structure of the ternary signaling complex of a TGF-beta superfamily member. *Proceedings of the National Academy of Sciences of*

the United States of America, *103*(20), 7643–7648. http://dx.doi.org/10.1073/pnas.0602558103.
Annes, J. P., Munger, J. S., & Rifkin, D. B. (2003). Making sense of latent TGFbeta activation. *Journal of Cell Science*, *116*(Pt 2), 217–224.
Aono, A., Hazama, M., Notoya, K., Taketomi, S., Yamasaki, H., Tsukuda, R., et al. (1995). Potent ectopic bone-inducing activity of bone morphogenetic protein-4/7 heterodimer. *Biochemical and Biophysical Research Communications*, *210*(3), 670–677. http://dx.doi.org/10.1006/bbrc.1995.1712.
Archdeacon, P., & Detwiler, R. K. (2008). Bone morphogenetic protein 7 (BMP7): A critical role in kidney development and a putative modulator of kidney injury. *Advances in Chronic Kidney Disease*, *15*(3), 314–320. http://dx.doi.org/10.1053/j.ackd.2008.04.011.
Argintar, E., Edwards, S., & Delahay, J. (2011). Bone morphogenetic proteins in orthopaedic trauma surgery. *Injury*, *42*(8), 730–734. http://dx.doi.org/10.1016/j.injury.2010.11.016.
Ariizumi, T., & Asashima, M. (1995). Control of the embryonic body plan by activin during amphibian development. *Zoological Science*, *12*(5), 509–521.
Asashima, M., Ariizumi, T., & Malacinski, G. M. (2000). In vitro control of organogenesis and body patterning by activin during early amphibian development. *Comparative Biochemistry and Physiology. Part B, Biochemistry & Molecular Biology*, *126*(2), 169–178.
Baarends, W. M., van Helmond, M. J., Post, M., van der Schoot, P. J., Hoogerbrugge, J. W., de Winter, J. P., et al. (1994). A novel member of the transmembrane serine/threonine kinase receptor family is specifically expressed in the gonads and in mesenchymal cells adjacent to the mullerian duct. *Development*, *120*(1), 189–197.
Babitt, J. L., Zhang, Y., Samad, T. A., Xia, Y., Tang, J., Campagna, J. A., et al. (2005). Repulsive guidance molecule (RGMa), a DRAGON homologue, is a bone morphogenetic protein co-receptor. *The Journal of Biological Chemistry*, *280*(33), 29820–29827. http://dx.doi.org/10.1074/jbc.M503511200.
Baek, S. J., Kim, K. S., Nixon, J. B., Wilson, L. C., & Eling, T. E. (2001). Cyclooxygenase inhibitors regulate the expression of a TGF-beta superfamily member that has proapoptotic and antitumorigenic activities. *Molecular Pharmacology*, *59*(4), 901–908.
Baloh, R. H., Tansey, M. G., Lampe, P. A., Fahrner, T. J., Enomoto, H., Simburger, K. S., et al. (1998). Artemin, a novel member of the GDNF ligand family, supports peripheral and central neurons and signals through the GFRalpha3-RET receptor complex. *Neuron*, *21*(6), 1291–1302.
Bandyopadhyay, A., Tsuji, K., Cox, K., Harfe, B. D., Rosen, V., & Tabin, C. J. (2006). Genetic analysis of the roles of BMP2, BMP4, and BMP7 in limb patterning and skeletogenesis. *PLoS Genetics*, *2*(12), e216. http://dx.doi.org/10.1371/journal.pgen.0020216.
Barakat, B., Itman, C., Mendis, S. H., & Loveland, K. L. (2012). Activins and inhibins in mammalian testis development: New models, new insights. *Molecular and Cellular Endocrinology*, *359*(1–2), 66–77. http://dx.doi.org/10.1016/j.mce.2012.02.018.
Beck, C. W., Christen, B., Barker, D., & Slack, J. M. (2006). Temporal requirement for bone morphogenetic proteins in regeneration of the tail and limb of Xenopus tadpoles. *Mechanisms of Development*, *123*(9), 674–688. http://dx.doi.org/10.1016/j.mod.2006.07.001.
Benazet, J. D., Bischofberger, M., Tiecke, E., Goncalves, A., Martin, J. F., Zuniga, A., et al. (2009). A self-regulatory system of interlinked signaling feedback loops controls mouse limb patterning. *Science*, *323*(5917), 1050–1053. http://dx.doi.org/10.1126/science.1168755.
Bessa, P. C., Cerqueira, M. T., Rada, T., Gomes, M. E., Neves, N. M., Nobre, A., et al. (2009). Expression, purification and osteogenic bioactivity of recombinant human BMP-4, -9, -10, -11 and -14. *Protein Expression and Purification*, *63*(2), 89–94. http://dx.doi.org/10.1016/j.pep.2008.09.014.

Bilezikjian, L. M., Blount, A. L., Donaldson, C. J., & Vale, W. W. (2006). Pituitary actions of ligands of the TGF-beta family: Activins and inhibins. *Reproduction, 132*(2), 207–215. http://dx.doi.org/10.1530/rep.1.01073.

Bisgrove, B. W., Essner, J. J., & Yost, H. J. (1999). Regulation of midline development by antagonism of lefty and nodal signaling. *Development, 126*(14), 3253–3262.

Blobe, G. C., Liu, X., Fang, S. J., How, T., & Lodish, H. F. (2001). A novel mechanism for regulating transforming growth factor beta (TGF-beta) signaling. Functional modulation of type III TGF-beta receptor expression through interaction with the PDZ domain protein, GIPC. *The Journal of Biological Chemistry, 276*(43), 39608–39617. http://dx.doi.org/10.1074/jbc.M106831200.

Blobe, G. C., Schiemann, W. P., Pepin, M. C., Beauchemin, M., Moustakas, A., Lodish, H. F., et al. (2001). Functional roles for the cytoplasmic domain of the type III transforming growth factor beta receptor in regulating transforming growth factor beta signaling. *The Journal of Biological Chemistry, 276*(27), 24627–24637. http://dx.doi.org/10.1074/jbc.M100188200.

Bocharov, E. V., Korzhnev, D. M., Blommers, M. J., Arvinte, T., Orekhov, V. Y., Billeter, M., et al. (2002). Dynamics-modulated biological activity of transforming growth factor beta3. *The Journal of Biological Chemistry, 277*(48), 46273–46279. http://dx.doi.org/10.1074/jbc.M206274200.

Bootcov, M. R., Bauskin, A. R., Valenzuela, S. M., Moore, A. G., Bansal, M., He, X. Y., et al. (1997). MIC-1, a novel macrophage inhibitory cytokine, is a divergent member of the TGF-beta superfamily. *Proceedings of the National Academy of Sciences of the United States of America, 94*(21), 11514–11519.

Bresnahan, P. A., Leduc, R., Thomas, L., Thorner, J., Gibson, H. L., Brake, A. J., et al. (1990). Human fur gene encodes a yeast KEX2-like endoprotease that cleaves pro-beta-NGF in vivo. *The Journal of Cell Biology, 111*(6 Pt 2), 2851–2859.

Brown, D. A., Breit, S. N., Buring, J., Fairlie, W. D., Bauskin, A. R., Liu, T., et al. (2002). Concentration in plasma of macrophage inhibitory cytokine-1 and risk of cardiovascular events in women: A nested case–control study. *Lancet, 359*(9324), 2159–2163. http://dx.doi.org/10.1016/S0140-6736(02)09093-1.

Brown, D. A., Hance, K. W., Rogers, C. J., Sansbury, L. B., Albert, P. S., Murphy, G., et al. (2012). Serum macrophage inhibitory cytokine-1 (MIC-1/GDF15): A potential screening tool for the prevention of colon cancer? *Cancer Epidemiology, Biomarkers & Prevention, 21*(2), 337–346. http://dx.doi.org/10.1158/1055-9965.EPI-11-0786.

Brown, D. A., Ward, R. L., Buckhaults, P., Liu, T., Romans, K. E., Hawkins, N. J., et al. (2003). MIC-1 serum level and genotype: Associations with progress and prognosis of colorectal carcinoma. *Clinical Cancer Research, 9*(7), 2642–2650.

Brown, M. A., Zhao, Q., Baker, K. A., Naik, C., Chen, C., Pukac, L., et al. (2005). Crystal structure of BMP-9 and functional interactions with pro-region and receptors. *The Journal of Biological Chemistry, 280*(26), 25111–25118. http://dx.doi.org/10.1074/jbc.M503328200.

Buijs, J. T., van der Horst, G., van den Hoogen, C., Cheung, H., de Rooij, B., Kroon, J., et al. (2012). The BMP2/7 heterodimer inhibits the human breast cancer stem cell subpopulation and bone metastases formation. *Oncogene, 31*(17), 2164–2174. http://dx.doi.org/10.1038/onc.2011.400.

Butler, S. J., & Dodd, J. (2003). A role for BMP heterodimers in roof plate-mediated repulsion of commissural axons. *Neuron, 38*(3), 389–401.

Calvanese, L., Falcigno, L., Maglione, C., Marasco, D., Ruggiero, A., Squeglia, F., et al. (2014). Structural and binding properties of the PASTA domain of PonA2, a key penicillin binding protein from Mycobacterium tuberculosis. *Biopolymers, 101*(7), 712–719. http://dx.doi.org/10.1002/bip.22447.

Calvanese, L., Marasco, D., Doti, N., Saporito, A., D'Auria, G., Paolillo, L., et al. (2010). Structural investigations on the Nodal-Cripto binding: A theoretical and experimental approach. *Biopolymers, 93*(11), 1011–1021. http://dx.doi.org/10.1002/bip.21517.

Carcamo, J., Weis, F. M., Ventura, F., Wieser, R., Wrana, J. L., Attisano, L., et al. (1994). Type I receptors specify growth-inhibitory and transcriptional responses to transforming growth factor beta and activin. *Molecular and Cellular Biology, 14*(6), 3810–3821.

Cash, J. N., Angerman, E. B., Kattamuri, C., Nolan, K., Zhao, H., Sidis, Y., et al. (2012). Structure of myostatin.follistatin-like 3: N-terminal domains of follistatin-type molecules exhibit alternate modes of binding. *The Journal of Biological Chemistry, 287*(2), 1043–1053. http://dx.doi.org/10.1074/jbc.M111.270801.

Cash, J. N., Rejon, C. A., McPherron, A. C., Bernard, D. J., & Thompson, T. B. (2009). The structure of myostatin:follistatin 288: Insights into receptor utilization and heparin binding. *The EMBO Journal, 28*(17), 2662–2676. http://dx.doi.org/10.1038/emboj.2009.205.

Chacko, B. M., Qin, B., Correia, J. J., Lam, S. S., de Caestecker, M. P., & Lin, K. (2001). The L3 loop and C-terminal phosphorylation jointly define Smad protein trimerization. *Nature Structural Biology, 8*(3), 248–253. http://dx.doi.org/10.1038/84995.

Chang, H., Brown, C. W., & Matzuk, M. M. (2002). Genetic analysis of the mammalian transforming growth factor-beta superfamily. *Endocrine Reviews, 23*(6), 787–823. http://dx.doi.org/10.1210/er.2002-0003.

Cheifetz, S., Like, B., & Massague, J. (1986). Cellular distribution of type I and type II receptors for transforming growth factor-beta. *The Journal of Biological Chemistry, 261*(21), 9972–9978.

Chen, C., & Shen, M. M. (2004). Two modes by which Lefty proteins inhibit nodal signaling. *Current Biology, 14*(7), 618–624. http://dx.doi.org/10.1016/j.cub.2004.02.042.

Cheng, H., Jiang, W., Phillips, F. M., Haydon, R. C., Peng, Y., Zhou, L., et al. (2003). Osteogenic activity of the fourteen types of human bone morphogenetic proteins (BMPs). *The Journal of Bone and Joint Surgery. American Volume, 85-A*(8), 1544–1552.

Cheng, S. K., Olale, F., Bennett, J. T., Brivanlou, A. H., & Schier, A. F. (2003). EGF-CFC proteins are essential coreceptors for the TGF-beta signals Vg1 and GDF1. *Genes & Development, 17*(1), 31–36. http://dx.doi.org/10.1101/gad.1041203.

Clackson, T., & Wells, J. A. (1995). A hot spot of binding energy in a hormone-receptor interface. *Science, 267*(5196), 383–386.

Conley, B. A., Koleva, R., Smith, J. D., Kacer, D., Zhang, D., Bernabeu, C., et al. (2004). Endoglin controls cell migration and composition of focal adhesions: Function of the cytosolic domain. *The Journal of Biological Chemistry, 279*(26), 27440–27449. http://dx.doi.org/10.1074/jbc.M312561200.

Constam, D. B. (2014). Regulation of TGFbeta and related signals by precursor processing. *Seminars in Cell & Developmental Biology, 32*, 85–97. http://dx.doi.org/10.1016/j.semcdb.2014.01.008.

Cook, R. W., Thompson, T. B., Jardetzky, T. S., & Woodruff, T. K. (2004). Molecular biology of inhibin action. *Seminars in Reproductive Medicine, 22*(3), 269–276. http://dx.doi.org/10.1055/S-2004-831902.

Correia, J. J., Chacko, B. M., Lam, S. S., & Lin, K. (2001). Sedimentation studies reveal a direct role of phosphorylation in Smad3:Smad4 homo- and hetero-trimerization. *Biochemistry, 40*(5), 1473–1482.

Craik, D. J. (2009). Circling the enemy: Cyclic proteins in plant defence. *Trends in Plant Science, 14*(6), 328–335. http://dx.doi.org/10.1016/j.tplants.2009.03.003.

Craik, D. J., Daly, N. L., & Waine, C. (2001). The cystine knot motif in toxins and implications for drug design. *Toxicon, 39*(1), 43–60.

Crotwell, P. L., Sommervold, A. R., & Mabee, P. M. (2004). Expression of bmp2a and bmp2b in late-stage zebrafish median fin development. *Gene Expression Patterns, 5*(2), 291–296. http://dx.doi.org/10.1016/j.modgep.2004.07.001.
Cui, Y., Hackenmiller, R., Berg, L., Jean, F., Nakayama, T., Thomas, G., et al. (2001). The activity and signaling range of mature BMP-4 is regulated by sequential cleavage at two sites within the prodomain of the precursor. *Genes & Development, 15*(21), 2797–2802. http://dx.doi.org/10.1101/gad.940001.
Daluiski, A., Engstrand, T., Bahamonde, M. E., Gamer, L. W., Agius, E., Stevenson, S. L., et al. (2001). Bone morphogenetic protein-3 is a negative regulator of bone density. *Nature Genetics, 27*(1), 84–88. http://dx.doi.org/10.1038/83810.
Danesh, S. M., Villasenor, A., Chong, D., Soukup, C., & Cleaver, O. (2009). BMP and BMP receptor expression during murine organogenesis. *Gene Expression Patterns, 9*(5), 255–265. http://dx.doi.org/10.1016/j.gep.2009.04.002.
Daniels, L. B., Clopton, P., Laughlin, G. A., Maisel, A. S., & Barrett-Connor, E. (2011). Growth-differentiation factor-15 is a robust, independent predictor of 11-year mortality risk in community-dwelling older adults: The Rancho Bernardo study. *Circulation, 123*(19), 2101–2110. http://dx.doi.org/10.1161/CIRCULATIONAHA.110.979740.
Dathe, K., Kjaer, K. W., Brehm, A., Meinecke, P., Nurnberg, P., Neto, J. C., et al. (2009). Duplications involving a conserved regulatory element downstream of BMP2 are associated with brachydactyly type A2. *American Journal of Human Genetics, 84*(4), 483–492. http://dx.doi.org/10.1016/j.ajhg.2009.03.001.
de Kretser, D. M., & Robertson, D. M. (1989). The isolation and physiology of inhibin and related proteins. *Biology of Reproduction, 40*(1), 33–47.
Dong, J., Albertini, D. F., Nishimori, K., Kumar, T. R., Lu, N., & Matzuk, M. M. (1996). Growth differentiation factor-9 is required during early ovarian folliculogenesis. *Nature, 383*(6600), 531–535. http://dx.doi.org/10.1038/383531a0.
Dube, J. L., Wang, P., Elvin, J., Lyons, K. M., Celeste, A. J., & Matzuk, M. M. (1998). The bone morphogenetic protein 15 gene is X-linked and expressed in oocytes. *Molecular Endocrinology, 12*(12), 1809–1817. http://dx.doi.org/10.1210/mend.12.12.0206.
Dubois, C. M., Blanchette, F., Laprise, M. H., Leduc, R., Grondin, F., & Seidah, N. G. (2001). Evidence that furin is an authentic transforming growth factor-beta1-converting enzyme. *The American Journal of Pathology, 158*(1), 305–316.
Dubois, C. M., Laprise, M. H., Blanchette, F., Gentry, L. E., & Leduc, R. (1995). Processing of transforming growth factor beta 1 precursor by human furin convertase. *The Journal of Biological Chemistry, 270*(18), 10618–10624.
Dudley, A. T., Lyons, K. M., & Robertson, E. J. (1995). A requirement for bone morphogenetic protein-7 during development of the mammalian kidney and eye. *Genes & Development, 9*(22), 2795–2807.
Ebisawa, T., Tada, K., Kitajima, I., Tojo, K., Sampath, T. K., Kawabata, M., et al. (1999). Characterization of bone morphogenetic protein-6 signaling pathways in osteoblast differentiation. *Journal of Cell Science, 112*(Pt 20), 3519–3527.
Fainsod, A., Steinbeisser, H., & De Robertis, E. M. (1994). On the function of BMP-4 in patterning the marginal zone of the Xenopus embryo. *The EMBO Journal, 13*(21), 5015–5025.
Fairlie, W. D., Moore, A. G., Bauskin, A. R., Russell, P. K., Zhang, H. P., & Breit, S. N. (1999). MIC-1 is a novel TGF-beta superfamily cytokine associated with macrophage activation. *Journal of Leukocyte Biology, 65*(1), 2–5.
Fajardo, R. J., Manoharan, R. K., Pearsall, R. S., Davies, M. V., Marvell, T., Monnell, T. E., et al. (2010). Treatment with a soluble receptor for activin improves bone mass and structure in the axial and appendicular skeleton of female cynomolgus macaques (Macaca fascicularis). *Bone, 46*(1), 64–71. http://dx.doi.org/10.1016/j.bone.2009.09.018.

Fang, J., Wang, S. Q., Smiley, E., & Bonadio, J. (1997). Genes coding for mouse activin beta C and beta E are closely linked and exhibit a liver-specific expression pattern in adult tissues. *Biochemical and Biophysical Research Communications*, *231*(3), 655–661. http://dx.doi.org/10.1006/bbrc.1997.6162.

Feldman, B., Gates, M. A., Egan, E. S., Dougan, S. T., Rennebeck, G., Sirotkin, H. I., et al. (1998). Zebrafish organizer development and germ-layer formation require nodal-related signals. *Nature*, *395*(6698), 181–185. http://dx.doi.org/10.1038/26013.

Feng, X. H., & Derynck, R. (1997). A kinase subdomain of transforming growth factor-beta (TGF-beta) type I receptor determines the TGF-beta intracellular signaling specificity. *The EMBO Journal*, *16*(13), 3912–3923. http://dx.doi.org/10.1093/emboj/16.13.3912.

Feng, X. H., & Derynck, R. (2005). Specificity and versatility in tgf-beta signaling through Smads. *Annual Review of Cell and Developmental Biology*, *21*, 659–693. http://dx.doi.org/10.1146/annurev.cellbio.21.022404.142018.

Fujii, M., Takeda, K., Imamura, T., Aoki, H., Sampath, T. K., Enomoto, S., et al. (1999). Roles of bone morphogenetic protein type I receptors and Smad proteins in osteoblast and chondroblast differentiation. *Molecular Biology of the Cell*, *10*(11), 3801–3813.

Gallea, S., Lallemand, F., Atfi, A., Rawadi, G., Ramez, V., Spinella-Jaegle, S., et al. (2001). Activation of mitogen-activated protein kinase cascades is involved in regulation of bone morphogenetic protein-2-induced osteoblast differentiation in pluripotent C2C12 cells. *Bone*, *28*(5), 491–498.

Galliher, A. J., & Schiemann, W. P. (2007). Src phosphorylates Tyr284 in TGF-beta type II receptor and regulates TGF-beta stimulation of p38 MAPK during breast cancer cell proliferation and invasion. *Cancer Research*, *67*(8), 3752–3758. http://dx.doi.org/10.1158/0008-5472.CAN-06-3851.

Gamer, L. W., Cox, K. A., Small, C., & Rosen, V. (2001). Gdf11 is a negative regulator of chondrogenesis and myogenesis in the developing chick limb. *Developmental Biology*, *229*(2), 407–420. http://dx.doi.org/10.1006/dbio.2000.9981.

Ghosh-Choudhury, N., Abboud, S. L., Chandrasekar, B., & Ghosh Choudhury, G. (2003). BMP-2 regulates cardiomyocyte contractility in a phosphatidylinositol 3 kinase-dependent manner. *FEBS Letters*, *544*(1–3), 181–184.

Gilboa, L., Nohe, A., Geissendorfer, T., Sebald, W., Henis, Y. I., & Knaus, P. (2000). Bone morphogenetic protein receptor complexes on the surface of live cells: A new oligomerization mode for serine/threonine kinase receptors. *Molecular Biology of the Cell*, *11*(3), 1023–1035.

Godin, R. E., Takaesu, N. T., Robertson, E. J., & Dudley, A. T. (1998). Regulation of BMP7 expression during kidney development. *Development*, *125*(17), 3473–3482.

Gold, E., Jetly, N., O'Bryan, M. K., Meachem, S., Srinivasan, D., Behuria, S., et al. (2009). Activin C antagonizes activin A in vitro and overexpression leads to pathologies in vivo. *The American Journal of Pathology*, *174*(1), 184–195. http://dx.doi.org/10.2353/ajpath.2009.080296.

Goldman, D. C., Donley, N., & Christian, J. L. (2009). Genetic interaction between Bmp2 and Bmp4 reveals shared functions during multiple aspects of mouse organogenesis. *Mechanisms of Development*, *126*(3–4), 117–127. http://dx.doi.org/10.1016/j.mod.2008.11.008.

Gordon, J., Patel, S. R., Mishina, Y., & Manley, N. R. (2010). Evidence for an early role for BMP4 signaling in thymus and parathyroid morphogenesis. *Developmental Biology*, *339*(1), 141–154. http://dx.doi.org/10.1016/j.ydbio.2009.12.026.

Graff, J. M., Bansal, A., & Melton, D. A. (1996). Xenopus Mad proteins transduce distinct subsets of signals for the TGF beta superfamily. *Cell*, *85*(4), 479–487.

Grainger, D. J., Wakefield, L., Bethell, H. W., Farndale, R. W., & Metcalfe, J. C. (1995). Release and activation of platelet latent TGF-beta in blood clots during dissolution with plasmin. *Nature Medicine*, *1*(9), 932–937.

Gray, P. C., Greenwald, J., Blount, A. L., Kunitake, K. S., Donaldson, C. J., Choe, S., et al. (2000). Identification of a binding site on the type II activin receptor for activin and inhibin. *The Journal of Biological Chemistry, 275*(5), 3206–3212.

Greenwald, J., Fischer, W. H., Vale, W. W., & Choe, S. (1999). Three-finger toxin fold for the extracellular ligand-binding domain of the type II activin receptor serine kinase. *Nature Structural Biology, 6*(1), 18–22. http://dx.doi.org/10.1038/4887.

Greenwald, J., Groppe, J., Gray, P., Wiater, E., Kwiatkowski, W., Vale, W., et al. (2003). The BMP7/ActRII extracellular domain complex provides new insights into the cooperative nature of receptor assembly. *Molecular Cell, 11*(3), 605–617.

Greenwald, J., Vega, M. E., Allendorph, G. P., Fischer, W. H., Vale, W., & Choe, S. (2004). A flexible activin explains the membrane-dependent cooperative assembly of TGF-beta family receptors. *Molecular Cell, 15*(3), 485–489. http://dx.doi.org/10.1016/j.molcel.2004.07.011.

Greiser, J. S., Stross, C., Heinrich, P. C., Behrmann, I., & Hermanns, H. M. (2002). Orientational constraints of the gp130 intracellular juxtamembrane domain for signaling. *The Journal of Biological Chemistry, 277*(30), 26959–26965. http://dx.doi.org/10.1074/jbc.M204113200.

Griffith, D. L., Keck, P. C., Sampath, T. K., Rueger, D. C., & Carlson, W. D. (1996). Three-dimensional structure of recombinant human osteogenic protein 1: Structural paradigm for the transforming growth factor beta superfamily. *Proceedings of the National Academy of Sciences of the United States of America, 93*(2), 878–883.

Gritsman, K., Zhang, J., Cheng, S., Heckscher, E., Talbot, W. S., & Schier, A. F. (1999). The EGF-CFC protein one-eyed pinhead is essential for nodal signaling. *Cell, 97*(1), 121–132.

Groppe, J., Greenwald, J., Wiater, E., Rodriguez-Leon, J., Economides, A. N., Kwiatkowski, W., et al. (2002). Structural basis of BMP signalling inhibition by the cystine knot protein Noggin. *Nature, 420*(6916), 636–642. http://dx.doi.org/10.1038/nature01245.

Groppe, J., Hinck, C. S., Samavarchi-Tehrani, P., Zubieta, C., Schuermann, J. P., Taylor, A. B., et al. (2008). Cooperative assembly of TGF-beta superfamily signaling complexes is mediated by two disparate mechanisms and distinct modes of receptor binding. *Molecular Cell, 29*(2), 157–168. http://dx.doi.org/10.1016/j.molcel.2007.11.039.

Gruber, C. W., Cemazar, M., Anderson, M. A., & Craik, D. J. (2007). Insecticidal plant cyclotides and related cystine knot toxins. *Toxicon, 49*(4), 561–575. http://dx.doi.org/10.1016/j.toxicon.2006.11.018.

Gumienny, T. L., & Savage-Dunn, C. (2013). TGF-beta signaling in C. elegans. *WormBook*, 1–34. http://dx.doi.org/10.1895/wormbook.1.22.2.

Guo, X., & Wang, X. F. (2009). Signaling cross-talk between TGF-beta/BMP and other pathways. *Cell Research, 19*(1), 71–88. http://dx.doi.org/10.1038/cr.2008.302.

Hamada, H., Meno, C., Watanabe, D., & Saijoh, Y. (2002). Establishment of vertebrate left-right asymmetry. *Nature Reviews. Genetics, 3*(2), 103–113. http://dx.doi.org/10.1038/nrg732.

Harrington, A. E., Morris-Triggs, S. A., Ruotolo, B. T., Robinson, C. V., Ohnuma, S., & Hyvonen, M. (2006). Structural basis for the inhibition of activin signalling by follistatin. *The EMBO Journal, 25*(5), 1035–1045. http://dx.doi.org/10.1038/sj.emboj.7601000.

Harrison, C. A., Gray, P. C., Fischer, W. H., Donaldson, C., Choe, S., & Vale, W. (2004). An activin mutant with disrupted ALK4 binding blocks signaling via type II receptors. *The Journal of Biological Chemistry, 279*(27), 28036–28044. http://dx.doi.org/10.1074/jbc.M402782200.

Hart, P. J., Deep, S., Taylor, A. B., Shu, Z., Hinck, C. S., & Hinck, A. P. (2002). Crystal structure of the human TbetaR2 ectodomain—TGF-beta3 complex. *Nature Structural Biology, 9*(3), 203–208. http://dx.doi.org/10.1038/nsb766.

Harth, S. (2010). Molecular recognition in BMP ligand-receptor interactions. (PhD Dissertation), University of Wuerzburg, Wuerzburg. Retrieved from http://nbn-resolving.de/urn:nbn:de:bvb:20-opus-52797.

Harth, S., Kotzsch, A., Hu, J., Sebald, W., & Mueller, T. D. (2010). A selection fit mechanism in BMP receptor IA as a possible source for BMP ligand-receptor promiscuity. *PloS One*, *5*(9), 1–13, http://dx.doi.org/10.1371/journal.pone.0013049.

Hartung, A., Bitton-Worms, K., Rechtman, M. M., Wenzel, V., Boergermann, J. H., Hassel, S., et al. (2006). Different routes of bone morphogenic protein (BMP) receptor endocytosis influence BMP signaling. *Molecular and Cellular Biology*, *26*(20), 7791–7805. http://dx.doi.org/10.1128/MCB.00022-06.

Hassel, S., Eichner, A., Yakymovych, M., Hellman, U., Knaus, P., & Souchelnytskyi, S. (2004). Proteins associated with type II bone morphogenetic protein receptor (BMPR-II) and identified by two-dimensional gel electrophoresis and mass spectrometry. *Proteomics*, *4*(5), 1346–1358. http://dx.doi.org/10.1002/pmic.200300770.

Hassel, S., Yakymovych, M., Hellman, U., Ronnstrand, L., Knaus, P., & Souchelnytskyi, S. (2006). Interaction and functional cooperation between the serine/threonine kinase bone morphogenetic protein type II receptor with the tyrosine kinase stem cell factor receptor. *Journal of Cellular Physiology*, *206*(2), 457–467. http://dx.doi.org/10.1002/jcp.20480.

Hazama, M., Aono, A., Ueno, N., & Fujisawa, Y. (1995). Efficient expression of a heterodimer of bone morphogenetic protein subunits using a baculovirus expression system. *Biochemical and Biophysical Research Communications*, *209*(3), 859–866. http://dx.doi.org/10.1006/bbrc.1995.1578.

He, X., Chow, D., Martick, M. M., & Garcia, K. C. (2001). Allosteric activation of a spring-loaded natriuretic peptide receptor dimer by hormone. *Science*, *293*(5535), 1657–1662. http://dx.doi.org/10.1126/science.1062246.

Hedger, M. P., Winnall, W. R., Phillips, D. J., & de Kretser, D. M. (2011). The regulation and functions of activin and follistatin in inflammation and immunity. *Vitamins and Hormones*, *85*, 255–297. http://dx.doi.org/10.1016/B978-0-12-385961-7.00013-5.

Heinecke, K., Seher, A., Schmitz, W., Mueller, T. D., Sebald, W., & Nickel, J. (2009). Receptor oligomerization and beyond: A case study in bone morphogenetic proteins. *BMC Biology*, 7, 59. http://dx.doi.org/10.1186/1741-7007-7-59.

Helder, M. N., Ozkaynak, E., Sampath, K. T., Luyten, F. P., Latin, V., Oppermann, H., et al. (1995). Expression pattern of osteogenic protein-1 (bone morphogenetic protein-7) in human and mouse development. *The Journal of Histochemistry and Cytochemistry*, *43*(10), 1035–1044.

Hillger, F., Herr, G., Rudolph, R., & Schwarz, E. (2005). Biophysical comparison of BMP-2, ProBMP-2, and the free pro-peptide reveals stabilization of the pro-peptide by the mature growth factor. *The Journal of Biological Chemistry*, *280*(15), 14974–14980. http://dx.doi.org/10.1074/jbc.M414413200.

Hinck, A. P. (2012). Structural studies of the TGF-betas and their receptors—Insights into evolution of the TGF-beta superfamily. *FEBS Letters*, *586*(14), 1860–1870. http://dx.doi.org/10.1016/j.febslet.2012.05.028.

Honda, J., Andou, H., Mannen, T., & Sugimoto, S. (2000). Direct refolding of recombinant human growth differentiation factor 5 for large-scale production process. *Journal of Bioscience and Bioengineering*, *89*(6), 582–589.

Huang, T., Schor, S. L., & Hinck, A. P. (2014). Biological activity differences between TGF-beta1 and TGF-beta3 correlate with differences in the rigidity and arrangement of their component monomers. *Biochemistry*, *53*(36), 5737–5749. http://dx.doi.org/10.1021/bi500647d.

Huse, M., Muir, T. W., Xu, L., Chen, Y. G., Kuriyan, J., & Massague, J. (2001). The TGF beta receptor activation process: An inhibitor- to substrate-binding switch. *Molecular Cell*, *8*(3), 671–682.

Irish, V. F., & Gelbart, W. M. (1987). The decapentaplegic gene is required for dorsal-ventral patterning of the Drosophila embryo. *Genes & Development, 1*(8), 868–879.

Israel, D. I., Nove, J., Kerns, K. M., Kaufman, R. J., Rosen, V., Cox, K. A., et al. (1996). Heterodimeric bone morphogenetic proteins show enhanced activity in vitro and in vivo. *Growth Factors, 13*(3–4), 291–300.

Itoh, F., Asao, H., Sugamura, K., Heldin, C. H., ten Dijke, P., & Itoh, S. (2001). Promoting bone morphogenetic protein signaling through negative regulation of inhibitory Smads. *The EMBO Journal, 20*(15), 4132–4142. http://dx.doi.org/10.1093/emboj/20.15.4132.

Jena, N., Martin-Seisdedos, C., McCue, P., & Croce, C. M. (1997). BMP7 null mutation in mice: Developmental defects in skeleton, kidney, and eye. *Experimental Cell Research, 230*(1), 28–37.

Josso, N. (1990). Anti-Mullerian hormone: Hormone or growth factor? *Progress in Growth Factor Research, 2*(3), 169–179.

Josso, N., di Clemente, N., & Gouedard, L. (2001). Anti-Mullerian hormone and its receptors. *Molecular and Cellular Endocrinology, 179*(1–2), 25–32.

Josso, N., Lamarre, I., Picard, J. Y., Berta, P., Davies, N., Morichon, N., et al. (1993). Anti-mullerian hormone in early human development. *Early Human Development, 33*(2), 91–99.

Josso, N., Racine, C., di Clemente, N., Rey, R., & Xavier, F. (1998). The role of anti-Mullerian hormone in gonadal development. *Molecular and Cellular Endocrinology, 145*(1–2), 3–7.

Kaartinen, V., Voncken, J. W., Shuler, C., Warburton, D., Bu, D., Heisterkamp, N., et al. (1995). Abnormal lung development and cleft palate in mice lacking TGF-beta 3 indicates defects of epithelial-mesenchymal interaction. *Nature Genetics, 11*(4), 415–421. http://dx.doi.org/10.1038/ng1295-415.

Kaivo-Oja, N., Mottershead, D. G., Mazerbourg, S., Myllymaa, S., Duprat, S., Gilchrist, R. B., et al. (2005). Adenoviral gene transfer allows Smad-responsive gene promoter analyses and delineation of type I receptor usage of transforming growth factor-beta family ligands in cultured human granulosa luteal cells. *The Journal of Clinical Endocrinology and Metabolism, 90*(1), 271–278. http://dx.doi.org/10.1210/jc.2004-1288.

Kaps, C., Hoffmann, A., Zilberman, Y., Pelled, G., Haupl, T., Sittinger, M., et al. (2004). Distinct roles of BMP receptors type IA and IB in osteo-/chondrogenic differentiation in mesenchymal progenitors (C3H10T1/2). *Biofactors, 20*(2), 71–84.

Keller, S., Nickel, J., Zhang, J. L., Sebald, W., & Mueller, T. D. (2004). Molecular recognition of BMP-2 and BMP receptor IA. *Nature Structural & Molecular Biology, 11*(5), 481–488. http://dx.doi.org/10.1038/nsmb756.

Keski-Oja, J., Koli, K., & von Melchner, H. (2004). TGF-beta activation by traction? *Trends in Cell Biology, 14*(12), 657–659. http://dx.doi.org/10.1016/j.tcb.2004.10.003.

Kiefer, S. M., & Saling, P. (2002). Proteolytic processing of human zona pellucida proteins. *Biology of Reproduction, 66*(2), 407–414.

Kim, R. Y., Robertson, E. J., & Solloway, M. J. (2001). Bmp6 and Bmp7 are required for cushion formation and septation in the developing mouse heart. *Developmental Biology, 235*(2), 449–466. http://dx.doi.org/10.1006/dbio.2001.0284.

Kimchi, A., Wang, X. F., Weinberg, R. A., Cheifetz, S., & Massague, J. (1988). Absence of TGF-beta receptors and growth inhibitory responses in retinoblastoma cells. *Science, 240*(4849), 196–199.

King, J. A., Marker, P. C., Seung, K. J., & Kingsley, D. M. (1994). BMP5 and the molecular, skeletal, and soft-tissue alterations in short ear mice. *Developmental Biology, 166*(1), 112–122. http://dx.doi.org/10.1006/dbio.1994.1300.

Kingsley, D. M., Bland, A. E., Grubber, J. M., Marker, P. C., Russell, L. B., Copeland, N. G., et al. (1992). The mouse short ear skeletal morphogenesis locus is associated with defects in a bone morphogenetic member of the TGF beta superfamily. *Cell, 71*(3), 399–410.

Kirsch, T., Nickel, J., & Sebald, W. (2000). BMP-2 antagonists emerge from alterations in the low-affinity binding epitope for receptor BMPR-II. *The EMBO Journal, 19*(13), 3314–3324. http://dx.doi.org/10.1093/emboj/19.13.3314.

Kishimoto, Y., Lee, K. H., Zon, L., Hammerschmidt, M., & Schulte-Merker, S. (1997). The molecular nature of zebrafish swirl: BMP2 function is essential during early dorsoventral patterning. *Development, 124*(22), 4457–4466.

Klages, J., Kotzsch, A., Coles, M., Sebald, W., Nickel, J., Muller, T., et al. (2008). The solution structure of BMPR-IA reveals a local disorder-to-order transition upon BMP-2 binding. *Biochemistry, 47*(46), 11930–11939. http://dx.doi.org/10.1021/bi801059j.

Kosaki, K., Bassi, M. T., Kosaki, R., Lewin, M., Belmont, J., Schauer, G., et al. (1999). Characterization and mutation analysis of human LEFTY A and LEFTY B, homologues of murine genes implicated in left-right axis development. *American Journal of Human Genetics, 64*(3), 712–721. http://dx.doi.org/10.1086/302289.

Koster, M., Plessow, S., Clement, J. H., Lorenz, A., Tiedemann, H., & Knochel, W. (1991). Bone morphogenetic protein 4 (BMP-4), a member of the TGF-beta family, in early embryos of Xenopus laevis: Analysis of mesoderm inducing activity. *Mechanisms of Development, 33*(3), 191–199.

Kotzbauer, P. T., Lampe, P. A., Heuckeroth, R. O., Golden, J. P., Creedon, D. J., Johnson, E. M., Jr., et al. (1996). Neurturin, a relative of glial-cell-line-derived neurotrophic factor. *Nature, 384*(6608), 467–470. http://dx.doi.org/10.1038/384467a0.

Kotzsch, A., Nickel, J., Sebald, W., & Mueller, T. D. (2009). Purification, crystallization and preliminary data analysis of ligand-receptor complexes of growth and differentiation factor 5 (GDF5) and BMP receptor IB (BRIB). *Acta Crystallographica. Section F, Structural Biology and Crystallization Communications, 65*(Pt 8), 779–783. http://dx.doi.org/10.1107/S1744309109024142.

Kotzsch, A., Nickel, J., Seher, A., Sebald, W., & Muller, T. D. (2009). Crystal structure analysis reveals a spring-loaded latch as molecular mechanism for GDF-5-type I receptor specificity. *The EMBO Journal, 28*(7), 937–947. http://dx.doi.org/10.1038/emboj.2009.37.

Kuhfahl, S., & Schwarz, E. (2014). A core domain of the BMP2 proregion is sufficient for the biogenesis of mature homodimeric growth factor. *Biological Chemistry, 396*(3), 215–223. http://dx.doi.org/10.1515/hsz-2014-0220.

Kunnapuu, J., Tauscher, P. M., Tiusanen, N., Nguyen, M., Loytynoja, A., Arora, K., et al. (2014). Cleavage of the Drosophila screw prodomain is critical for a dynamic BMP morphogen gradient in embryogenesis. *Developmental Biology, 389*(2), 149–159. http://dx.doi.org/10.1016/j.ydbio.2014.02.007.

Kusumoto, K., Bessho, K., Fujimura, K., Akioka, J., Ogawa, Y., & Iizuka, T. (1997). Comparison of ectopic osteoinduction in vivo by recombinant human BMP-2 and recombinant Xenopus BMP-4/7 heterodimer. *Biochemical and Biophysical Research Communications, 239*(2), 575–579. http://dx.doi.org/10.1006/bbrc.1997.7497.

Kutty, G., Kutty, R. K., Samuel, W., Duncan, T., Jaworski, C., & Wiggert, B. (1998). Identification of a new member of transforming growth factor-beta superfamily in Drosophila: The first invertebrate activin gene. *Biochemical and Biophysical Research Communications, 246*(3), 644–649. http://dx.doi.org/10.1006/bbrc.1998.8678.

Kwong, F. N., Hoyland, J. A., Evans, C. H., & Freemont, A. J. (2009). Regional and cellular localisation of BMPs and their inhibitors' expression in human fractures. *International Orthopaedics, 33*(1), 281–288. http://dx.doi.org/10.1007/s00264-008-0691-z.

Lach-Trifilieff, E., Minetti, G. C., Sheppard, K., Ibebunjo, C., Feige, J. N., Hartmann, S., et al. (2014). An antibody blocking activin type II receptors induces strong skeletal muscle hypertrophy and protects from atrophy. *Molecular and Cellular Biology, 34*(4), 606–618. http://dx.doi.org/10.1128/MCB.01307-13.

Laitinen, M., Vuojolainen, K., Jaatinen, R., Ketola, I., Aaltonen, J., Lehtonen, E., et al. (1998). A novel growth differentiation factor-9 (GDF-9) related factor is co-expressed with GDF-9 in mouse oocytes during folliculogenesis. *Mechanisms of Development, 78*(1–2), 135–140.

Lang, R. A. (2004). Pathways regulating lens induction in the mouse. *The International Journal of Developmental Biology, 48*(8–9), 783–791. http://dx.doi.org/10.1387/ijdb.041903rl.

Laverty, H. G., Wakefield, L. M., Occleston, N. L., O'Kane, S., & Ferguson, M. W. (2009). TGF-beta3 and cancer: A review. *Cytokine & Growth Factor Reviews, 20*(4), 305–317. http://dx.doi.org/10.1016/j.cytogfr.2009.07.002.

Leask, A., & Abraham, D. J. (2004). TGF-beta signaling and the fibrotic response. *The FASEB Journal, 18*(7), 816–827. http://dx.doi.org/10.1096/fj.03-1273rev.

Lee, M. K., Pardoux, C., Hall, M. C., Lee, P. S., Warburton, D., Qing, J., et al. (2007). TGF-beta activates Erk MAP kinase signalling through direct phosphorylation of ShcA. *The EMBO Journal, 26*(17), 3957–3967. http://dx.doi.org/10.1038/sj.emboj.7601818.

Letterio, J. J., & Roberts, A. B. (1996). Transforming growth factor-beta1-deficient mice: Identification of isoform-specific activities in vivo. *Journal of Leukocyte Biology, 59*(6), 769–774.

Lewis, K. A., Gray, P. C., Blount, A. L., MacConell, L. A., Wiater, E., Bilezikjian, L. M., et al. (2000). Betaglycan binds inhibin and can mediate functional antagonism of activin signalling. *Nature, 404*(6776), 411–414. http://dx.doi.org/10.1038/35006129.

Li, J. Z., Li, H., Sasaki, T., Holman, D., Beres, B., Dumont, R. J., et al. (2003). Osteogenic potential of five different recombinant human bone morphogenetic protein adenoviral vectors in the rat. *Gene Therapy, 10*(20), 1735–1743. http://dx.doi.org/10.1038/sj.gt.3302075.

Lin, L. F., Doherty, D. H., Lile, J. D., Bektesh, S., & Collins, F. (1993). GDNF: A glial cell line-derived neurotrophic factor for midbrain dopaminergic neurons. *Science, 260*(5111), 1130–1132.

Ling, N., Ying, S. Y., Ueno, N., Shimasaki, S., Esch, F., Hotta, M., et al. (1986a). A homodimer of the beta-subunits of inhibin A stimulates the secretion of pituitary follicle stimulating hormone. *Biochemical and Biophysical Research Communications, 138*(3), 1129–1137.

Ling, N., Ying, S. Y., Ueno, N., Shimasaki, S., Esch, F., Hotta, M., et al. (1986b). Pituitary FSH is released by a heterodimer of the beta-subunits from the two forms of inhibin. *Nature, 321*(6072), 779–782. http://dx.doi.org/10.1038/321779a0.

Liu, F., Hata, A., Baker, J. C., Doody, J., Carcamo, J., Harland, R. M., et al. (1996). A human Mad protein acting as a BMP-regulated transcriptional activator. *Nature, 381*(6583), 620–623. http://dx.doi.org/10.1038/381620a0.

Livnah, O., Johnson, D. L., Stura, E. A., Farrell, F. X., Barbone, F. P., You, Y., et al. (1998). An antagonist peptide-EPO receptor complex suggests that receptor dimerization is not sufficient for activation. *Nature Structural Biology, 5*(11), 993–1004. http://dx.doi.org/10.1038/2965.

Livnah, O., Stura, E. A., Middleton, S. A., Johnson, D. L., Jolliffe, L. K., & Wilson, I. A. (1999). Crystallographic evidence for preformed dimers of erythropoietin receptor before ligand activation. *Science, 283*(5404), 987–990.

Lo, R. S., Chen, Y. G., Shi, Y., Pavletich, N. P., & Massague, J. (1998). The L3 loop: A structural motif determining specific interactions between SMAD proteins and TGF-beta receptors. *The EMBO Journal, 17*(4), 996–1005. http://dx.doi.org/10.1093/emboj/17.4.996.

Lo, P. C., & Frasch, M. (1999). Sequence and expression of myoglianin, a novel Drosophila gene of the TGF-beta superfamily. *Mechanisms of Development, 86*(1–2), 171–175.

Lorda-Diez, C. I., Montero, J. A., Garcia-Porrero, J. A., & Hurle, J. M. (2010). Tgfbeta2 and 3 are coexpressed with their extracellular regulator Ltbp1 in the early limb bud and modulate mesodermal outgrowth and BMP signaling in chicken embryos. *BMC Developmental Biology, 10*, 69. http://dx.doi.org/10.1186/1471-213X-10-69.

Lotinun, S., Pearsall, R. S., Davies, M. V., Marvell, T. H., Monnell, T. E., Ucran, J., et al. (2010). A soluble activin receptor type IIA fusion protein (ACE-011) increases bone mass via a dual anabolic-antiresorptive effect in Cynomolgus monkeys. *Bone, 46*(4), 1082–1088. http://dx.doi.org/10.1016/j.bone.2010.01.370.

Lotinun, S., Pearsall, R. S., Horne, W. C., & Baron, R. (2012). Activin receptor signaling: A potential therapeutic target for osteoporosis. *Current Molecular Pharmacology, 5*(2), 195–204.

Luo, G., Hofmann, C., Bronckers, A. L., Sohocki, M., Bradley, A., & Karsenty, G. (1995). BMP-7 is an inducer of nephrogenesis, and is also required for eye development and skeletal patterning. *Genes & Development, 9*(22), 2808–2820.

Luyten, F. P., Cunningham, N. S., Ma, S., Muthukumaran, N., Hammonds, R. G., Nevins, W. B., et al. (1989). Purification and partial amino acid sequence of osteogenin, a protein initiating bone differentiation. *The Journal of Biological Chemistry, 264*(23), 13377–13380.

Mace, P. D., Cutfield, J. F., & Cutfield, S. M. (2006). High resolution structures of the bone morphogenetic protein type II receptor in two crystal forms: Implications for ligand binding. *Biochemical and Biophysical Research Communications, 351*(4), 831–838. http://dx.doi.org/10.1016/j.bbrc.2006.10.109.

Mahlawat, P., Ilangovan, U., Biswas, T., Sun, L. Z., & Hinck, A. P. (2012). Structure of the Alk1 extracellular domain and characterization of its bone morphogenetic protein (BMP) binding properties. *Biochemistry, 51*(32), 6328–6341. http://dx.doi.org/10.1021/bi300942x.

Martens, J. W., de Winter, J. P., Timmerman, M. A., McLuskey, A., van Schaik, R. H., Themmen, A. P., et al. (1997). Inhibin interferes with activin signaling at the level of the activin receptor complex in Chinese hamster ovary cells. *Endocrinology, 138*(7), 2928–2936. http://dx.doi.org/10.1210/endo.138.7.5250.

Massague, J. (2000). How cells read TGF-beta signals. *Nature Reviews. Molecular Cell Biology, 1*(3), 169–178. http://dx.doi.org/10.1038/35043051.

Massague, J., & Wotton, D. (2000). Transcriptional control by the TGF-beta/Smad signaling system. *The EMBO Journal, 19*(8), 1745–1754. http://dx.doi.org/10.1093/emboj/19.8.1745.

Mazerbourg, S., Klein, C., Roh, J., Kaivo-Oja, N., Mottershead, D. G., Korchynskyi, O., et al. (2004). Growth differentiation factor-9 signaling is mediated by the type I receptor, activin receptor-like kinase 5. *Molecular Endocrinology, 18*(3), 653–665. http://dx.doi.org/10.1210/me.2003-0393.

McNatty, K. P., Juengel, J. L., Reader, K. L., Lun, S., Myllymaa, S., Lawrence, S. B., et al. (2005a). Bone morphogenetic protein 15 and growth differentiation factor 9 co-operate to regulate granulosa cell function. *Reproduction, 129*(4), 473–480. http://dx.doi.org/10.1530/rep.1.0511.

McNatty, K. P., Juengel, J. L., Reader, K. L., Lun, S., Myllymaa, S., Lawrence, S. B., et al. (2005b). Bone morphogenetic protein 15 and growth differentiation factor 9 co-operate to regulate granulosa cell function in ruminants. *Reproduction, 129*(4), 481–487. http://dx.doi.org/10.1530/rep.1.00517.

McPherron, A. C., Lawler, A. M., & Lee, S. J. (1997). Regulation of skeletal muscle mass in mice by a new TGF-beta superfamily member. *Nature, 387*(6628), 83–90. http://dx.doi.org/10.1038/387083a0.

McPherron, A. C., Lawler, A. M., & Lee, S. J. (1999). Regulation of anterior/posterior patterning of the axial skeleton by growth/differentiation factor 11. *Nature Genetics, 22*(3), 260–264. http://dx.doi.org/10.1038/10320.

McPherron, A. C., & Lee, S. J. (1997). Double muscling in cattle due to mutations in the myostatin gene. *Proceedings of the National Academy of Sciences of the United States of America, 94*(23), 12457–12461.

Mercola, M. (2003). Left-right asymmetry: Nodal points. *Journal of Cell Science, 116*(Pt 16), 3251–3257. http://dx.doi.org/10.1242/jcs.00668.

Merwin, J. R., Newman, W., Beall, L. D., Tucker, A., & Madri, J. (1991). Vascular cells respond differentially to transforming growth factors beta 1 and beta 2 in vitro. *The American Journal of Pathology, 138*(1), 37–51.

Mi, L. Z., Brown, C. T., Gao, Y., Tian, Y., Le, V. Q., Walz, T., et al. (2015). Structure of bone morphogenetic protein 9 procomplex. *Proceedings of the National Academy of Sciences of the United States of America, 112*(12), 3710–3715. http://dx.doi.org/10.1073/pnas.1501303112.

Mickiewicz, A. L., & Kordower, J. H. (2011). GDNF family ligands: A potential future for Parkinson's disease therapy. *CNS & Neurological Disorders Drug Targets, 10*(6), 703–711.

Mikic, B., Entwistle, R., Rossmeier, K., & Bierwert, L. (2008). Effect of GDF-7 deficiency on tail tendon phenotype in mice. *Journal of Orthopaedic Research, 26*(6), 834–839. http://dx.doi.org/10.1002/jor.20581.

Mikic, B., Rossmeier, K., & Bierwert, L. (2009). Identification of a tendon phenotype in GDF6 deficient mice. *The anatomical record: Advances in integrative anatomy and evolutionary biology, 292*(3), 396–400. http://dx.doi.org/10.1002/ar.20852.

Milbrandt, J., de Sauvage, F. J., Fahrner, T. J., Baloh, R. H., Leitner, M. L., Tansey, M. G., et al. (1998). Persephin, a novel neurotrophic factor related to GDNF and neurturin. *Neuron, 20*(2), 245–253.

Molloy, S. S., Thomas, L., VanSlyke, J. K., Stenberg, P. E., & Thomas, G. (1994). Intracellular trafficking and activation of the furin proprotein convertase: Localization to the TGN and recycling from the cell surface. *The EMBO Journal, 13*(1), 18–33.

Moore, R. K., Otsuka, F., & Shimasaki, S. (2003). Molecular basis of bone morphogenetic protein-15 signaling in granulosa cells. *The Journal of Biological Chemistry, 278*(1), 304–310. http://dx.doi.org/10.1074/jbc.M207362200.

Mottershead, D. G., Harrison, C. A., Mueller, T. D., Stanton, P. G., Gilchrist, R. B., & McNatty, K. P. (2013). Growth differentiation factor 9:bone morphogenetic protein 15 (GDF9:BMP15) synergism and protein heterodimerization. *Proceedings of the National Academy of Sciences of the United States of America, 110*(25), E2257. http://dx.doi.org/10.1073/pnas.1303459110.

Mottershead, D. G., Pulkki, M. M., Muggalla, P., Pasternack, A., Tolonen, M., Myllymaa, S., et al. (2008). Characterization of recombinant human growth differentiation factor-9 signaling in ovarian granulosa cells. *Molecular and Cellular Endocrinology, 283*(1–2), 58–67. http://dx.doi.org/10.1016/j.mce.2007.11.007.

Mueller, T. D., & Nickel, J. (2012). Promiscuity and specificity in BMP receptor activation. *FEBS Letters, 586*(14), 1846–1859. http://dx.doi.org/10.1016/j.febslet.2012.02.043.

Muenster, U., Harrison, C. A., Donaldson, C., Vale, W., & Fischer, W. H. (2005). An activin-A/C chimera exhibits activin and myostatin antagonistic properties. *The Journal of Biological Chemistry, 280*(44), 36626–36632. http://dx.doi.org/10.1074/jbc.M507236200.

Nakamura, K., Shirai, T., Morishita, S., Uchida, S., Saeki-Miura, K., & Makishima, F. (1999). p38 mitogen-activated protein kinase functionally contributes to chondrogenesis induced by growth/differentiation factor-5 in ATDC5 cells. *Experimental Cell Research, 250*(2), 351–363. http://dx.doi.org/10.1006/excr.1999.4535.

Namiki, M., Akiyama, S., Katagiri, T., Suzuki, A., Ueno, N., Yamaji, N., et al. (1997). A kinase domain-truncated type I receptor blocks bone morphogenetic protein-2-induced signal transduction in C2C12 myoblasts. *The Journal of Biological Chemistry, 272*(35), 22046–22052.

Nawshad, A., Lagamba, D., Polad, A., & Hay, E. D. (2005). Transforming growth factor-beta signaling during epithelial-mesenchymal transformation: Implications for embryogenesis and tumor metastasis. *Cells, Tissues, Organs, 179*(1–2), 11–23. http://dx.doi.org/10.1159/000084505.

Neave, B., Holder, N., & Patient, R. (1997). A graded response to BMP-4 spatially coordinates patterning of the mesoderm and ectoderm in the zebrafish. *Mechanisms of Development, 62*(2), 183–195.

Nelsen, S. M., & Christian, J. L. (2009). Site-specific cleavage of BMP4 by furin, PC6, and PC7. *The Journal of Biological Chemistry, 284*(40), 27157–27166. http://dx.doi.org/10.1074/jbc.M109.028506.

Newfeld, S. J., Wisotzkey, R. G., & Kumar, S. (1999). Molecular evolution of a developmental pathway: Phylogenetic analyses of transforming growth factor-beta family ligands, receptors and Smad signal transducers. *Genetics, 152*(2), 783–795.

Nguyen, M., Parker, L., & Arora, K. (2000). Identification of Maverick, a novel member of the TGF-beta superfamily in Drosophila. *Mechanisms of Development, 95*(1–2), 201–206.

Nicholls, P. K., Harrison, C. A., Gilchrist, R. B., Farnworth, P. G., & Stanton, P. G. (2009). Growth differentiation factor 9 is a germ cell regulator of Sertoli cell function. *Endocrinology, 150*(5), 2481–2490. http://dx.doi.org/10.1210/en.2008-1048.

Nickel, J., Kotzsch, A., Sebald, W., & Mueller, T. D. (2005). A single residue of GDF-5 defines binding specificity to BMP receptor IB. *Journal of Molecular Biology, 349*(5), 933–947. http://dx.doi.org/10.1016/j.jmb.2005.04.015.

Nicks, K. M., Perrien, D. S., Akel, N. S., Suva, L. J., & Gaddy, D. (2009). Regulation of osteoblastogenesis and osteoclastogenesis by the other reproductive hormones, Activin and Inhibin. *Molecular and Cellular Endocrinology, 310*(1–2), 11–20. http://dx.doi.org/10.1016/j.mce.2009.07.001.

Nikaido, M., Tada, M., Saji, T., & Ueno, N. (1997). Conservation of BMP signaling in zebrafish mesoderm patterning. *Mechanisms of Development, 61*(1–2), 75–88.

Nishimatsu, S., & Thomsen, G. H. (1998). Ventral mesoderm induction and patterning by bone morphogenetic protein heterodimers in Xenopus embryos. *Mechanisms of Development, 74*(1–2), 75–88.

Nishitoh, H., Ichijo, H., Kimura, M., Matsumoto, T., Makishima, F., Yamaguchi, A., et al. (1996). Identification of type I and type II serine/threonine kinase receptors for growth/differentiation factor-5. *The Journal of Biological Chemistry, 271*(35), 21345–21352.

Nohe, A., Hassel, S., Ehrlich, M., Neubauer, F., Sebald, W., Henis, Y. I., et al. (2002). The mode of bone morphogenetic protein (BMP) receptor oligomerization determines different BMP-2 signaling pathways. *The Journal of Biological Chemistry, 277*(7), 5330–5338. http://dx.doi.org/10.1074/jbc.M102750200.

Nolan, K., Kattamuri, C., Luedeke, D. M., Deng, X., Jagpal, A., Zhang, F., et al. (2013). Structure of protein related to Dan and Cerberus: Insights into the mechanism of bone morphogenetic protein antagonism. *Structure, 21*(8), 1417–1429. http://dx.doi.org/10.1016/j.str.2013.06.005.

O'Connor, M. B., Umulis, D., Othmer, H. G., & Blair, S. S. (2006). Shaping BMP morphogen gradients in the Drosophila embryo and pupal wing. *Development, 133*(2), 183–193. http://dx.doi.org/10.1242/dev.02214.

Ogawa, H., Qiu, Y., Ogata, C. M., & Misono, K. S. (2004). Crystal structure of hormone-bound atrial natriuretic peptide receptor extracellular domain: Rotation mechanism for transmembrane signal transduction. *The Journal of Biological Chemistry, 279*(27), 28625–28631. http://dx.doi.org/10.1074/jbc.M313222200.

Oh, S. P., Seki, T., Goss, K. A., Imamura, T., Yi, Y., Donahoe, P. K., et al. (2000). Activin receptor-like kinase 1 modulates transforming growth factor-beta 1 signaling in the regulation of angiogenesis. *Proceedings of the National Academy of Sciences of the United States of America, 97*(6), 2626–2631.

Okada, Y., Nonaka, S., Tanaka, Y., Saijoh, Y., Hamada, H., & Hirokawa, N. (1999). Abnormal nodal flow precedes situs inversus in iv and inv mice. *Molecular Cell, 4*(4), 459–468.

Ottemann, K. M., Xiao, W., Shin, Y. K., & Koshland, D. E., Jr. (1999). A piston model for transmembrane signaling of the aspartate receptor. *Science, 285*(5434), 1751–1754.

Parker, L., Ellis, J. E., Nguyen, M. Q., & Arora, K. (2006). The divergent TGF-beta ligand Dawdle utilizes an activin pathway to influence axon guidance in Drosophila. *Development, 133*(24), 4981–4991. http://dx.doi.org/10.1242/dev.02673.

Pauklin, S., & Vallier, L. (2015). Activin/Nodal signalling in stem cells. *Development, 142*(4), 607–619. http://dx.doi.org/10.1242/dev.091769.

Pearce, K. H., Jr., Cunningham, B. C., Fuh, G., Teeri, T., & Wells, J. A. (1999). Growth hormone binding affinity for its receptor surpasses the requirements for cellular activity. *Biochemistry, 38*(1), 81–89. http://dx.doi.org/10.1021/bi9817008.

Pearsall, R. S., Canalis, E., Cornwall-Brady, M., Underwood, K. W., Haigis, B., Ucran, J., et al. (2008). A soluble activin type IIA receptor induces bone formation and improves skeletal integrity. *Proceedings of the National Academy of Sciences of the United States of America, 105*(19), 7082–7087. http://dx.doi.org/10.1073/pnas.0711263105.

Pelton, R. W., Johnson, M. D., Perkett, E. A., Gold, L. I., & Moses, H. L. (1991). Expression of transforming growth factor-beta 1, -beta 2, and -beta 3 mRNA and protein in the murine lung. *American Journal of Respiratory Cell and Molecular Biology, 5*(6), 522–530. http://dx.doi.org/10.1165/ajrcmb/5.6.522.

Peng, J., Li, Q., Wigglesworth, K., Rangarajan, A., Kattamuri, C., Peterson, R. T., et al. (2013). Growth differentiation factor 9:bone morphogenetic protein 15 heterodimers are potent regulators of ovarian functions. *Proceedings of the National Academy of Sciences of the United States of America, 110*(8), E776–E785. http://dx.doi.org/10.1073/pnas.1218020110.

Peng, C., & Mukai, S. T. (2000). Activins and their receptors in female reproduction. *Biochemistry and Cell Biology, 78*(3), 261–279.

Perr, H. A., Ye, J., & Gitelman, S. E. (1999). Smooth muscle expresses bone morphogenetic protein (Vgr-1/BMP-6) in human fetal intestine. *Biology of the Neonate, 75*(3), 210–214. http://dx.doi.org/10.1159/000014097.

Phillips, D. J., de Kretser, D. M., & Hedger, M. P. (2009). Activin and related proteins in inflammation: Not just interested bystanders. *Cytokine & Growth Factor Reviews, 20*(2), 153–164. http://dx.doi.org/10.1016/j.cytogfr.2009.02.007.

Piek, E., Afrakhte, M., Sampath, K., van Zoelen, E. J., Heldin, C. H., & ten Dijke, P. (1999). Functional antagonism between activin and osteogenic protein-1 in human embryonal carcinoma cells. *Journal of Cellular Physiology, 180*(2), 141–149. http://dx.doi.org/10.1002/(SICI)1097-4652(199908)180:2<141::AID-JCP1>3.0.CO;2-I.

Pizette, S., & Niswander, L. (1999). BMPs negatively regulate structure and function of the limb apical ectodermal ridge. *Development, 126*(5), 883–894.

Poorgholi Belverdi, M., Krause, C., Guzman, A., & Knaus, P. (2012). Comprehensive analysis of TGF-beta and BMP receptor interactomes. *European Journal of Cell Biology, 91*(4), 287–293. http://dx.doi.org/10.1016/j.ejcb.2011.05.004.

Pulkki, M. M., Myllymaa, S., Pasternack, A., Lun, S., Ludlow, H., Al-Qahtani, A., et al. (2011). The bioactivity of human bone morphogenetic protein-15 is sensitive to C-terminal modification: Characterization of the purified untagged processed mature region. *Molecular and Cellular Endocrinology, 332*(1–2), 106–115. http://dx.doi.org/10.1016/j.mce.2010.10.002.

Rebbapragada, A., Benchabane, H., Wrana, J. L., Celeste, A. J., & Attisano, L. (2003). Myostatin signals through a transforming growth factor beta-like signaling pathway to block adipogenesis. *Molecular and Cellular Biology, 23*(20), 7230–7242.

Reddi, A. H. (2005). BMPs: From bone morphogenetic proteins to body morphogenetic proteins. *Cytokine & Growth Factor Reviews, 16*(3), 249–250. http://dx.doi.org/10.1016/j.cytogfr.2005.04.003.

Reissmann, E., Jornvall, H., Blokzijl, A., Andersson, O., Chang, C., Minchiotti, G., et al. (2001). The orphan receptor ALK7 and the activin receptor ALK4 mediate signaling by Nodal proteins during vertebrate development. *Genes & Development, 15*(15), 2010–2022. http://dx.doi.org/10.1101/gad.201801.

Roberts, A. B., Kondaiah, P., Rosa, F., Watanabe, S., Good, P., Danielpour, D., et al. (1990). Mesoderm induction in Xenopus laevis distinguishes between the various TGF-beta isoforms. *Growth Factors, 3*(4), 277–286.

Roberts, A. B., Sporn, M. B., Assoian, R. K., Smith, J. M., Roche, N. S., Wakefield, L. M., et al. (1986). Transforming growth factor type beta: Rapid induction of fibrosis and angiogenesis in vivo and stimulation of collagen formation in vitro. *Proceedings of the National Academy of Sciences of the United States of America, 83*(12), 4167–4171.

Sahni, V., Mukhopadhyay, A., Tysseling, V., Hebert, A., Birch, D., McGuire, T. L., et al. (2010). BMPR1a and BMPR1b signaling exert opposing effects on gliosis after spinal cord injury. *The Journal of Neuroscience, 30*(5), 1839–1855. http://dx.doi.org/10.1523/JNEUROSCI.4459-09.2010.

Saitoh, M., Nishitoh, H., Amagasa, T., Miyazono, K., Takagi, M., & Ichijo, H. (1996). Identification of important regions in the cytoplasmic juxtamembrane domain of type I receptor that separate signaling pathways of transforming growth factor-beta. *The Journal of Biological Chemistry, 271*(5), 2769–2775.

Sako, D., Grinberg, A. V., Liu, J., Davies, M. V., Castonguay, R., Maniatis, S., et al. (2010). Characterization of the ligand binding functionality of the extracellular domain of activin receptor type IIb. *The Journal of Biological Chemistry, 285*(27), 21037–21048. http://dx.doi.org/10.1074/jbc.M110.114959.

Sakuma, R., Ohnishi Yi, Y., Meno, C., Fujii, H., Juan, H., Takeuchi, J., et al. (2002). Inhibition of Nodal signalling by Lefty mediated through interaction with common receptors and efficient diffusion. *Genes to Cells*, 7(4), 401–412.

Samad, T. A., Rebbapragada, A., Bell, E., Zhang, Y., Sidis, Y., Jeong, S. J., et al. (2005). DRAGON, a bone morphogenetic protein co-receptor. *The Journal of Biological Chemistry, 280*(14), 14122–14129. http://dx.doi.org/10.1074/jbc.M410034200.

Sampath, T. K., Coughlin, J. E., Whetstone, R. M., Banach, D., Corbett, C., Ridge, R. J., et al. (1990). Bovine osteogenic protein is composed of dimers of OP-1 and BMP-2A, two members of the transforming growth factor-beta superfamily. *The Journal of Biological Chemistry, 265*(22), 13198–13205.

Sampath, T. K., & Reddi, A. H. (1981). Dissociative extraction and reconstitution of extracellular matrix components involved in local bone differentiation. *Proceedings of the National Academy of Sciences of the United States of America, 78*(12), 7599–7603.

Sampath, K., Rubinstein, A. L., Cheng, A. M., Liang, J. O., Fekany, K., Solnica-Krezel, L., et al. (1998). Induction of the zebrafish ventral brain and floorplate requires cyclops/nodal signalling. *Nature, 395*(6698), 185–189. http://dx.doi.org/10.1038/26020.

Sanford, L. P., Ormsby, I., Gittenberger-de Groot, A. C., Sariola, H., Friedman, R., Boivin, G. P., et al. (1997). TGFbeta2 knockout mice have multiple developmental defects that are non-overlapping with other TGFbeta knockout phenotypes. *Development, 124*(13), 2659–2670.

Saremba, S., Nickel, J., Seher, A., Kotzsch, A., Sebald, W., & Mueller, T. D. (2008). Type I receptor binding of bone morphogenetic protein 6 is dependent on N-glycosylation of the ligand. *The FEBS Journal, 275*(1), 172–183. http://dx.doi.org/10.1111/j.1742-4658.2007.06187.x.

Sariola, H., & Saarma, M. (2003). Novel functions and signalling pathways for GDNF. *Journal of Cell Science, 116*(Pt 19), 3855–3862. http://dx.doi.org/10.1242/jcs.00786.

Savage, C., Das, P., Finelli, A. L., Townsend, S. R., Sun, C. Y., Baird, S. E., et al. (1996). Caenorhabditis elegans genes sma-2, sma-3, and sma-4 define a conserved family of transforming growth factor beta pathway components. *Proceedings of the National Academy of Sciences of the United States of America, 93*(2), 790–794.

Savage-Dunn, C. (2005). TGF-beta signaling. *WormBook*, 1–12. http://dx.doi.org/10.1895/wormbook.1.22.1.

Scheufler, C., Sebald, W., & Hulsmeyer, M. (1999). Crystal structure of human bone morphogenetic protein-2 at 2.7 A resolution. *Journal of Molecular Biology, 287*(1), 103–115. http://dx.doi.org/10.1006/jmbi.1999.2590.

Schmid, B., Furthauer, M., Connors, S. A., Trout, J., Thisse, B., Thisse, C., et al. (2000). Equivalent genetic roles for bmp7/snailhouse and bmp2b/swirl in dorsoventral pattern formation. *Development, 127*(5), 957–967.

Schmidt, J. E., Suzuki, A., Ueno, N., & Kimelman, D. (1995). Localized BMP-4 mediates dorsal/ventral patterning in the early Xenopus embryo. *Developmental Biology, 169*(1), 37–50. http://dx.doi.org/10.1006/dbio.1995.1124.

Schramm, C., Huber, S., Protschka, M., Czochra, P., Burg, J., Schmitt, E., et al. (2004). TGFbeta regulates the CD4+CD25+ T-cell pool and the expression of Foxp3 in vivo. *International Immunology, 16*(9), 1241–1249. http://dx.doi.org/10.1093/intimm/dxh126.

Schreuder, H., Liesum, A., Pohl, J., Kruse, M., & Koyama, M. (2005). Crystal structure of recombinant human growth and differentiation factor 5: Evidence for interaction of the type I and type II receptor-binding sites. *Biochemical and Biophysical Research Communications, 329*(3), 1076–1086. http://dx.doi.org/10.1016/j.bbrc.2005.02.078.

Schuelke, M., Wagner, K. R., Stolz, L. E., Hubner, C., Riebel, T., Komen, W., et al. (2004). Myostatin mutation associated with gross muscle hypertrophy in a child. *The New England Journal of Medicine, 350*(26), 2682–2688. http://dx.doi.org/10.1056/NEJMoa040933.

Seemann, P., Schwappacher, R., Kjaer, K. W., Krakow, D., Lehmann, K., Dawson, K., et al. (2005). Activating and deactivating mutations in the receptor interaction site of GDF5 cause symphalangism or brachydactyly type A2. *The Journal of Clinical Investigation, 115*(9), 2373–2381. http://dx.doi.org/10.1172/JCI25118.

Seidah, N. G., Mayer, G., Zaid, A., Rousselet, E., Nassoury, N., Poirier, S., et al. (2008). The activation and physiological functions of the proprotein convertases. *The International Journal of Biochemistry & Cell Biology, 40*(6–7), 1111–1125. http://dx.doi.org/10.1016/j.biocel.2008.01.030.

Sengle, G., Charbonneau, N. L., Ono, R. N., Sasaki, T., Alvarez, J., Keene, D. R., et al. (2008). Targeting of bone morphogenetic protein growth factor complexes to fibrillin. *The Journal of Biological Chemistry, 283*(20), 13874–13888. http://dx.doi.org/10.1074/jbc.M707820200.

Sengle, G., Ono, R. N., Sasaki, T., & Sakai, L. Y. (2011). Prodomains of transforming growth factor beta (TGFbeta) superfamily members specify different functions: Extracellular matrix interactions and growth factor bioavailability. *The Journal of Biological Chemistry, 286*(7), 5087–5099. http://dx.doi.org/10.1074/jbc.M110.188615.

Serpe, M., & O'Connor, M. B. (2006). The metalloprotease tolloid-related and its TGF-beta-like substrate Dawdle regulate Drosophila motoneuron axon guidance. *Development, 133*(24), 4969–4979. http://dx.doi.org/10.1242/dev.02711.

Settle, S. H., Jr., Rountree, R. B., Sinha, A., Thacker, A., Higgins, K., & Kingsley, D. M. (2003). Multiple joint and skeletal patterning defects caused by single and double mutations in the mouse Gdf6 and Gdf5 genes. *Developmental Biology*, *254*(1), 116–130.

Shi, M., Zhu, J., Wang, R., Chen, X., Mi, L., Walz, T., et al. (2011). Latent TGF-beta structure and activation. *Nature*, *474*(7351), 343–349. http://dx.doi.org/10.1038/nature10152.

Shimmi, O., Umulis, D., Othmer, H., & O'Connor, M. B. (2005). Facilitated transport of a Dpp/Scw heterodimer by Sog/Tsg leads to robust patterning of the Drosophila blastoderm embryo. *Cell*, *120*(6), 873–886. http://dx.doi.org/10.1016/j.cell.2005.02.009.

Shnaper, S., Desbaillets, I., Brown, D. A., Murat, A., Migliavacca, E., Schluep, M., et al. (2009). Elevated levels of MIC-1/GDF15 in the cerebrospinal fluid of patients are associated with glioblastoma and worse outcome. *International Journal of Cancer*, *125*(11), 2624–2630. http://dx.doi.org/10.1002/ijc.24639.

Shull, M. M., Ormsby, I., Kier, A. B., Pawlowski, S., Diebold, R. J., Yin, M., et al. (1992). Targeted disruption of the mouse transforming growth factor-beta 1 gene results in multifocal inflammatory disease. *Nature*, *359*(6397), 693–699. http://dx.doi.org/10.1038/359693a0.

Smeland, E. B., Blomhoff, H. K., Holte, H., Ruud, E., Beiske, K., Funderud, S., et al. (1987). Transforming growth factor type beta (TGF beta) inhibits G1 to S transition, but not activation of human B lymphocytes. *Experimental Cell Research*, *171*(1), 213–222.

Solloway, M. J., Dudley, A. T., Bikoff, E. K., Lyons, K. M., Hogan, B. L., & Robertson, E. J. (1998). Mice lacking Bmp6 function. *Developmental Genetics*, *22*(4), 321–339. http://dx.doi.org/10.1002/(SICI)1520-6408(1998)22:4<321::AID-DVG3>3.0.CO;2-8.

Solursh, M., Langille, R. M., Wood, J., & Sampath, T. K. (1996). Osteogenic protein-1 is required for mammalian eye development. *Biochemical and Biophysical Research Communications*, *218*(2), 438–443.

Sparrow, D. B., Kotecha, S., Towers, N., & Mohun, T. J. (1998). Xenopus eHAND: A marker for the developing cardiovascular system of the embryo that is regulated by bone morphogenetic proteins. *Mechanisms of Development*, *71*(1–2), 151–163.

Stamler, R., Keutmann, H. T., Sidis, Y., Kattamuri, C., Schneyer, A., & Thompson, T. B. (2008). The structure of FSTL3.activin A complex. Differential binding of N-terminal domains influences follistatin-type antagonist specificity. *The Journal of Biological Chemistry*, *283*(47), 32831–32838. http://dx.doi.org/10.1074/jbc.M801266200.

Storm, E. E., Huynh, T. V., Copeland, N. G., Jenkins, N. A., Kingsley, D. M., & Lee, S. J. (1994). Limb alterations in brachypodism mice due to mutations in a new member of the TGF beta-superfamily. *Nature*, *368*(6472), 639–643. http://dx.doi.org/10.1038/368639a0.

Strelau, J., Schober, A., Sullivan, A., Schilling, L., & Unsicker, K. (2003). Growth/differentiation factor-15 (GDF-15), a novel member of the TGF-beta superfamily, promotes survival of lesioned mesencephalic dopaminergic neurons in vitro and in vivo and is induced in neurons following cortical lesioning. *Journal of Neural Transmission. Supplementum*, *65*, 197–203.

Strelau, J., Sullivan, A., Bottner, M., Lingor, P., Falkenstein, E., Suter-Crazzolara, C., et al. (2000). Growth/differentiation factor-15/macrophage inhibitory cytokine-1 is a novel trophic factor for midbrain dopaminergic neurons in vivo. *The Journal of Neuroscience*, *20*(23), 8597–8603.

Sun, P. D., & Davies, D. R. (1995). The cystine-knot growth-factor superfamily. *Annual Review of Biophysics and Biomolecular Structure*, *24*, 269–291. http://dx.doi.org/10.1146/annurev.bb.24.060195.001413.

Swencki-Underwood, B., Mills, J. K., Vennarini, J., Boakye, K., Luo, J., Pomerantz, S., et al. (2008). Expression and characterization of a human BMP-7 variant with improved biochemical properties. *Protein Expression and Purification, 57*(2), 312–319. http://dx.doi.org/10.1016/j.pep.2007.09.016.

Thisse, B., Wright, C. V. E., & Thisse, C. (2000). Activin- and Nodal-related factors control antero-posterior patterning of the zebrafish embryo. *Nature, 403*(6768), 425–428. http://dx.doi.org/10.1038/35000200.

Thompson, T. B., Lerch, T. F., Cook, R. W., Woodruff, T. K., & Jardetzky, T. S. (2005). The structure of the follistatin:Activin complex reveals antagonism of both type I and type II receptor binding. *Developmental Cell, 9*(4), 535–543. http://dx.doi.org/10.1016/j.devcel.2005.09.008.

Thompson, T. B., Woodruff, T. K., & Jardetzky, T. S. (2003). Structures of an ActRIIB: Activin A complex reveal a novel binding mode for TGF-beta ligand:receptor interactions. *The EMBO Journal, 22*(7), 1555–1566. http://dx.doi.org/10.1093/emboj/cdg156.

Ting, C. Y., Herman, T., Yonekura, S., Gao, S., Wang, J., Serpe, M., et al. (2007). Tiling of r7 axons in the Drosophila visual system is mediated both by transduction of an activin signal to the nucleus and by mutual repulsion. *Neuron, 56*(5), 793–806. http://dx.doi.org/10.1016/j.neuron.2007.09.033.

Tovar, Y. R. L. B., Ramirez-Jarquin, U. N., Lazo-Gomez, R., & Tapia, R. (2014). Trophic factors as modulators of motor neuron physiology and survival: Implications for ALS therapy. *Frontiers in Cellular Neuroscience, 8*, 61. http://dx.doi.org/10.3389/fncel.2014.00061.

Townson, S. A., Martinez-Hackert, E., Greppi, C., Lowden, P., Sako, D., Liu, J., et al. (2012). Specificity and structure of a high affinity activin receptor-like kinase 1 (ALK1) signaling complex. *The Journal of Biological Chemistry, 287*(33), 27313–27325. http://dx.doi.org/10.1074/jbc.M112.377960.

Vale, W., Rivier, C., Hsueh, A., Campen, C., Meunier, H., Bicsak, T., et al. (1988). Chemical and biological characterization of the inhibin family of protein hormones. *Recent Progress in Hormone Research, 44*, 1–34.

Vallier, L., Reynolds, D., & Pedersen, R. A. (2004). Nodal inhibits differentiation of human embryonic stem cells along the neuroectodermal default pathway. *Developmental Biology, 275*(2), 403–421. http://dx.doi.org/10.1016/j.ydbio.2004.08.031.

Veverka, V., Henry, A. J., Slocombe, P. M., Ventom, A., Mulloy, B., Muskett, F. W., et al. (2009). Characterization of the structural features and interactions of sclerostin: Molecular insight into a key regulator of Wnt-mediated bone formation. *The Journal of Biological Chemistry, 284*(16), 10890–10900. http://dx.doi.org/10.1074/jbc.M807994200.

Vitt, U. A., Mazerbourg, S., Klein, C., & Hsueh, A. J. (2002). Bone morphogenetic protein receptor type II is a receptor for growth differentiation factor-9. *Biology of Reproduction, 67*(2), 473–480.

von Einem, S., Schwarz, E., & Rudolph, R. (2010). A novel TWO-STEP renaturation procedure for efficient production of recombinant BMP-2. *Protein Expression and Purification, 73*(1), 65–69. http://dx.doi.org/10.1016/j.pep.2010.03.009.

Wakefield, L. M., Smith, D. M., Broz, S., Jackson, M., Levinson, A. D., & Sporn, M. B. (1989). Recombinant TGF-beta 1 is synthesized as a two-component latent complex that shares some structural features with the native platelet latent TGF-beta 1 complex. *Growth Factors, 1*(3), 203–218.

Wang, X. (2013). Structural studies of GDNF family ligands with their receptors-Insights into ligand recognition and activation of receptor tyrosine kinase RET. *Biochimica et Biophysica Acta, 1834*(10), 2205–2212. http://dx.doi.org/10.1016/j.bbapap.2012.10.008.

Wang, E. A., Rosen, V., Cordes, P., Hewick, R. M., Kriz, M. J., Luxenberg, D. P., et al. (1988). Purification and characterization of other distinct bone-inducing factors. *Proceedings of the National Academy of Sciences of the United States of America, 85*(24), 9484–9488.

Weber, D., Kotzsch, A., Nickel, J., Harth, S., Seher, A., Mueller, U., et al. (2007). A silent H-bond can be mutationally activated for high-affinity interaction of BMP-2 and activin type IIB receptor. *BMC Structural Biology*, 7, 6. http://dx.doi.org/10.1186/1472-6807-7-6.

Wei, Z., Salmon, R. M., Upton, P. D., Morrell, N. W., & Li, W. (2014). Regulation of bone morphogenetic protein 9 (BMP9) by redox-dependent proteolysis. *The Journal of Biological Chemistry*, *289*(45), 31150–31159. http://dx.doi.org/10.1074/jbc.M114.579771.

Weidauer, S. E., Schmieder, P., Beerbaum, M., Schmitz, W., Oschkinat, H., & Mueller, T. D. (2009). NMR structure of the Wnt modulator protein Sclerostin. *Biochemical and Biophysical Research Communications*, *380*(1), 160–165. http://dx.doi.org/10.1016/j.bbrc.2009.01.062.

Wharton, K. A., Cook, J. M., Torres-Schumann, S., de Castro, K., Borod, E., & Phillips, D. A. (1999). Genetic analysis of the bone morphogenetic protein-related gene, gbb, identifies multiple requirements during Drosophila development. *Genetics*, *152*(2), 629–640.

Wieser, R., Wrana, J. L., & Massague, J. (1995). GS domain mutations that constitutively activate T beta R-I, the downstream signaling component in the TGF-beta receptor complex. *The EMBO Journal*, *14*(10), 2199–2208.

Wiklund, F. E., Bennet, A. M., Magnusson, P. K., Eriksson, U. K., Lindmark, F., Wu, L., et al. (2010). Macrophage inhibitory cytokine-1 (MIC-1/GDF15): A new marker of all-cause mortality. *Aging Cell*, *9*(6), 1057–1064. http://dx.doi.org/10.1111/j.1474-9726.2010.00629.x.

Willis, S. A., Zimmerman, C. M., Li, L. I., & Mathews, L. S. (1996). Formation and activation by phosphorylation of activin receptor complexes. *Molecular Endocrinology*, *10*(4), 367–379. http://dx.doi.org/10.1210/mend.10.4.8721982.

Winnier, G., Blessing, M., Labosky, P. A., & Hogan, B. L. (1995). Bone morphogenetic protein-4 is required for mesoderm formation and patterning in the mouse. *Genes & Development*, *9*(17), 2105–2116.

Wise, R. J., Barr, P. J., Wong, P. A., Kiefer, M. C., Brake, A. J., & Kaufman, R. J. (1990). Expression of a human proprotein processing enzyme: Correct cleavage of the von Willebrand factor precursor at a paired basic amino acid site. *Proceedings of the National Academy of Sciences of the United States of America*, *87*(23), 9378–9382.

Wu, Q., Kanata, K., Saba, R., Deng, C. X., Hamada, H., & Saga, Y. (2013). Nodal/activin signaling promotes male germ cell fate and suppresses female programming in somatic cells. *Development*, *140*(2), 291–300. http://dx.doi.org/10.1242/dev.087882.

Xia, Y., Yu, P. B., Sidis, Y., Beppu, H., Bloch, K. D., Schneyer, A. L., et al. (2007). Repulsive guidance molecule RGMa alters utilization of bone morphogenetic protein (BMP) type II receptors by BMP2 and BMP4. *The Journal of Biological Chemistry*, *282*(25), 18129–18140. http://dx.doi.org/10.1074/jbc.M701679200.

Xu, J., Wang, A. H., Oses-Prieto, J., Makhijani, K., Katsuno, Y., Pei, M., et al. (2013). Arginine methylation initiates BMP-induced Smad signaling. *Molecular Cell*, *51*(1), 5–19. http://dx.doi.org/10.1016/j.molcel.2013.05.004.

Yamaguchi, K., Nagai, S., Ninomiya-Tsuji, J., Nishita, M., Tamai, K., Irie, K., et al. (1999). XIAP, a cellular member of the inhibitor of apoptosis protein family, links the receptors to TAB1-TAK1 in the BMP signaling pathway. *The EMBO Journal*, *18*(1), 179–187. http://dx.doi.org/10.1093/emboj/18.1.179.

Ye, F., Sun, T., Luo, H., Ding, H., Chen, K., Shen, X., et al. (2006). Quantitative characterization of 15-deoxy-delta(12,14)-prostaglandin J2 in regulating EGFPSmad2 translocation in CHO cells through PPARgamma/TGFbeta/Smad2 pathway. *Cellular Physiology and Biochemistry*, *18*(1–3), 143–150. http://dx.doi.org/10.1159/000095183.

Yeh, L. C., Falcon, W. E., Garces, A., Lee, J. C., & Lee, J. C. (2012). A host-guest relationship in bone morphogenetic protein receptor-II defines specificity in ligand-receptor recognition. *Biochemistry*, *51*(35), 6968–6980. http://dx.doi.org/10.1021/bi3003023.

Yi, S. E., Daluiski, A., Pederson, R., Rosen, V., & Lyons, K. M. (2000). The type I BMP receptor BMPRIB is required for chondrogenesis in the mouse limb. *Development*, *127*(3), 621–630.

Yi, J. Y., Shin, I., & Arteaga, C. L. (2005). Type I transforming growth factor beta receptor binds to and activates phosphatidylinositol 3-kinase. *The Journal of Biological Chemistry*, *280*(11), 10870–10876. http://dx.doi.org/10.1074/jbc.M413223200.

Yoshimura, A., & Muto, G. (2011). TGF-beta function in immune suppression. *Current Topics in Microbiology and Immunology*, *350*, 127–147. http://dx.doi.org/10.1007/82_2010_87.

Zhang, H., & Bradley, A. (1996). Mice deficient for BMP2 are nonviable and have defects in amnion/chorion and cardiac development. *Development*, *122*(10), 2977–2986.

Zhang, J. L., Qiu, L. Y., Kotzsch, A., Weidauer, S., Patterson, L., Hammerschmidt, M., et al. (2008). Crystal structure analysis reveals how the Chordin family member crossveinless 2 blocks BMP-2 receptor binding. *Developmental Cell*, *14*(5), 739–750. http://dx.doi.org/10.1016/j.devcel.2008.02.017.

Zhao, G. Q. (2003). Consequences of knocking out BMP signaling in the mouse. *Genesis*, *35*(1), 43–56. http://dx.doi.org/10.1002/gene.10167.

Zhu, C. C., Boone, J. Q., Jensen, P. A., Hanna, S., Podemski, L., Locke, J., et al. (2008). Drosophila activin- and the activin-like product Dawdle function redundantly to regulate proliferation in the larval brain. *Development*, *135*(3), 513–521. http://dx.doi.org/10.1242/dev.010876.

Zou, Z., & Sun, P. D. (2004). Overexpression of human transforming growth factor-beta1 using a recombinant CHO cell expression system. *Protein Expression and Purification*, *37*(2), 265–272. http://dx.doi.org/10.1016/j.pep.2003.06.001.

Zou, H., Wieser, R., Massague, J., & Niswander, L. (1997). Distinct roles of type I bone morphogenetic protein receptors in the formation and differentiation of cartilage. *Genes & Development*, *11*(17), 2191–2203.

Zuniga, J. E., Groppe, J. C., Cui, Y., Hinck, C. S., Contreras-Shannon, V., Pakhomova, O. N., et al. (2005). Assembly of TbetaRI:TbetaRII:TGFbeta ternary complex in vitro with receptor extracellular domains is cooperative and isoform-dependent. *Journal of Molecular Biology*, *354*(5), 1052–1068. http://dx.doi.org/10.1016/j.jmb.2005.10.014.

Zuniga, J. E., Ilangovan, U., Mahlawat, P., Hinck, C. S., Huang, T., Groppe, J. C., et al. (2011). The TbetaR-I pre-helix extension is structurally ordered in the unbound form and its flanking prolines are essential for binding. *Journal of Molecular Biology*, *412*(4), 601–618. http://dx.doi.org/10.1016/j.jmb.2011.07.046.

CHAPTER TWO

The Bone Morphogenetic Proteins and Their Antagonists

Barbara Mulloy, Chris C. Rider[1]
Centre for Biomedical Sciences, School of Biological Sciences, Royal Holloway, University of London, Egham, Surrey, United Kingdom
[1]Corresponding author: e-mail address: c.rider@rhul.ac.uk

Contents

Abstract

The bone morphogenetic proteins (BMPs) and the growth and differentiation factors comprise a single family of some 20 homologous, dimeric cytokines which share the cystine-knot domain typical of the TGF-β superfamily. They control the differentiation and activity of a range of cell types, including many outside bone and cartilage. They serve as developmental morphogens, but are also important in chronic pathologies, including tissue fibrosis and cancer. One mechanism for enabling tight spatiotemporal control of their activities is through a number of antagonist proteins, including Noggin, Follistatin, Chordin, Twisted gastrulation (TSG), and the seven members of the Cerberus and Dan family. These antagonists are secreted proteins that bind selectively to particular BMPs with high affinity, thereby blocking receptor engagement and signaling. Most of these antagonists also possess a TGF-β cystine-knot domain. Here, we discuss current knowledge and understanding of the structures and activities of the BMPs and their antagonists, with a particular focus on the latter proteins. Recent advances in structural

Vitamins and Hormones, Volume 99
ISSN 0083-6729
http://dx.doi.org/10.1016/bs.vh.2015.06.004

biology of BMP antagonists have begun the process of elucidating the molecular basis of their activity, displaying a surprising variety between the modes of action of these closely related proteins. We also discuss the interactions of the antagonists with the glycosaminoglycan heparan sulfate, which is found ubiquitously on cell surfaces and in the extracellular matrix.

1. BONE MORPHOGENETIC PROTEINS

The bone morphogenetic proteins (BMPs) are small, secreted proteins that can act as morphogens. They were first identified by their ability to drive differentiation of cells toward bone and cartilage, and later found to be involved in many other developmental processes (Brazil, Church, Surae, Godson, & Martin, 2015; Carreira, Alves, Zambuzzi, Sogayar, & Granjeiro, 2014; Rider & Mulloy, 2010). With one exception (BMP1; a member of the astacin family of metalloproteinases (Sarras, 1996)), they are structurally members of the TGF-β superfamily of cystine-knot proteins, as are nearly all of the BMP antagonist proteins as described below. Both BMPs and BMP antagonists are among the most ancient members of the TGF-β superfamily, traceable back to the early metazoan animals (Herpin et al., 2005; Roch & Sherwood, 2014). The BMPs and the growth and differentiation factors (GDFs) are a joint family of cytokines, being separated only by nomenclature; they are neither phylogenetically nor functionally distinct (Carreira et al., 2014; Rider & Mulloy, 2010). Currently, over 20 BMP/GDF proteins have been described (Carreira et al., 2014). Signaling and regulation of the BMPs have been extensively studied and are the subjects of recent reviews (Bragdon et al., 2011; Brazil et al., 2015). In summary, BMP signaling is initiated by cell-surface assembly of receptor complex involving type I and type II TGF-β superfamily receptor polypeptides, leading to phosphorylation of one or more of the SMAD family of signal transduction proteins.

BMPs are synthesized as large precursor proteins, including an N-terminal signal sequence for secretion, a prodomain, and the mature cytokine polypeptide. The prodomain is cleaved from the mature cytokine, but may remain associated with the C-terminal BMP after cleavage, targeting the active protein to sites in the extracellular matrix (ECM) by binding to fibrillin or to heparan sulfate (HS) proteoglycans (Sengle, Ono, Sasaki, & Sakai, 2011). Mature, active BMPs are homo- or heterodimers: there is evidence that heterodimers can be more potent than homodimers in cell differentiation (Valera, Isaacs, Kawakami, Izpisúa Belmonte, & Choe, 2010) and bone regeneration (Sun, Wang, Zheng, Fan, & Gu, 2012) models. Both dimerization and prodomain association are important in the complex extracellular regulation of BMP activity, with many questions awaiting clarification (Bragdon et al., 2011).

In terms of solved three-dimensional structures, since our earlier survey (Rider & Mulloy, 2010), only one new BMP crystal structure has been added to the Protein Data Bank (PDB), a 1.90 Å resolution structure of the dimer of BMP-9/GDF-2 (PDB: 4MPL; Wei, Salmon, Upton, Morrell, & Li, 2014). This BMP can dimerize with or without a covalent disulfide linkage between the monomers; the noncovalent dimer being more susceptible to protease cleavage. This crystal structure, as is usual for other solved BMP structures, was determined for only the C-terminal, cystine-knot domain, without involvement of the prodomain.

Very recently, one member of the BMP/GDF family, GDF-11, also known as BMP-11, has risen to prominence in the field of tissue regeneration. As BMP-11, it was originally identified as a developmental morphogen inducing dorsal mesoderm and neural tissue in *Xenopus* embryos (Gamer et al., 1999). However, investigations of the limited cellular regeneration of tissues in the adult rodent now implicate GDF-11 as a factor able to reverse the effects of aging in heart (Loffredo et al., 2013), skeletal muscle (Sinha et al., 2014), and brain (Katsimpardi et al., 2014). This makes it an exciting potential agent in future therapies for age-related degenerative conditions.

2. BMP ANTAGONIST PROTEINS

An unusual feature of the biology of BMPs is that there are a number of endogenous high-affinity BMP antagonist proteins, which function by binding directly to BMPs thereby blocking their subsequent binding to their cell-surface receptors and inhibiting downstream intracellular signaling. The expression of such proteins thus represents a mechanism for modulating BMP activities. There are a number of such BMP antagonist proteins, which as detailed below, can be grouped into various families on the basis of shared structural homology. As with the BMPs themselves, expression of the various antagonists is under tight spatiotemporal control and therefore, the consequences of gene knockout of each antagonist generates a differing phenotype, as we have previously summarized elsewhere (Rider & Mulloy, 2010).

2.1 Chordin Family

Chordin was first identified as a protein with a prominent role in dorsal–ventral patterning secreted by the Spemann organizer in amphibian embryos. A characteristic structural feature of chordin is the presence of four cysteine-rich repeats, which are 58–75 residues long, each containing 10 cysteines with characteristic spacings (Garcia, Coffinier, Larrain, Oelgeschlager, & De Robertis, 2002). These repeats are homologous with domains in a number of ECM proteins, including von Willebrand factor. Chordin is now seen to be the prototypic member of a family of proteins, characterized by the possession of multiple copies of this cysteine-rich repeat. These are chordin-like 1/Crim 1, chordin-like 2, crossveinless 2, brolin, brolin-like (Miwa et al., 2009), and kielin/chordin-like 1 (KCP) (Lin et al., 2005) also referred to as Crim 2.

The cysteine-rich domains of chordin, in particular the first and third, mediate binding to BMPs. However, possession of such domains is not evidence that a protein is a BMP antagonist, as for instance KCP functions as a promoter rather than an antagonist of BMP-4 and BMP-7 activity (Lin et al., 2005). It achieves this at least in part through binding to and suppressing the signaling activity of TGF-β and activin A (Lin, Patel, Wang, & Dressler, 2006).

The three-dimensional transmission electron microscopy (TEM) structure of chordin resembles a horseshoe, and a mode of interaction with BMP involving cooperation between the terminal cysteine-rich domains has been proposed (Troilo et al., 2014).

2.2 Follistatin and Follistatin-Like Proteins

Follistatin is a multidomain, secreted glycoprotein, with a molecular weight in the 31–39 kD range, due to alternative splicing. It was first identified in ovarian fluid as a high-affinity antagonist of the TGF-β superfamily cytokine activin, giving rise to a role in the regulation of follicle-stimulating hormone secretion (Welt, Sidis, Keutmann, & Schneyer, 2002). Since then, high-affinity interactions with various BMPs have been identified, and there has been a corresponding increase in the number of tissues and developmental processes recognized to be regulated by follistatin, as has been previously reviewed (Rider & Mulloy, 2010). One such example is that follistatin is now known to be an antagonist of GDF-8/myostatin (Sidis et al., 2006). Since myostatin suppresses muscle growth, follistatin, in particular the follistatin 288 isoform, acts to increase muscle mass (Cash, Rejon,

McPherron, Bernard, & Thompson, 2009), an activity which may prove useful in the treatment of muscle-wasting conditions. This activity has also not been lost on the muscle-building community.

A further instance of the counteracting activities of GDF-8 and follistatin lies in the field of obesity research and concerns adipose tissue differentiation. Unlike white adipose tissue which merely stores triglyceride reserves, brown adipose tissue is active in consuming fatty acids for thermogenesis. As recently reviewed by Singh, Braga, and Pervin (2014), GDF-8 inhibits brown adipose tissue formation, whereas follistatin, which is highly expressed in brown adipose tissue, promotes its differentiation. The extent to which these opposing activities arise from direct antagonism and consequent inhibition of canonical GDF-8-induced SMAD phosphorylation or via indirect mechanisms currently remains uncertain.

Follistatin can now be regarded as the prototypic member of a subfamily of BMP antagonists, recognizable through a shared domain homology (Sylva, Moorman, & van den Hoff, 2013). The four follistatin-like proteins, FSTLs, would initially appear to be BMP antagonists with different roles in embryogenesis. Thus, FSTL1 gene knockout mice show abnormalities in skeletogenesis and ureter development (Sylva et al., 2013), but the perinatal mortality of such animals actually results from respiratory failure due to abnormalities in the trachea and lung. These appear to arise from FSTL1 acting as a high-affinity antagonist, K_d 7 nM, of BMP-4 (Geng et al., 2011). By contrast, FSTL3 acts to limit testicular size (Oldknow et al., 2013), primarily due to inhibition of activin activity. It also regulates pancreatic islet number and size, and consequently glucose metabolism and fat deposition (Mukherjee et al., 2007). FSTL3 has been crystallized with myostatin (PDB: 3SEK), in a study that demonstrated differences in ligand binding between the various N-terminal domains of the FST family members (Cash et al., 2012). Since FSTL4, also known as SPIG1, antagonizes the expression and activity of the cytokine brain-derived neurotrophic factor (Suzuki et al., 2014), it remains currently uncertain as to whether it and also FSTL5 are actually BMP antagonists.

2.3 Noggin

Noggin was also first identified as a dorsalizing factor promoting neural tissue formation secreted by the Spemann organizer. It is a 67-kD homodimeric glycoprotein, which was the first BMP antagonist to have its crystallographic structure resolved at high resolution (PDB ID: 1M4U). Moreover, this is a structure of Noggin complexed with BMP-7 (Groppe et al., 2002). It has therefore served as a paradigm for the structure of BMP antagonists and their mode of BMP binding. The Noggin structure, as shown in Fig. 1, shares with that of its BMP ligands, the TGF-β cytokine superfamily fold of a cystine-knot domain, containing two β-strand finger-like loops. In the case of Noggin, this is preceded by a lengthy sequence containing several short α-helices region. Noggin exists as a dimer of identical subunits and resembles a flat "butterfly-like" shape with the helical regions of the two subunits in close contact with each other as the body of the butterfly, and the β-strands of each subunit projecting outward as wings in parallel with each other, and at some 45° away from the axis of symmetry of the dimer. In the Noggin–BMP-7 complex, the two proteins face toward each other and make contact through the tips of their β-strand loops. Since the loops of Noggin are longer than those of BMP-7, they extend over them, obscuring the binding sites on BMP-7 for its type II receptor proteins which are also in the tips of the BMP-7 β-strand loops (Greenwald et al., 2003). As the binding of

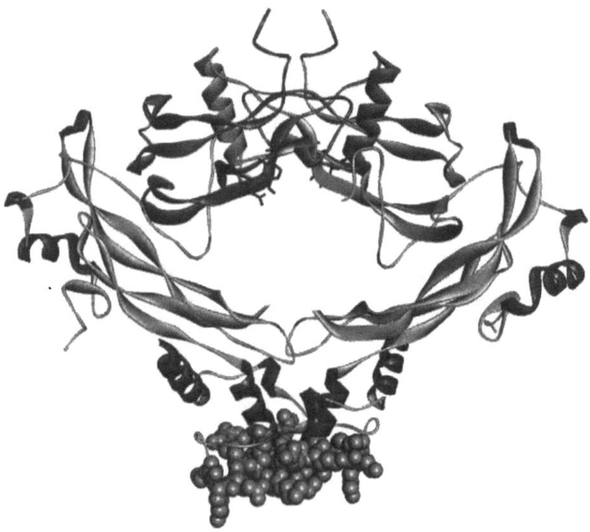

Figure 1 The crystal structure of the BMP antagonist Noggin complexed with BMP-7 (PDB: 1M4U). The BMP-7 dimer is represented as a magenta-colored ribbon at the top of the diagram, and the Noggin dimer as a ribbon colored by secondary structure: red, α-helices; aquamarine, β-strands; and, green, turns. The heparin-binding site, in amino acids 133–144, which encompasses a cluster of eight basic arginine and lysine residues (at the bottom of the diagram), has been represented in CPK space-filling format and colored yellow. (See the color plate.)

BMP-7 to type II receptors is a key step in generating the BMP-receptor signaling complex, these crystallographic insights provide a detailed structural basis for the mode of action of the BMP antagonists.

Noggin binds with varying affinities to several BMPs, including BMPs -2, -4, -7, and -14, but not to others, as recently summarized elsewhere (Krause, Guzman, & Knaus, 2011). In the cases of BMPs -7 and -14 (GDF-5), substitution of single amino acids, lysine 60 and asparagine 64 respectively, in the β-strand loops are sufficient to substantially affect their binding to Noggin (Seemann et al., 2009; Song et al., 2010). Homozygous gene deletion of Noggin in mice is perinatally lethal with major defects in neural tube closure, limb malformations, and deficient somite development (McMahon et al., 1998). In mice heterozygous for Noggin gene deletion, various skeletal abnormalities arise due to cartilage hyperplasia and failures to initiate joint formation (Brunet, McMahon, McMahon, & Harland, 1998). All these abnormalities may be attributed to inadequate attenuation of BMP activity in the affected tissues. Moreover, overexpression of Noggin in osteoblasts under the control of the osteocalcin promoter gives rise to pronounced osteopenia, due to significantly reduced bone mineralization and density, with resulting spontaneous fractures of the long bones (Devlin et al., 2003). Noggin antagonism of BMP activity is also important in the development of numerous organs including pituitary (Davis & Camper, 2007), heart (Choi, Stottmann, Yang, Meyers, & Klingensmith, 2007), prostate (Cook et al., 2007), thymus, and parathyroid (Patel, Gordon, Mahbub, Blackburn, & Manley, 2006).

Noggin has various activities of relevance to potential future regenerative strategies for neurological disorders and injury. Treatment of embryonic stem cell cultures with Noggin suppresses nonneural cell differentiation and enhances the production of neural progenitors (Gerrard, Rodgers, & Cui, 2005), including those which can give rise to astrocytes, oligodendrocytes, dopaminergic neurons (Chiba, Lee, Zhou, & Freed, 2008; Itsykson et al., 2005; Izrael et al., 2007), and serotonergic neurons (Shimada et al., 2012). In the hippocampus, a brain region in which unusually new neurons are added throughout adult life, Noggin has been shown to promote this activity (Bonaguidi et al., 2008). Noggin is expressed in reactive astrocytes surrounding brain stab wound injuries, and the inhibition of Noggin activity by infusion of a blocking antibody resulted in the potential conversion of inhibitory oligodendrocyte precursors into type II astrocytes (Hampton et al., 2007). Administration of recombinant Noggin has been reported to improve brain recovery from ischemic injury by promoting microglia activation (Shin, Lim, Jeong, Kang, & Park, 2014).

Beyond regulating the fate of neural cells, Noggin has been shown to be a potent inducer of adipocyte formation in mesenchymal stem cell cultures. Interestingly, this effect has been reported to be independent of BMP activity, raising the question as to whether Noggin can regulate additional cell signaling pathways (Sawant et al., 2012).

Noggin is also implicated in cancer progression, with its BMP antagonist activity resulting in various associations with the osteolytic metastatic colonization of bone. The forced overexpression of Noggin inhibits the BMP-2 and -4 stimulated growth of experimental osteolytic metastases of prostate cancer (Feeley, Krenek, et al., 2006) and also inhibits the osteoblastic component of mixed osteoblastic/osteolytic lung cancer metastases (Feeley, Liu, et al., 2006). In human prostate and mammary cancer cell lines, Noggin expression is constitutive in osteolytic but not in osteoblastic cells lines (Schwaninger et al., 2007). In breast cancer, Noggin is overexpressed in bone metastases, where it promotes colonization and osteolytic tumor growth (Tarragona et al., 2012).

In addition to binding BMPs, Noggin also binds strongly to heparin and HS glycosaminoglycans, the latter being widely distributed on cell surfaces and in the ECM. Such binding is likely to restrict the diffusion of Noggin away from its sites of secretion in the tissues (Paine-Saunders, Viviano, Economides, & Saunders, 2002). The binding of Noggin to heparin was established by these workers as being within the "wrist" or "body" region of Noggin (see Fig. 1), since deletion of a 12-residue sequence, amino acids 133–144, which encompasses a cluster of eight basic arginine and lysine residues is sufficient to abolish this interaction. This deletion mutant exerts comparable BMP-4 antagonist activity to wild-type Noggin, thus indicating that heparin/HS binding and BMP-4 binding are unlinked activities. This is in keeping with the considerable physical distance between the respective binding sites within the three-dimensional structure of the Noggin protein. Thus, binding to cell-surface HS serves to capture secreted Noggin, without apparently affecting its affinity for its BMP ligands (Paine-Saunders et al., 2002). Recently, a novel point mutation in Noggin resulting in the substitution of Arg136 with cysteine was identified in a Japanese family suffering from hereditary proximal symphalangism and conductive hearing loss, in which there is a fusion of joints in the fingers and toes, and between the auditory ossicles (Masuda et al., 2014). This mutation lies within the heparin-binding site and is predicted to reduce affinity for the glycosaminoglycan. Thus, this study implies that interaction with heparin and HS is essential for normal spatially localized functioning of Noggin.

2.4 Twisted Gastrulation

Twisted gastrulation (TSG) is a phylogenetically conserved, secreted protein which clearly functions as a BMP antagonist in the morphogenesis of several embryonic organs and tissues including the forebrain (Sun et al., 2010; Wills, Harland, & Khokha, 2006) and mammary gland (Forsman et al., 2013). Moreover, by acting as an antagonist of the protective activity of BMP-7 in renal podocytes, it exacerbates induced injury in these cells in an experimental model of the early events underlying chronic kidney disease (Yamada et al., 2014). However, in a number of other developmental contexts, a relatively recent example being in the genesis of osteoclasts (Sotillo Rodriguez et al., 2009), TSG is seen to promote BMP-2 activity. One explanation for these apparently opposing activities of TSG lies in its well-established interactions with chordin and its homologues, chordin-like 2 and crossveinless 2 (Ikeya et al., 2008; Larrain et al., 2001; Xie & Fisher, 2005; Zakin, Metzinger, Chang, Coffinier, & De Robertis, 2008). TSG cooperates with chordin by promoting its antagonist activity. However, TSG also promotes chordin degradation by the metalloprotease Tolloid. Thus, depending on the extracellular context, TSG can possess a duality of function, either antagonizing or promoting BMP activity (Larrain et al., 2001). This complex regulation implies that fine control of chordin–BMP interactions is of considerable developmental importance (Plouhinec, Zakin, Moriyama, & De Robertis, 2013). In addition, there is evidence that TSG enhances BMP activity in a chordin-independent manner (Xie & Fisher, 2005).

3. CERBERUS AND DAN FAMILY BMP ANTAGONISTS

Seven BMP antagonists can be classified as members of the Cerberus and Dan (CAN) family due to their close sequence homology. They each possess a single TGF-β cytokine superfamily domain. As previously discussed by Avsian-Kretchmer and Hsueh (2004), this domain like that of the BMPs themselves contains a ring composed of eight amino acids which includes four of the conserved cysteines: cysteine 2 bridging with cysteine 5 and cysteine 3 with cysteine 6. The disulfide bridge between cysteines 1 and 4 passes this ring, in order to create a three-dimensional knot structure. However, as we have previously discussed elsewhere (Rider & Mulloy, 2010), the CAN family members possess an additional pair of cysteines which stabilize the protein fold by forming a disulfide bridge between the tips of the two β-strand loops. Avsian-Kretchmer and Hsueh (2004) have elucidated the phylogenetic relationship between the CAN family members, and with other BMP antagonists. The seven CAN family members are detailed below.

3.1 Cerberus

Cerberus is yet another BMP antagonist that was first isolated as a factor secreted by the *Xenopus* Spemann organizer which functions as an inhibitor of signaling by BMPs, and also of the TGF-β superfamily cytokine Nodal, as well as Wnts (Piccolo et al., 1999). Microinjection of Cerberus mRNA into *Xenopus* embryos causes the ectopic induction of heads, and duplicate hearts and livers. However mouse Cerberus, sometimes referred to as Cerberus-like or Cerberus 1, is not a Wnt antagonist, and is also not essential in head development (Belo et al., 2000). The mammalian protein, however, does retain a role in initiating cardiogenesis. Its expression within the presumptive foregut endoderm in *Xenopus* embryos is induced by Nodal, and triggers the differentiation of the nearby cardiac mesoderm (Foley, Korol, Timmer, & Mercola, 2007). More recently in mouse kidney development, Cerberus has been shown to regulate the spatial organization within the ureteric tree, at least in part by antagonizing BMP-2 and -4 activity (Chi et al., 2011). In the chick embryo, Cerberus interacts with Nodal during left–right axis determination through complex interactions which appear to involve not only the induction of expression of Cerberus by Nodal on the left-side of the mesoderm (Tavares, Andrade, Silva, & Belo, 2007) but also the functioning of Cerberus in this process as a BMP coagonist rather than antagonist (Yu et al., 2008). Birds express Caronte, a close homologue of Cerberus which is also involved in left–right asymmetrical morphogenesis (Yokouchi, Vogan, Pearse, & Tabin, 1999). A mammalian counterpart of Caronte has yet to be described.

3.2 Coco

Coco is also referred to as Dante and Cerberus-like 2 (Cerl2), the latter name arising from its close homology with Cerberus. Like Cerberus, Coco has both Nodal and BMP antagonist activities. In mouse embryos, injection of Coco mRNA causes defects in gastrulation and mesoderm formation, consistent with antagonism of Nodal activity (Marques et al., 2004). Homozygous inactivation of the gene encoding Coco gives rise to a 35% perinatal mortality due to left–right polarity defects in the heart. Animals apparently unaffected by such lateral abnormalities still show other cardiac malformations (Araújo, Marques, & Belo, 2014).

In a significant study of the mechanisms by which metastatic tumor cells can overcome organ-specific barriers to enable the growth of distant tumor colonies, Gao et al. (2012) performed a screen for cDNAs mediating lung metastasis in breast cancer cell lines. This led to the identification of a truncated variant of Coco as being a single sequence able to confer lung metastatic activity. Coco was shown by these workers to be expressed as a secreted but cell-surface-retained protein which appeared to promote colony expansion by neutralizing the high localized BMP bioactivity within lung. Coco had no effect on metastasis to bone or brain, both of which have low tissue BMP expression.

3.3 Dan

Dan (*D*ifferential screening-selected gene *a*berrative in *n*euroblastoma) was originally identified in a search for tumor suppressor genes, rather than the investigation of embryonic amphibian development. It is also known as NBL1 (*n*euro*bl*astoma suppressor of tumorgenicity 1). Homozygous deletion of Dan expression in mice did not result in a defective phenotype, although combination of Dan deletion with heterozygous deletion of Noggin resulted in animals with a selective defect in the patterning of the vertebrae (Dionne, Skarnes, & Harland, 2001). This study also suggested that Dan might function as an antagonist of GDFs -5, -6, and -7, rather than of BMPs -2 and -4. However, luciferase reporter assays conducted by Hung, Wu, Wang, and Luo (2012) have demonstrated that Dan functions as a potent BMP-2 and -4 antagonist. More recently, similar studies by Nolan et al. (2015) have compared the IC_{50} values of Dan with BMP-2, BMP-7, and GDF-5, with those for *P*rotein *R*elated to *D*an and *C*erberus (PRDC) (see below). Dan provided IC_{50}s in the 100 nM range with the two BMPs, significantly higher than the low nM values obtained with PRDC. Dan is also a markedly less efficient antagonist of GDF-5 (IC_{50} 10 μM) compared to PRDC (IC_{50} 92 nM). Overall, although these studies require some resolution, they indicate that the various BMP antagonists show considerable differences in their specificities and affinities for the different BMPs/GDFs.

The Dan transcript is widely distributed in different organs (Hung et al., 2012), but within mouse forebrain, Dan expression shows highly specific spatiotemporal patterning (Kim & Pleasure, 2003). Dan has been shown to regulate the development of ovarian follicles (Hung et al., 2012), the migration of neurons through the corpus callosum joining the brain cortical hemispheres (Choe, Siegenthaler, & Pleasure, 2012), and in birds, the morphogenesis of the inner ear (Yamanishi, Katsu, Funahashi, Yumoto, & Yokouchi, 2007).

The homodimerization of Dan is discussed in Section 3.5. A high-resolution structure for Dan is now available (4X1J; Nolan et al., 2015).

3.4 Gremlin

Gremlin, often referred to as gremlin-1 and previously named drm (*d*own-*r*egulated by v-*mos*), has an essential role in skeletal development and homeostasis, and in kidney development. In some mouse strains, homozygous gremlin gene deletion causes perinatal mortality with an absence of kidneys and an extensive skeletal phenotype with limb and digit abnormalities (Khokha, Hsu, Brunet, Dionne, & Harland, 2003; Michos et al., 2004). On a more resilient genetic background with reduced penetrance of kidney absence, allowing some 50% of animals to survive the perinatal period, shortened limbs with missing digits and reduced bone mineral densities are observed in 1-month-old pups (Canalis, Parker, & Zanotti, 2012). However, the conditional overexpression of gremlin under the control of the osteocalcin promoter which is active in the osteoblastic lineage also gives rise to reduced limb bone length and bone mineral densities (Gazzerro et al., 2005). Overall, these findings are consistent with physiological antagonism of BMP activity and indicate that proper skeletal development depends on the correct spatiotemporal expression of gremlin. Similarly, as recently reviewed elsewhere, the fine control of the balance between BMP activity and gremlin appears to be essential for normal organogenesis of the kidney (Nishinakamura & Sakaguchi, 2014). Gremlin is a more potent antagonist of BMPs -2 and -4 compared to BMP-7 (Church et al., 2014; Sudo, Avsian-Kretchmer, Wang, & Hsueh, 2004). Where both BMP-4 and gremlin are expressed in the same cell, they are able to interact intracellularly (Sun et al., 2006). Beyond its function as a BMP antagonist, gremlin acts as a high-affinity agonist of VEGF receptor 2, promoting angiogenesis (Mitola et al., 2010). Moreover, there is one report that gremlin binds to Slit proteins to inhibit monocyte chemotaxis (Chen et al., 2004).

As well as its important roles in normal physiological development, gremlin is reexpressed in chronic fibrotic diseases. This is well established in the kidney where gremlin is reexpressed in diabetic nephropathy, with direct correlation to fibrosis score (Dolan et al., 2005). Its expression is also upregulated at fibrotic sites in chronic renal allograft nephropathy (Carvajal et al., 2008). Gremlin appears to have a causative role in renal fibrosis, since single allelic gene deletion of gremlin attenuates experimental diabetic nephropathy (Roxburgh et al., 2009), whereas the overexpression of gremlin in kidney tubules increases the level of damage and fibrosis on exposure to folic acid, a commonly used inducer of renal fibrosis (Droguett et al., 2014). Investigations of the pathogenesis of diabetic nephropathy have established that exposure of renal cell cultures to supra-physiological glucose concentrations results in profibrotic changes. Such glucose concentrations induce gremlin expression in cultures of renal mesangial cells (Zhang et al., 2010) and immortalized podocytes (Li et al., 2013), consistent with the view that gremlin is a key signaling molecule in the development of renal fibrosis.

As with chronic kidney fibrosis, idiopathic pulmonary fibrosis is a progressive disease with no curative therapy and ultimately leads to organ failure. In both diseases, reduced BMP activity is thought to lead to myofibroblast accumulation in part via epithelial to mesenchymal transitions. In idiopathic pulmonary fibrosis, gremlin is upregulated (Myllärniemi et al., 2008), and experimental transient overexpression of gremlin induced in the lung by an adenoviral vector resulted in transient profibrotic cellular changes within the alveoli (Farkas et al., 2011). Gremlin, via its antagonism of BMP activity in the lung, is also seen to play a central role in development of pulmonary hypertension (Cahill et al., 2012).

Gremlin overexpression has been found in various carcinomas, being localized with the tumor-associated stroma (Sneddon et al., 2006). More recently, gremlin has been shown to be expressed in cancer-associated fibroblasts at the invasion fronts of colorectal cancer (Karagiannis et al., 2014). Such studies support the view that gremlin is involved in tumor cell–stromal cell interactions and may promote tumor invasion. Gremlin is also overexpressed in malignant mesothelioma (Wang et al., 2012) and in lung adenocarcinoma, but not in squamous cell carcinoma of the lung (Mulvihill et al., 2012). Thus, overall, gremlin evokes considerable interest in molecular pathology as it appears to be important in the etiology of a number of chronic diseases.

3.5 PRDC/Gremlin 2

PRDC, Protein Related to Dan and Cerberus, is also known as gremlin 2, due to its close homology to gremlin. PRDC was originally identified in a gene trap investigation of mouse embryonic development, where it was shown to be expressed in certain neurons in the developing spinal chord (Minabe-Saegusa, Saegusa, Tsukahara, & Noguchi, 1998). PRDC preferentially antagonizes BMP-2 and -4 activity compared to that of other BMPs (Sudo et al., 2004). It is widely expressed in a variety of tissues, but its expression in the ovarian granulosa cells implicates it in the regulation of ovarian follicle development. The finding that PRDC binds not only to BMP-4 but also to anti-Mullerian hormone substantiates such a hypothesis (Nilsson, Larsen, & Skinner, 2014).

PRDC is also expressed in mouse embryonic preosteoblasts, where its forced overexpression inhibits BMP-2 driven osteoblast differentiation (Ideno et al., 2009). In Zebrafish embryos, PRDC shows complex expression patterns during development in a variety of locations including eye, pharyngeal arches, somites, swim bladder (Müller et al., 2006), and the facial skeleton (Zuniga, Rippen, Alexander, Schilling, & Crump, 2011). In the chick, PRDC has been shown to modulate BMP-4 activity in the development of the neurons of cranial sensory ganglia (Kriebitz et al., 2009). Finally, in murine cardiac development, PRDC promotes the differentiation of atrial cardiomyocytes and the establishment of cardiac rhythm (Tanwar et al., 2014).

Recent investigation of the structure of PRDC has revealed an unexpected mode of dimerization. PRDC along with Cerberus, Dan, and gremlin, but not Coco, sclerostin, and USAG-1 (*U*terine *s*ensitization-*a*ssociated *g*ene-1), have in common with the BMPs a cysteine residue within their cystine-knot domains which is not involved in intrachain disulfide bridges (Nolan et al., 2013). In the BMPs, such cysteines give rise to an interchain disulfide bridge which is the basis of an offset face-to-face dimerization of the polypeptides (see Fig. 1). A similar disulfide-bridged dimer is formed in the case of Noggin (Groppe et al., 2002), albeit using unpaired cysteines at the carboxyterminal end of the cystine-knot domain. This dimerization is central to the ability of Noggin to bind its BMP ligands with high affinity in a dimer–dimer interaction (see Fig. 1). It was therefore envisaged that such disulfide-bridged dimers would be widespread throughout the CAN family, however evidence for CAN protein dimers had been lacking. Kattamuri et al. (2012) established using size-exclusion chromatography, sedimentation ultracentrifugation, and glutaraldehyde cross-linking, that PRDC and also Dan form noncovalent dimers in which the unpaired cysteines play no role. The existence of Dan as a noncovalently linked dimer has also been confirmed by cross-linking studies elsewhere (Hung et al., 2012).

The crystallographic structure of PRDC (Nolan et al., 2013) supports the existence of this novel dimer as an overlapping head-to-tail, elongated "banana"-like arched structure, involving extensive contact between the β-strands of the subunits (Fig. 2A and B). As has been discussed elsewhere (Hinck & Huang, 2013), it is of fundamental interest that proteins of high primary and secondary structural homology can form two very different dimer

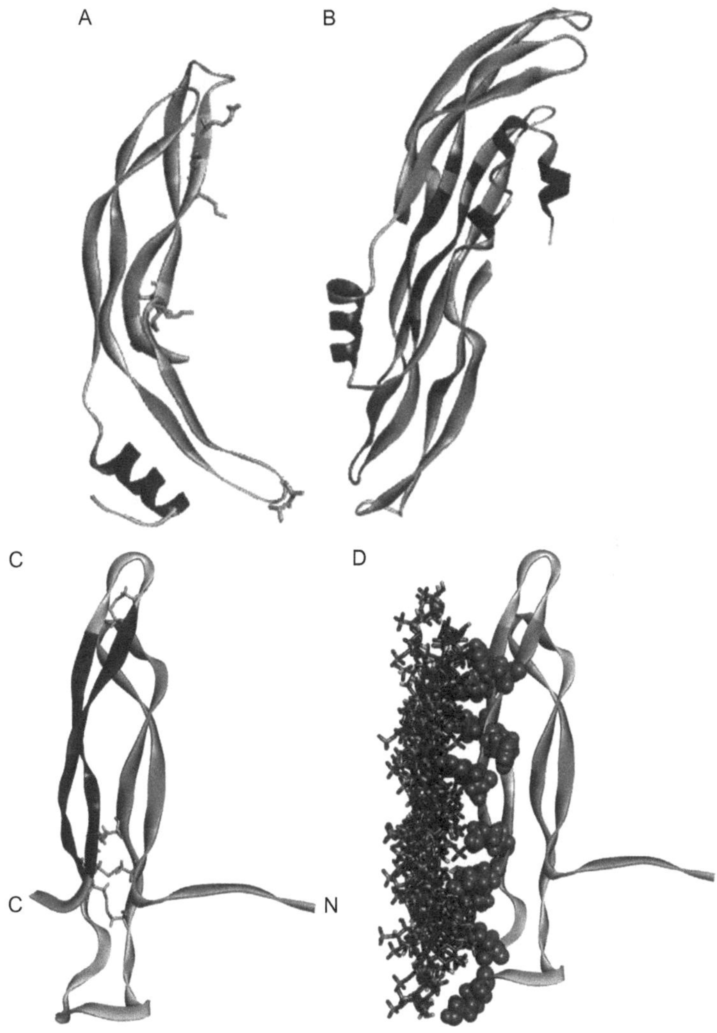

Figure 2 See legend on next page.

associations. Very recently, a crystallographic structure of Dan (4X1J) shows a dimer with a highly similar architecture to that of PRDC (Nolan et al., 2015).

3.6 Sclerostin

The *SOST* gene encoding sclerostin was first identified in searches for the mutation underlying rare inherited diseases of sclerotic overgrowth of the skeleton resulting in excessive bone densities (Balemans et al., 2001; Brunkow et al., 2001). Although initially characterized as a high-affinity antagonist of BMPs -5 and -6 (Winkler et al., 2003), sclerostin also antagonizes Wnt signaling (Li et al., 2005; Semënov, Tamai, & He, 2005). Later studies have argued that both modes of signaling may be relevant in different cellular contexts (Krause et al., 2010). Irrespective of its mode of action, sclerostin appeared to influence only bone deposition. This selective activity appears to be largely dependent on its restricted expression, which axiomatically is described as occurring only in osteocytes, a mature form of osteoblasts. However, there is recent evidence for sclerostin expression outside bone, particularly in cartilage (Hernandez, Whitty, Wardale, & Henson, 2014; Roudier et al., 2013). Moreover, localized overexpression of sclerostin is associated with aortic valve calcification (Koos et al., 2013). Nonetheless, inhibition of sclerostin activity has been perceived as an attractive therapeutic target for the treatment of osteoporosis, a condition common in postmenopausal women. A number of experimental studies of sclerostin-neutralizing monoclonal antibodies have established their effectiveness in improving bone density, bone volume, and trabecular thickness, as has been systematically reviewed elsewhere (Das & Sakthiswary, 2013). Moreover, in a phase II clinical trial in postmenopausal women with low bone mineral densities, the humanized sclerostin-neutralizing monoclonal

Figure 2 Ribbon representations of DAN family members, colored red for α-helices, aquamarine for β-strands, and green for turns. (A) Monomer of PRDC with possible heparin-binding residues in yellow stick representation. (B) Dimer of PRDC, taken from the crystal structure PDB: 4JPH; with heparin-binding site colored yellow, dimer interface purple, and BMP-binding site orange. (C) Ribbon representation of sclerostin structure, one of the ensemble of NMR structures determined by Veverka et al. (2009). N- and C-termini are marked and colors distinguish the β-strand fingers 1 (on the right: orange) and 3 (on the left: red); the disorganized loop 2 is colored white. Cystine bridges are shown in yellow. (D) The same sclerostin polypeptide structure (β-strands colored aquamarine) docked with a heparin oligosaccharide (on the left: stick representation); the results of multiple docking calculations are superimposed. The sidechains of contact arginines and lysines are shown in green space-fill. (See the color plate.)

antibody romosozumab, significantly increased bone mineral density in the lumbar spine, hip, and femoral neck with only mild adverse effects (McClung et al., 2014).

The three-dimensional structure of sclerostin has been solved for the human protein using NMR (Veverka et al., 2009). As may be seen in Fig. 2C, sclerostin conforms to the TGF-β cytokine superfamily fold, with three loops extending out from the central cysteine knot, of which loops 1 and 3 are β-strand fingers. Loop 2, compared to its counterparts in other CAN family members is relatively disorganized and flexible. Through the analysis of changes in the NMR backbone amide signals occurring on the binding of the Fab fragment of an antisclerostin antibody which blocks sclerostin's effect on Wnt signaling, loop 2 was shown to be essential in this activity (Veverka et al., 2009), a finding recently confirmed elsewhere by mutagenesis (Boschert et al., 2013). The existence of a dimeric form of sclerostin in tissue extracts and cultured cell lysates has been demonstrated (Hernandez et al., 2014). Interestingly, although both sclerostin and Noggin are BMP antagonists, they bind each other with high affinity (K_d 3 nM) and when complexed together, neutralize each other's antagonist activities (Winkler et al., 2004).

3.7 USAG-1/Wise

The final member of the CAN family was identified as USAG-1, a tightly regulated gene expressed in the rat uterine endometrium (Simmons & Kennedy, 2002), and independently as Wise, a Wnt inhibitor in the early *Xenopus* embryo, involved in the anterior–posterior axis and neural tube development (Itasaki et al., 2003). Homozygous gene knockout of *USAG1* results in mice with the limited phenotype of supernumerary teeth (Murashima-Suginami et al., 2008). This arises from a lack of USAG-1 inhibition of BMP-7 activity in the odontogenic epithelium (Kiso et al., 2014). USAG-1 is also expressed in renal tubules and *USAG1* knockout mice are more resistant than wild type to experimentally induced renal injury (Yanagita et al., 2006), and to a progressive hereditary kidney disease, Alport syndrome (Tanaka et al., 2010). Within human kidney, USAG-1 is abundantly expressed by podocytes in the glomerulus (Turk et al., 2009). USAG-1 is considered to be an antagonist of BMP-7, the latter remaining highly expressed in the adult kidney beyond the developmental stages. Since USAG-1 and sclerostin share a close phylogenetic relationship within the CAN family (Avsian-Kretchmer & Hsueh, 2004), it is unsurprising that both are dual function BMP and Wnt antagonists. Moreover, as in sclerostin, the high-affinity-binding site for the Wnt coreceptor LRP6, through which Wnt signaling inhibition is exercised, occurs via loop 2 of the predicted USAG-1 structure (Lintern, Guidato, Rowe, Saldanha, & Itasaki, 2009).

4. BMP-ANTAGONISTS BINDING TO HEPARIN/HS

An important property common for many BMP antagonists, as indeed it is among many of the BMPs themselves, is binding to the sulfated polysaccharides, heparin, and HS. Although heparin is localized within the cytoplasmic granules of mast cells, HS is widely distributed, being found in the extracellular matrices and cell surfaces in all tissues. The binding of a secreted protein to HS will restrict its diffusion away from its cellular sources. Moreover, there are known instances where such an interaction can protect the protein involved from degradation, and also enhance its biological activity through a coreceptor role (Mulloy & Rider, 2006). As we have previously reviewed elsewhere, chordin and follistatin are well established as heparin/HS-binding proteins (Rider & Mulloy, 2010). The binding of Noggin to heparin/HS has been discussed above.

Among the CAN family, we previously used the then known high-resolution structures of the antagonists, or where unavailable, homology models, in molecular dynamic docking calculations with heparin oligosaccharide structures to predicted that all CAN antagonists, except Dan and Cerberus, would bind at physiological pH and ionic strength (Rider & Mulloy, 2010, Supplementary material). The experimental evidence supporting these interactions is now being accumulated although it is as yet not comprehensive. In the case of sclerostin, the heparin-binding site has been mapped through changes in the backbone amide NMR signals when the protein is incubated with the glycosaminoglycan (Veverka et al., 2009). This binding site involves a linear array of basic residues, arginines and lysines, running down one side of the β-strand loops (Fig. 2D) which interact with the highly acidic sulfate groups of heparin. In addition, a supporting network of hydrogen bonds between neighboring amino acids and the glycosaminoglycan has been postulated. Similar binding also occurs with cell-surface HS, as the forced overexpression of sclerostin in an osteoblastic cell line results in the protein accumulating in the conditioned supernatant only on the addition of exogenous heparin, which would displace the protein from retention on the cell surface. Moreover, this cell-surface retention was reduced on alanine replacement of some of the key arginines and lysines of the heparin-binding site (Veverka et al., 2009).

High-affinity binding to heparin has also been observed for gremlin, which required the application of 1 M NaCl for its elution from a preparative heparin-Sepharose column (Stabile et al., 2007). Moreover, the same laboratory has since used surface plasmon resonance to determine the binding of gremlin to heparin as having a low dissociation constant, K_d 20 nM (Chiodelli et al., 2011). They further showed that HS, but not other glycosaminoglycans can compete with this binding, and that gremlin bound to cell-surface HS.

Other members of the CAN family have been less well studied in this regard, but PRDC is retained on affinity columns of immobilized heparin requiring 0.66 M salt for its elution; moreover, heparin inhibits the binding of PRDC to BMP-2 (Nolan et al., 2013). The forced overexpression of USAG-1 in HEK293 cells shows binding to the cell surface, and addition of chlorate, an inhibitor of the sulfation of heparin/HS chains reduces this retention (Lintern et al., 2009). Finally in the case of Coco, pericellular accumulation on the surfaces of metastatic breast cancer cells has been noted (Gao et al., 2012).

It has been established that HS is essential for the interaction of gremlin with VEGF receptor-2 (Chiodelli et al., 2011). Otherwise, there is little insight into how the binding of BMP antagonists to HS might affect their biological activities, beyond restricting their diffusion through the tissues, thereby establishing localized concentration gradients. However, given their functions as morphogen antagonists during development, this localization alone will be important in the spatiotemporal regulation of their activities. What it does mean in the case of the CAN family is that these small proteins, with a monomeric M_r 20–25 kD, and a relatively simple protein fold may possess up to four specific binding sites on their surfaces; one for their BMP ligands; one for any secondary ligand (such as the Wnt receptor for sclerostin and USAG-1, and the VEGF receptor for gremlin); and, in addition for at least for some of them, a heparin/HS-binding site, and a dimerization site for binding to a second subunit.

REFERENCES

Araújo, A. C., Marques, S., & Belo, J. A. (2014). Targeted inactivation of Cerberus like-2 leads to left ventricular cardiac hyperplasia and systolic dysfunction in the mouse. *PLoS One*, *9*, e102716. http://dx.doi.org/10.1371/journal.pone.0102716.

Avsian-Kretchmer, O., & Hsueh, A. J. (2004). Comparative genomic analysis of the eight-membered ring cystine knot-containing bone morphogenetic protein antagonists. *Molecular Endocrinology*, *18*, 1–12.

Balemans, W., Ebeling, M., Patel, N., Van, H. E., Olson, P., Dioszegi, M., et al. (2001). Increased bone density in sclerosteosis is due to the deficiency of a novel secreted protein (SOST). *Human Molecular Genetics*, *10*, 537–543.

Belo, J. A., Bachiller, D., Agius, E., Kemp, C., Borges, A. C., Marques, S., et al. (2000). Cerberus-like is a secreted BMP and nodal antagonist not essential for mouse development. *Genesis*, *26*, 265–270.

Bonaguidi, M. A., Peng, C. Y., McGuire, T., Falciglia, G., Gobeske, K. T., Czeisler, C., et al. (2008). Noggin expands neural stem cells in the adult hippocampus. *The Journal of Neuroscience*, *28*, 9194–9204.

Boschert, V., van Dinther, M., Weidauer, S., van Pee, K., Muth, E. M., Ten Dijke, P., et al. (2013). Mutational analysis of sclerostin shows importance of the flexible loop and the cystine-knot for Wnt-signaling inhibition. *PLoS One*, *8*, e81710. http://dx.doi.org/10.1371/journal.pone.0081710.

Bragdon, B., Moseychuk, O., Saldanha, S., King, D., Julian, J., & Nohe, A. (2011). Bone morphogenetic proteins: A critical review. *Cellular Signalling*, *23*, 609–620.

Brazil, D. P., Church, R. H., Surae, S., Godson, G., & Martin, F. (2015). BMP signalling: Agony and antagony in the family. *Trends in Cell Biology*, *25*, 249–264.

Brunet, L. J., McMahon, J. A., McMahon, A. P., & Harland, R. M. (1998). Noggin, cartilage morphogenesis, and joint formation in the mammalian skeleton. *Science*, *280*, 1455–1457.

Brunkow, M. E., Gardner, J. C., Van Ness, J., Paeper, B. W., Kovacevich, B. R., Proll, S., et al. (2001). Bone dysplasia sclerosteosis results from loss of the SOST gene product, a novel cystine knot-containing protein. *American Journal of Human Genetics*, *68*, 577–589.

Cahill, E., Costello, C. M., Rowan, S. C., Harkin, S., Howell, K., Leonard, M. O., et al. (2012). Gremlin plays a key role in the pathogenesis of pulmonary hypertension. *Circulation*, *125*, 920–930.

Canalis, E., Parker, K., & Zanotti, S. (2012). Gremlin1 is required for skeletal development and postnatal skeletal homeostasis. *Journal of Cellular Physiology*, *227*, 269–277.

Carreira, A. C., Alves, G. G., Zambuzzi, W. F., Sogayar, M. C., & Granjeiro, J. M. (2014). Bone morphogenetic proteins: Structure, biological function and therapeutic applications. *Archives of Biochemistry and Biophysics*, *561*, 64–73.

Carvajal, G., Droguett, A., Burgos, M. E., Aros, C., Ardiles, L., Flores, C., et al. (2008). Gremlin: A novel mediator of epithelial mesenchymal transition and fibrosis in chronic allograft nephropathy. *Transplantation Proceedings*, *40*, 734–739.

Cash, J. N., Angerman, E. B., Kattamuri, C., Nolan, K., Zhao, H., Sidis, Y., et al. (2012). Structure of myostatin·follistatin-like 3: N-terminal domains of follistatin-type molecules exhibit alternate modes of binding. *The Journal of Biological Chemistry*, *287*, 1043–1053.

Cash, J. N., Rejon, C. A., McPherron, A. C., Bernard, D. J., & Thompson, T. B. (2009). The structure of myostatin:follistatin 288: Insights into receptor utilization and heparin binding. *The EMBO Journal*, *28*, 2662–2676.

Chen, B., Blair, D. G., Plisov, S., Vasiliev, G., Perantoni, A. O., Chen, Q., et al. (2004). Bone morphogenetic protein antagonists Drm/Gremlin and Dan interact with Slits and act as negative regulators of monocyte chemotaxis. *Journal of Immunology*, *173*, 5914–5917.

Chi, L., Saarela, U., Railo, A., Prunskaite-Hyyryläinen, R., Skovorodkin, I., Anthony, S., et al. (2011). A secreted BMP antagonist, Cer1, fine tunes the spatial organization of the ureteric bud tree during mouse kidney development. *PLoS One*, *6*, e27676. http://dx.doi.org/10.1371/journal.pone.0027676.

Chiba, S., Lee, Y. M., Zhou, W., & Freed, C. R. (2008). Noggin enhances dopamine neuron production from human embryonic stem cells and improves behavioral outcome after transplantation into Parkinsonian rats. *Stem Cells*, *26*, 2810–2820.

Chiodelli, P., Mitola, S., Ravelli, C., Oreste, P., Rusnati, M., & Presta, M. (2011). Heparan sulfate proteoglycans mediate the angiogenic activity of the vascular endothelial growth factor receptor-2 agonist gremlin. *Arteriosclerosis, Thrombosis, and Vascular Biology, 31*, e116–e127. http://dx.doi.org/10.1161/ATVBAHA.111.235184.

Choe, Y., Siegenthaler, J. A., & Pleasure, S. J. (2012). A cascade of morphogenic signaling initiated by the meninges controls corpus callosum formation. *Neuron, 73*, 698–712.

Choi, M., Stottmann, R. W., Yang, Y. P., Meyers, E. N., & Klingensmith, J. (2007). The bone morphogenetic protein antagonist noggin regulates mammalian cardiac morphogenesis. *Circulation Research, 100*, 220–228.

Church, R. H., Krishnakumar, A., Urbanek, A., Geschwindner, S., Meneely, J., Bianchi, A., et al. (2014). Gremlin1 preferentially binds to Bone Morphogenetic Protein-2 (BMP-2) and BMP-4 over BMP-7. *The Biochemical Journal, 466*, 55–68.

Cook, C., Vezina, C. M., Allgeier, S. H., Shaw, A., Yu, M., Peterson, R. E., et al. (2007). Noggin is required for normal lobe patterning and ductal budding in the mouse prostate. *Developmental Biology, 312*, 217–230.

Das, S., & Sakthiswary, R. (2013). Bone metabolism and histomorphometric changes in murine models treated with sclerostin antibody: A systematic review. *Current Drug Targets, 14*, 1667–1674.

Davis, S. W., & Camper, S. A. (2007). Noggin regulates Bmp4 activity during pituitary induction. *Developmental Biology, 305*, 145–160.

Devlin, R. D., Du, Z., Pereira, R. C., Kimble, R. B., Economides, A. N., Jorgetti, V., et al. (2003). Skeletal overexpression of noggin results in osteopenia and reduced bone formation. *Endocrinology, 144*, 1972–1978.

Dionne, M. S., Skarnes, W. C., & Harland, R. M. (2001). Mutation and analysis of Dan, the founding member of the Dan family of transforming growth factor beta antagonists. *Molecular and Cellular Biology, 21*, 636–643.

Dolan, V., Murphy, M., Sadlier, D., Lappin, D., Doran, P., Godson, C., et al. (2005). Expression of gremlin, a bone morphogenetic protein antagonist, in human diabetic nephropathy. *American Journal of Kidney Diseases, 45*, 1034–1039.

Droguett, A., Krall, P., Burgos, M. E., Valderrama, G., Carpio, D., Ardiles, L., et al. (2014). Tubular overexpression of gremlin induces renal damage susceptibility in mice. *PLoS One, 9*, e101879. http://dx.doi.org/10.1371/journal.pone.0101879.

Farkas, L., Farkas, D., Gauldie, J., Warburton, D., Shi, W., & Kolb, M. (2011). Transient overexpression of Gremlin results in epithelial activation and reversible fibrosis in rat lungs. *American Journal of Respiratory Cell and Molecular Biology, 44*, 870–878.

Feeley, B. T., Krenek, L., Liu, N., Hsu, W. K., Gamradt, S. C., Schwarz, E. M., et al. (2006). Overexpression of noggin inhibits BMP-mediated growth of osteolytic prostate cancer lesions. *Bone, 38*, 154–166.

Feeley, B. T., Liu, N. Q., Conduah, A. H., Krenek, L., Roth, K., Dougall, W. C., et al. (2006). Mixed metastatic lung cancer lesions in bone are inhibited by noggin overexpression and Rank:Fc administration. *Journal of Bone and Mineral Research, 21*, 1571–1580.

Foley, A. C., Korol, O., Timmer, A. M., & Mercola, M. (2007). Multiple functions of Cerberus cooperate to induce heart downstream of Nodal. *Developmental Biology, 303*, 57–65.

Forsman, C. L., Ng, B. C., Heinze, R. K., Kuo, C., Sergi, C., Gopalakrishnan, R., et al. (2013). BMP-binding protein twisted gastrulation is required in mammary gland epithelium for normal ductal elongation and myoepithelial compartmentalization. *Developmental Biology, 373*, 95–106.

Gamer, L. W., Wolfman, N. M., Celeste, A. J., Hattersley, G., Hewick, R., & Rosen, V. (1999). A novel BMP expressed in developing mouse limb, spinal cord, and tail bud is a potent mesoderm inducer in Xenopus embryos. *Developmental Biology, 208*, 222–232.

Gao, H., Chakraborty, G., Lee-Lim, A. P., Mo, Q., Decker, M., Vonica, A., et al. (2012). The BMP inhibitor Coco reactivates breast cancer cells at lung metastatic sites. *Cell, 150*, 764–779.
Garcia, A. J., Coffinier, C., Larrain, J., Oelgeschlager, M., & De Robertis, E. M. (2002). Chordin-like CR domains and the regulation of evolutionarily conserved extracellular signaling systems. *Gene, 287*, 39–47.
Gazzerro, E., Pereira, R. C., Jorgetti, V., Olson, S., Economides, A. N., & Canalis, E. (2005). Skeletal overexpression of gremlin impairs bone formation and causes osteopenia. *Endocrinology, 146*, 655–665.
Geng, Y., Dong, Y., Yu, M., Zhang, L., Yan, X., Sun, J., et al. (2011). Follistatin-like 1 (Fstl1) is a bone morphogenetic protein (BMP) 4 signaling antagonist in controlling mouse lung development. *Proceedings of the National Academy of Sciences of the United States of America, 108*, 7058–7063.
Gerrard, L., Rodgers, L., & Cui, W. (2005). Differentiation of human embryonic stem cells to neural lineages in adherent culture by blocking bone morphogenetic protein signaling. *Stem Cells, 23*, 1234–1241.
Greenwald, J., Groppe, J., Gray, P., Wiater, E., Kwiatkowski, W., Vale, W., et al. (2003). The BMP7/ActRII extracellular domain complex provides new insights into the cooperative nature of receptor assembly. *Molecular Cell, 11*, 605–617.
Groppe, J., Greenwald, J., Wiater, E., Rodriguez-Leon, J., Economides, A. N., Kwiatkowski, W., et al. (2002). Structural basis of BMP signalling inhibition by the cystine knot protein noggin. *Nature, 420*, 636–642.
Hampton, D. W., Asher, R. A., Kondo, T., Steeves, J. D., Ramer, M. S., & Fawcett, J. W. (2007). A potential role for bone morphogenetic protein signalling in glial cell fate determination following adult central nervous system injury in vivo. *The European Journal of Neuroscience, 26*, 3024–3035.
Hernandez, P., Whitty, C., Wardale, J. R., & Henson, F. M. (2014). New insights into the location and form of sclerostin. *Biochemical and Biophysical Research Communications, 446*, 1108–1113.
Herpin, A., Lelong, C., Becker, T., Rosa, F., Favrel, P., & Cunningham, C. (2005). Structural and functional evidence for a singular repertoire of BMP receptor signal transducing proteins in the lophotrochozoan *Crassostrea gigas* suggests a shared ancestral BMP/activin pathway. *The FEBS Journal, 272*, 3424–3440.
Hinck, A. P., & Huang, T. (2013). TGF-β antagonists: Same knot, but different hold. *Structure, 21*, 1269–1270.
Hung, W. T., Wu, F. J., Wang, C. J., & Luo, C. W. (2012). DAN (NBL1) specifically antagonizes BMP2 and BMP4 and modulates the actions of GDF9, BMP2, and BMP4 in the rat ovary. *Biology of Reproduction, 86*, 1–9.
Ideno, H., Takanabe, R., Shimada, A., Imaizumi, K., Araki, R., Abe, M., et al. (2009). Protein related to DAN and cerberus (PRDC) inhibits osteoblastic differentiation and its suppression promotes osteogenesis in vitro. *Experimental Cell Research, 315*, 474–484.
Ikeya, M., Nosaka, T., Fukushima, K., Kawada, M., Furuta, Y., Kitamura, T., et al. (2008). Twisted gastrulation mutation suppresses skeletal defect phenotypes in Crossveinless 2 mutant mice. *Mechanisms of Development, 125*, 832–842.
Itasaki, N., Jones, C. M., Mercurio, S., Rowe, A., Domingos, P. M., Smith, J. C., et al. (2003). Wise, a context-dependent activator and inhibitor of Wnt signalling. *Development, 130*, 4295–4305.
Itsykson, P., Ilouz, N., Turetsky, T., Goldstein, R. S., Pera, M. F., Fishbein, I., et al. (2005). Derivation of neural precursors from human embryonic stem cells in the presence of noggin. *Molecular and Cellular Neurosciences, 30*, 24–36.
Izrael, M., Zhang, P., Kaufman, R., Shinder, V., Ella, R., Amit, M., et al. (2007). Human oligodendrocytes derived from embryonic stem cells: Effect of noggin on phenotypic differentiation in vitro and on myelination in vivo. *Molecular and Cellular Neurosciences, 34*, 310–323.

Karagiannis, G. S., Musrap, N., Saraon, P., Treacy, A., Schaeffer, D. F., Kirsch, R., et al. (2014). Bone morphogenetic protein antagonist gremlin-1 regulates colon cancer progression. *Biological Chemistry, 396*, 163–183.

Katsimpardi, L., Litterman, N. K., Schein, P. A., Miller, C. M., Loffredo, F. S., Wojtkiewicz, G. R., et al. (2014). Vascular and neurogenic rejuvenation of the aging mouse brain by young systemic factors. *Science, 344*, 630–634.

Kattamuri, C., Luedeke, D. M., Nolan, K., Rankin, S. A., Greis, K. D., Zorn, A. M., et al. (2012). Members of the DAN family are BMP antagonists that form highly stable noncovalent dimers. *Journal of Molecular Biology, 14*, 313–327.

Khokha, M. K., Hsu, D., Brunet, L. J., Dionne, M. S., & Harland, R. M. (2003). Gremlin is the BMP antagonist required for maintenance of Shh and Fgf signals during limb patterning. *Nature Genetics, 34*, 303–307.

Kim, A. S., & Pleasure, S. J. (2003). Expression of the BMP antagonist Dan during murine forebrain development. *Brain Research. Developmental Brain Research, 145*, 159–162.

Kiso, H., Takahashi, K., Saito, K., Togo, Y., Tsukamoto, H., Huang, B., et al. (2014). Interactions between BMP-7 and USAG-1 (uterine sensitization-associated gene-1) regulate supernumerary organ formations. *PLoS One, 9*, e96938. http://dx.doi.org/10.1371/journal.pone.0096938.

Koos, R., Brandenburg, V., Mahnken, A. H., Schneider, R., Dohmen, G., Autschbach, R., et al. (2013). Sclerostin as a potential novel biomarker for aortic valve calcification: An in-vivo and ex-vivo study. *The Journal of Heart Valve Disease, 22*, 317–325.

Krause, C., Guzman, A., & Knaus, P. (2011). Noggin. *The International Journal of Biochemistry & Cell Biology, 43*, 478–481.

Krause, C., Korchynskyi, O., de Rooij, K., Weidauer, S. E., de Gorter, D. J., van Bezooijen, R. L., et al. (2010). Distinct modes of inhibition by sclerostin on bone morphogenetic protein and Wnt signaling pathways. *The Journal of Biological Chemistry, 285*, 41614–41626.

Kriebitz, N. N., Kiecker, C., McCormick, L., Lumsden, A., Graham, A., & Bell, E. (2009). PRDC regulates placode neurogenesis in chick by modulating BMP signalling. *Developmental Biology, 336*, 280–292.

Larrain, J., Oelgeschlager, M., Ketpura, N. I., Reversade, B., Zakin, L., & De Robertis, E. M. (2001). Proteolytic cleavage of chordin as a switch for the dual activities of twisted gastrulation in BMP signaling. *Development, 128*, 4439–4447.

Li, G., Li, Y., Liu, S., Shi, Y., Chi, Y., Liu, G., et al. (2013). Gremlin aggravates hyperglycemia-induced podocyte injury by a TGFβ/SMAD dependent signaling pathway. *Journal of Cellular Biochemistry, 114*, 2101–2113.

Li, X., Zhang, Y., Kang, H., Liu, W., Liu, P., Zhang, J., et al. (2005). Sclerostin binds to LRP5/6 and antagonizes canonical Wnt signaling. *The Journal of Biological Chemistry, 280*, 19883–19887.

Lin, J., Patel, S. R., Cheng, X., Cho, E. A., Levitan, I., Ullenbruch, M., et al. (2005). Kielin/chordin-like protein, a novel enhancer of BMP signaling, attenuates renal fibrotic disease. *Nature Medicine, 11*, 387–393.

Lin, J., Patel, S. R., Wang, M., & Dressler, G. R. (2006). The cysteine-rich domain protein KCP is a suppressor of transforming growth factor beta/activin signaling in renal epithelia. *Molecular and Cellular Biology, 26*, 4577–4585.

Lintern, K. B., Guidato, S., Rowe, A., Saldanha, J. W., & Itasaki, N. (2009). Characterization of wise protein and its molecular mechanism to interact with both Wnt and BMP signals. *The Journal of Biological Chemistry, 284*, 23159–23168.

Loffredo, F. S., Steinhauser, M. L., Jay, S. M., Gannon, J., Pancoast, J. R., Yalamanchi, P., et al. (2013). Growth differentiation factor 11 is a circulating factor that reverses age-related cardiac hypertrophy. *Cell, 153*, 828–839.

Marques, S., Borges, A. C., Silva, A. C., Freitas, S., Cordenonsi, M., & Belo, J. A. (2004). The activity of the Nodal antagonist Cerl-2 in the mouse node is required for correct L/R body axis. *Genes & Development*, *18*, 2342–2347.
Masuda, S., Namba, K., Mutai, H., Usui, S., Miyanaga, Y., Kaneko, H., et al. (2014). A mutation in the heparin-binding site of noggin as a novel mechanism of proximal symphalangism and conductive hearing loss. *Biochemical and Biophysical Research Communications*, *447*, 496–502.
McClung, M. R., Grauer, A., Boonen, S., Bolognese, M. A., Brown, J. P., Diez-Perez, A., et al. (2014). Romosozumab in postmenopausal women with low bone mineral density. *The New England Journal of Medicine*, *370*, 412–420.
McMahon, J. A., Takada, S., Zimmerman, L. B., Fan, C. M., Harland, R. M., & McMahon, A. P. (1998). Noggin-mediated antagonism of BMP signaling is required for growth and patterning of the neural tube and somite. *Genes & Development*, *12*, 1438–1452.
Michos, O., Panman, L., Vintersten, K., Beier, K., Zeller, R., & Zuniga, A. (2004). Gremlin-mediated BMP antagonism induces the epithelial-mesenchymal feedback signaling controlling metanephric kidney and limb organogenesis. *Development*, *131*, 3401–3410.
Minabe-Saegusa, C., Saegusa, H., Tsukahara, M., & Noguchi, S. (1998). Sequence and expression of a novel mouse gene PRDC (protein related to DAN and cerberus) identified by a gene trap approach. *Development, Growth & Differentiation*, *40*, 343–353.
Mitola, S., Ravelli, C., Moroni, E., Salvi, V., Leali, D., Ballmer-Hofer, K., et al. (2010). Gremlin is a novel agonist of the major proangiogenic receptor VEGFR2. *Blood*, *116*, 3677–3680.
Miwa, H., Miyake, A., Kouta, Y., Shimada, A., Yamashita, Y., Nakayama, Y., et al. (2009). A novel neural-specific BMP antagonist, Brorin-like, of the Chordin family. *FEBS Letters*, *583*, 3643–3648.
Mukherjee, A., Sidis, Y., Mahan, A., Raher, M. J., Xia, Y., Rosen, E. D., et al. (2007). FSTL3 deletion reveals roles for TGF-beta family ligands in glucose and fat homeostasis in adults. *Proceedings of the National Academy of Sciences of the United States of America*, *104*, 1348–1353.
Müller, I. I., Knapik, E. W., & Hatzopoulos, A. K. (2006). Expression of the protein related to Dan and Cerberus gene—prdc—During eye, pharyngeal arch, somite, and swim bladder development in zebrafish. *Developmental Dynamics*, *235*, 2881–2888.
Mulloy, B., & Rider, C. C. (2006). Cytokines and proteoglycans: An introductory overview. *Biochemical Society Transactions*, *34*, 409–413.
Mulvihill, M. S., Kwon, Y. W., Lee, S., Fang, L. T., Choi, H., Ray, R., et al. (2012). Gremlin is overexpressed in lung adenocarcinoma and increases cell growth and proliferation in normal lung cells. *PLoS One*, *7*, e42264. http://dx.doi.org/10.1371/journal.pone.0042264.
Murashima-Suginami, A., Takahashi, K., Sakata, T., Tsukamoto, H., Sugai, M., Yanagita, M., et al. (2008). Enhanced BMP signaling results in supernumerary tooth formation in USAG-1 deficient mouse. *Biochemical and Biophysical Research Communications*, *369*, 1012–1016.
Myllärniemi, M., Lindholm, P., Ryynänen, M. J., Kliment, C. R., Salmenkivi, K., Keski-Oja, J., et al. (2008). Gremlin-mediated decrease in bone morphogenetic protein signaling promotes pulmonary fibrosis. *American Journal of Respiratory and Critical Care Medicine*, *177*, 321–329.
Nilsson, E. E., Larsen, G., & Skinner, M. K. (2014). Roles of Gremlin 1 and Gremlin 2 in regulating ovarian primordial to primary follicle transition. *Reproduction*, *147*, 865–874.
Nishinakamura, R., & Sakaguchi, M. (2014). BMP signaling and its modifiers in kidney development. *Pediatric Nephrology*, *29*, 681–686.

Nolan, K., Kattamuri, C., Luedeke, D. M., Angerman, E. B., Rankin, S. A., Stevens, M. L., et al. (2015). Structure of neuroblastoma suppressor of tumorigenicity 1 (NBL1): Insights for the functional variability across bone morphogenetic protein (BMP) antagonists. *The Journal of Biological Chemistry*, *290*, 4759–4771, pii: jbc.M114.628412.

Nolan, K., Kattamuri, C., Luedeke, D. M., Deng, X., Jagpal, A., Zhang, F., et al. (2013). Structure of protein related to Dan and Cerberus: Insights into the mechanism of bone morphogenetic protein antagonism. *Structure*, *21*, 1417–1429.

Oldknow, K. J., Seebacher, J., Goswami, T., Villen, J., Pitsillides, A. A., O'Shaughnessy, P. J., et al. (2013). Follistatin-like 3 (FSTL3) mediated silencing of transforming growth factor β (TGFβ) signaling is essential for testicular aging and regulating testis size. *Endocrinology*, *154*, 1310–1320.

Paine-Saunders, S., Viviano, B. L., Economides, A. N., & Saunders, S. (2002). Heparan sulfate proteoglycans retain Noggin at the cell surface: A potential mechanism for shaping bone morphogenetic protein gradients. *The Journal of Biological Chemistry*, *277*, 2089–2096.

Patel, S. R., Gordon, J., Mahbub, F., Blackburn, C. C., & Manley, N. R. (2006). Bmp4 and Noggin expression during early thymus and parathyroid organogenesis. *Gene Expression Patterns*, *6*, 794–799.

Piccolo, S., Agius, E., Leyns, L., Bhattacharyya, S., Grunz, H., Bouwmeester, T., et al. (1999). The head inducer Cerberus is a multifunctional antagonist of Nodal, BMP and Wnt signals. *Nature*, *397*, 707–710.

Plouhinec, J. L., Zakin, L., Moriyama, Y., & De Robertis, E. M. (2013). Chordin forms a self-organizing morphogen gradient in the extracellular space between ectoderm and mesoderm in the Xenopus embryo. *Proceedings of the National Academy of Sciences of the United States of America*, *110*, 20372–20379.

Rider, C. C., & Mulloy, B. (2010). Bone morphogenetic protein and growth differentiation factor cytokine families and their protein antagonists. *The Biochemical Journal*, *429*, 1–12.

Roch, G. J., & Sherwood, N. M. (2014). Glycoprotein hormones and their receptors emerged at the origin of metazoans. *Genome Biology and Evolution*, *6*, 1466–1479.

Roudier, M., Li, X., Niu, Q. T., Pacheco, E., Pretorius, J. K., Graham, K., et al. (2013). Sclerostin is expressed in articular cartilage but loss or inhibition does not affect cartilage remodeling during aging or following mechanical injury. *Arthritis and Rheumatism*, *65*, 721–731.

Roxburgh, S. A., Kattla, J. J., Curran, S. P., O'Meara, Y. M., Pollock, C. A., Goldschmeding, R., et al. (2009). Allelic depletion of grem1 attenuates diabetic kidney disease. *Diabetes*, *58*, 1641–1650.

Sarras, M. P. (1996). BMP-1 and the astacin family of metalloproteinases: A potential link between the extracellular matrix, growth factors and pattern formation. *Bioessays*, *18*, 439–442.

Sawant, A., Chanda, D., Isayeva, T., Tsuladze, G., Garvey, W. T., & Ponnazhagan, S. (2012). Noggin is novel inducer of mesenchymal stem cell adipogenesis: Implications for bone health and obesity. *The Journal of Biological Chemistry*, *287*, 12241–12249.

Schwaninger, R., Rentsch, C. A., Wetterwald, A., van der, H. G., van Bezooijen, R. L., van der, P. G., et al. (2007). Lack of noggin expression by cancer cells is a determinant of the osteoblast response in bone metastases. *The American Journal of Pathology*, *170*, 160–175.

Seemann, P., Brehm, A., König, J., Reissner, C., Stricker, S., Kuss, P., et al. (2009). Mutations in GDF5 reveal a key residue mediating BMP inhibition by noggin. *PLoS Genetics*, *5*(11), e1000747. http://dx.doi.org/10.1371/journal.pgen.1000747.

Semënov, M., Tamai, K., & He, X. (2005). SOST is a ligand for LRP5/LRP6 and a Wnt signaling inhibitor. *The Journal of Biological Chemistry*, *280*, 26770–26775.

Sengle, G., Ono, R. N., Sasaki, T., & Sakai, L. Y. (2011). Prodomains of transforming growth factor beta (TGFbeta) superfamily members specify different functions:

Extracellular matrix interactions and growth factor bioavailability. *The Journal of Biological Chemistry*, *286*, 5087–5099.

Shimada, T., Takai, Y., Shinohara, K., Yamasaki, A., Tominaga-Yoshino, K., Ogura, A., et al. (2012). A simplified method to generate serotonergic neurons from mouse embryonic stem and induced pluripotent stem cells. *Journal of Neurochemistry*, *122*, 81–93.

Shin, J. A., Lim, S. M., Jeong, S. I., Kang, J. L., & Park, E. M. (2014). Noggin improves ischemic brain tissue repair and promotes alternative activation of microglia in mice. *Brain, Behavior, and Immunity*, *40*, 143–154.

Sidis, Y., Mukherjee, A., Keutmann, H., Delbaere, A., Sadatsuki, M., & Schneyer, A. (2006). Biological activity of follistatin isoforms and follistatin-like-3 is dependent on differential cell surface binding and specificity for activin, myostatin, and bone morphogenetic proteins. *Endocrinology*, *147*, 3586–3597.

Simmons, D. G., & Kennedy, T. G. (2002). Uterine sensitization-associated gene-1: A novel gene induced within the rat endometrium at the time of uterine receptivity/sensitization for the decidual cell reaction. *Biology of Reproduction*, *67*, 1638–1645.

Singh, R., Braga, M., & Pervin, S. (2014). Regulation of brown adipocyte metabolism by myostatin/follistatin signaling. *Frontiers in Cell and Developmental Biology*. *2*. http://dx.doi.org/10.3389/fcell.2014.00060.

Sinha, M., Jang, Y. C., Oh, J., Khong, D., Wu, E. Y., Manohar, R., et al. (2014). Restoring systemic GDF11 levels reverses age-related dysfunction in mouse skeletal muscle. *Science*, *344*, 649–652.

Sneddon, J. B., Zhen, H. H., Montgomery, K., van de Rijn, M., Tward, A. D., West, R., et al. (2006). Bone morphogenetic protein antagonist gremlin 1 is widely expressed by cancer-associated stromal cells and can promote tumor cell proliferation. *Proceedings of the National Academy of Sciences of the United States of America*, *103*, 14842–14847.

Song, K., Krause, C., Shi, S., Patterson, M., Suto, R., Grgurevic, L., et al. (2010). Identification of a key residue mediating bone morphogenetic protein (BMP)-6 resistance to noggin inhibition allows for engineered BMPs with superior agonist activity. *The Journal of Biological Chemistry*, *285*, 12169–12180.

Sotillo Rodriguez, J. E., Mansky, K. C., Jensen, E. D., Carlson, A. E., Schwarz, T., Pham, L., et al. (2009). Enhanced osteoclastogenesis causes osteopenia in twisted gastrulation-deficient mice through increased BMP signaling. *Journal of Bone and Mineral Research*, *24*, 1917–1926.

Stabile, H., Mitola, S., Moroni, E., Belleri, M., Nicoli, S., Coltrini, D., et al. (2007). Bone morphogenic protein antagonist Drm/gremlin is a novel proangiogenic factor. *Blood*, *109*, 1834–1840.

Sudo, S., Avsian-Kretchmer, O., Wang, L. S., & Hsueh, A. J. (2004). Protein related to DAN and cerberus is a bone morphogenetic protein antagonist that participates in ovarian paracrine regulation. *The Journal of Biological Chemistry*, *279*, 23134–23141.

Sun, M., Forsman, C., Sergi, C., Gopalakrishnan, R., O'Connor, M. B., & Petryk, A. (2010). The expression of twisted gastrulation in postnatal mouse brain and functional implications. *Neuroscience*, *169*, 920–931.

Sun, P., Wang, J., Zheng, Y., Fan, Y., & Gu, Z. (2012). BMP2/7 heterodimer is a stronger inducer of bone regeneration in peri-implant bone defects model than BMP2 or BMP7 homodimer. *Dental Materials Journal*, *31*, 239–248.

Sun, J., Zhuang, F. F., Mullersman, J. E., Chen, H., Robertson, E. J., Warburton, D., et al. (2006). BMP4 activation and secretion are negatively regulated by an intracellular gremlin-BMP4 interaction. *The Journal of Biological Chemistry*, *281*, 29349–29356.

Suzuki, R., Matsumoto, M., Fujikawa, A., Kato, A., Kuboyama, K., Yonehara, K., et al. (2014). SPIG1 negatively regulates BDNF maturation. *The Journal of Neuroscience*, *34*, 3429–3442.

Sylva, M., Moorman, A. F., & van den Hoff, M. J. (2013). Follistatin-like 1 in vertebrate development. *Birth Defects Research. Part C, Embryo Today*, *99*, 61–69.

Tanaka, M., Asada, M., Higashi, A. Y., Nakamura, J., Oguchi, A., Tomita, M., et al. (2010). Loss of the BMP antagonist USAG-1 ameliorates disease in a mouse model of the progressive hereditary kidney disease Alport syndrome. *The Journal of Clinical Investigation*, *120*, 768–777.

Tanwar, V., Bylund, J. B., Hu, J., Yan, J., Walthall, J. M., Mukherjee, A., et al. (2014). Gremlin 2 promotes differentiation of embryonic stem cells to atrial fate by activation of the JNK signaling pathway. *Stem Cells*, *32*, 1774–1788.

Tarragona, M., Pavlovic, M., Arnal-Estapé, A., Urosevic, J., Morales, M., Guiu, M., et al. (2012). Identification of *NOG* as a specific breast cancer bone metastasis-supporting gene. *The Journal of Biological Chemistry*, *287*, 21346–21355.

Tavares, A. T., Andrade, S., Silva, A. C., & Belo, J. A. (2007). Cerberus is a feedback inhibitor of Nodal asymmetric signaling in the chick embryo. *Development*, *134*, 2051–2060.

Troilo, H., Zuk, A. V., Tunnicliffe, R. B., Wohl, A. P., Berry, R., Collins, R. F., et al. (2014). Nanoscale structure of the BMP antagonist chordin supports cooperative BMP binding. *Proceedings of the National Academy of Sciences*, *111*, 13063–13068.

Turk, T., Leeuwis, J. W., Gray, J., Torti, S. V., Lyons, K. M., Nguyen, T. Q., et al. (2009). BMP signaling and podocyte markers are decreased in human diabetic nephropathy in association with CTGF overexpression. *The Journal of Histochemistry and Cytochemistry*, *57*, 623–631.

Valera, E., Isaacs, M. J., Kawakami, Y., Izpisúa Belmonte, J. C., & Choe, S. (2010). BMP-2/6 heterodimer is more effective than BMP-2 or BMP-6 homodimers as inductor of differentiation of human embryonic stem cells. *PLoS One*, *5*, e11167. http://dx.doi.org/10.1371/journal.pone.0011167.

Veverka, V., Henry, A. J., Slocombe, P. M., Ventom, A., Mulloy, B., Muskett, F. W., et al. (2009). Characterization of the structural features and interactions of sclerostin: Molecular insight into a key regulator of Wnt-mediated bone formation. *The Journal of Biological Chemistry*, *284*, 10890–10900.

Wang, D. J., Zhi, X. Y., Zhang, S. C., Jiang, M., Liu, P., Han, X. P., et al. (2012). The bone morphogenetic protein antagonist Gremlin is overexpressed in human malignant mesothelioma. *Oncology Reports*, *27*, 58–64.

Wei, Z., Salmon, R. M., Upton, P. D., Morrell, N. W., & Li, W. (2014). Regulation of bone morphogenetic protein 9 (BMP9) by redox-dependent proteolysis. *The Journal of Biological Chemistry*, *289*, 31150–31159.

Welt, C., Sidis, Y., Keutmann, H., & Schneyer, A. (2002). Activins, inhibins, and follistatins: From endocrinology to signaling. A paradigm for the new millennium. *Experimental Biology and Medicine*, *227*, 724–752.

Wills, A., Harland, R. M., & Khokha, M. K. (2006). Twisted gastrulation is required for forebrain specification and cooperates with Chordin to inhibit BMP signaling during *X. tropicalis* gastrulation. *Developmental Biology*, *289*, 166–178.

Winkler, D. G., Sutherland, M. K., Geoghegan, J. C., Yu, C., Hayes, T., Skonier, J. E., et al. (2003). Osteocyte control of bone formation via sclerostin, a novel BMP antagonist. *The EMBO Journal*, *22*, 6267–6276.

Winkler, D. G., Yu, C., Geoghegan, J. C., Ojala, E. W., Skonier, J. E., Shpektor, D., et al. (2004). Noggin and sclerostin bone morphogenetic protein antagonists form a mutually inhibitory complex. *The Journal of Biological Chemistry*, *279*, 36293–36298.

Xie, J., & Fisher, S. (2005). Twisted gastrulation enhances BMP signaling through chordin dependent and independent mechanisms. *Development*, *132*, 383–391.

Yamada, S., Nakamura, J., Asada, M., Takase, M., Matsusaka, T., Iguchi, T., et al. (2014). Twisted gastrulation, a BMP antagonist, exacerbates podocyte injury. *PLoS One*, *9*, e89135. http://dx.doi.org/10.1371/journal.pone.0089135.

Yamanishi, T., Katsu, K., Funahashi, J., Yumoto, E., & Yokouchi, Y. (2007). Dan is required for normal morphogenesis and patterning in the developing chick inner ear. *Development, Growth & Differentiation*, *49*, 13–26.

Yanagita, M., Okuda, T., Endo, S., Tanaka, M., Takahashi, K., Sugiyama, F., et al. (2006). Uterine sensitization-associated gene-1 (USAG-1), a novel BMP antagonist expressed in the kidney, accelerates tubular injury. *The Journal of Clinical Investigation*, *116*, 70–79.

Yokouchi, Y., Vogan, K. J., Pearse, R. V., & Tabin, C. J. (1999). Antagonistic signaling by Caronte, a novel Cerberus-related gene, establishes left-right asymmetric gene expression. *Cell*, *98*, 573–583.

Yu, X., He, F., Zhang, T., Espinoza-Lewis, R. A., Lin, L., Yang, J., et al. (2008). Cerberus functions as a BMP agonist to synergistically induce nodal expression during left-right axis determination in the chick embryo. *Developmental Dynamics*, *237*, 3613–3623.

Zakin, L., Metzinger, C. A., Chang, E. Y., Coffinier, C., & De Robertis, E. M. (2008). Development of the vertebral morphogenetic field in the mouse: Interactions between Crossveinless-2 and Twisted Gastrulation. *Developmental Biology*, *323*, 6–18.

Zhang, Q., Shi, Y., Wada, J., Malakauskas, S. M., Liu, M., Ren, Y., et al. (2010). In vivo delivery of Gremlin siRNA plasmid reveals therapeutic potential against diabetic nephropathy by recovering bone morphogenetic protein-7. *PLoS One*, *5*, e11709. http://dx.doi.org/10.1371/journal.pone.0011709.

Zuniga, E., Rippen, M., Alexander, C., Schilling, T. F., & Crump, J. G. (2011). Gremlin 2 regulates distinct roles of BMP and endothelin 1 signaling in dorsoventral patterning of the facial skeleton. *Development*, *138*, 5147–5156.

CHAPTER THREE

BMP-7 Signaling and its Critical Roles in Kidney Development, the Responses to Renal Injury, and Chronic Kidney Disease

Scott R. Manson[1], Paul F. Austin, Qiusha Guo, Katelynn H. Moore
Department of Surgery, Division of Urology, Washington University School of Medicine, St. Louis Children's Hospital, St. Louis, Missouri, USA
[1]Corresponding author: e-mail address: mansons@wustl.edu

Contents

Abstract

Chronic kidney disease (CKD) is a significant health problem that most commonly results from congenital abnormalities in children and chronic renal injury in adults. The therapeutic potential of BMP-7 was first recognized nearly two decades ago with studies demonstrating its requirement for kidney development and ability to inhibit the

Vitamins and Hormones, Volume 99
ISSN 0083-6729
http://dx.doi.org/10.1016/bs.vh.2015.05.003

pathogenesis of renal injury in models of CKD. Since this time, our understanding of CKD has advanced considerably and treatment strategies have evolved with the identification of many additional signaling pathways, cell types, and pathologic processes that contribute to disease progression. The purpose of this review is to revisit the seminal studies that initially established the importance of BMP-7, highlight recent advances in BMP-7 research, and then integrate this knowledge with current research paradigms. We will provide an overview of the evolutionarily conserved roles of BMP proteins and the features that allow BMP signaling pathways to function as critical signaling nodes for controlling biological processes, including those related to CKD. We will discuss the multifaceted functions of BMP-7 during kidney development and the potential for alterations in BMP-7 signaling to result in congenital abnormalities and pediatric kidney disease. We will summarize the renal protective effects of recombinant BMP-7 in experimental models of CKD and then propose a model to describe the potential physiological role of endogenous BMP-7 in the innate repair mechanisms of the kidneys that respond to renal injury. Finally, we will highlight emerging clinical approaches for applying our knowledge of BMP-7 toward improving the treatment of patients with CKD.

1. INTRODUCTION

More than 31 million Americans (~10% of the population) are afflicted with chronic kidney disease (CKD) (Collins et al., 2014). While CKD has already reached pandemic proportions, its significance is only exacerbated by reports that the incidence of CKD continues to rise (Hamer & El Nahas, 2006). Patients with end-stage renal disease (ESRD) are faced with long-term dialysis and kidney transplantation. However, even these options for renal replacement are associated with considerable morbidity and mortality (Collins et al., 2014). Existing pharmacologic therapies are typically unable to prevent or even slow progression to ESRD and the most effective treatment option remains managing the underlying causes of CKD. Therefore, it is imperative that we develop a better understanding of both the conditions that cause CKD and the mechanisms that drive progression to ESRD (National Kidney Foundation, 2002). Achieving these goals may ultimately enable the development of improved therapies for patients with CKD.

Over the past two decades, studies have revealed important roles in the kidneys for the bone morphogenetic proteins (BMPs) that are members of the ubiquitous TGF-β superfamily of proteins. The BMP proteins were discovered in 1988 by isolating proteins from bone explants that are capable of stimulating *de novo* bone formation (Wozney et al., 1988). Since this time, nearly 20 BMP proteins have been characterized and their signaling pathways have been implicated in cellular processes that contribute to development, tissue homeostasis, and disease progression in numerous organ systems. While BMP proteins are broadly important, the most prominent role in the kidneys has been widely attributed to BMP-7. Indeed, BMP-7 is the only BMP family member that has been demonstrated to be either absolutely required for kidney development (Dudley, Lyons, & Robertson, 1995; Luo et al., 1995) or inhibit disease progression in experimental models of acute kidney injury (AKI) (Vukicevic et al., 1998) and CKD (Hruska et al., 2000). These findings strongly suggest that further defining the physiologic roles of BMP-7 in the kidneys and evaluating their importance in the pathogenesis and treatment of CKD are key steps in the development of renal protective therapies.

2. THE TGF-β/BMP PROTEIN SUPERFAMILY

The biological importance of BMP signaling can be traced back throughout much of evolutionary history. Indeed, the acquisition of mechanisms for cell-to-cell signaling was a landmark event in the evolution of metazoans. It was this property that allowed for the existence of distinct cell types that are capable of functioning independently while still coordinating their actions with other cells to maintain the framework of a multicellular organism (Pires-daSilva & Sommer, 2003). This is evident in humans by the gradients of growth factors that control morphogenic processes during development, the multitude of cytokines that coordinate inflammatory responses and tissue regeneration during wound healing, the mutations that disrupt these communication networks to drive tumorigenesis, and countless other processes. While primitive metazoans possessed only a limited number of signaling proteins, over time they evolved into superfamilies as increasing biological complexity demanded a wide array of signaling proteins to control tissue-specific functions. This included the TGF-β/BMP protein superfamily, which is one of only seven superfamilies of signaling proteins to have been conserved from early metazoans, spanning over 600 million years of evolution (Gerhart, 1999).

TGF-β/BMP proteins are linked by their signaling functions as secreted glycoproteins, significant homology within their receptor-binding domains, and characteristic cysteine-knot fold tertiary structure. To date, this superfamily includes 35 members organized into four homology subgroups: TGF-β proteins, BMP/GDF proteins, GDNF proteins, and activin/inhibin proteins. These subgroups emerged at differing points in evolutionary history as evidenced by the expression of BMP/GDF proteins in nematodes and flies, activins/inhibins in flies but not nematodes, and TGF-β and GDNF proteins in only vertebrates (Hinck, 2012). This strongly suggests that BMP/GDF proteins were the prototypical members of this superfamily, from which all other TGF-β/BMP proteins evolved. Indeed, BMP orthologs are required for body morphogenesis, dorsal–ventral patterning, and other critical processes in *Drosophila melanogaster* and *Caenorhabditis elegans* (Irish & Gelbart, 1987; Suzuki et al., 1999). This role has only expanded over evolutionary history as primitive functions have been conserved and new functions have emerged with the assimilation of additional BMP family members.

3. THE BMP-7 PROTEIN

Since the discovery of BMP proteins in humans (Wozney et al., 1988), thousands of manuscripts have described the important functions of BMP proteins, which encompass every organ system. However, it is largely BMP-7 that has captured the attention of nephrology researchers. BMP-7 was first identified in bovine as osteogenic protein-1 (OP-1) (Sampath et al., 1990) and soon thereafter a human ortholog was cloned (Ozkaynak et al., 1990) and found to be highly expressed in the kidneys (Ozkaynak, Schnegelsberg, & Oppermann, 1991). This was followed by seminal studies which demonstrated its requirement for kidney development (Dudley et al., 1995; Luo et al., 1995) and renal protective functions in models of AKI (Vukicevic et al., 1998) and CKD (Hruska et al., 2000).

The *Bmp7* gene is located at the 20q13.31 chromosomal locus and encodes a mature 4.0 kb mRNA transcript with seven exons, which is translated into a 431-amino-acid protein. BMP-7 shows the most significant homology to BMP-5, BMP-6, and BMP-8, particularly within the carboxy-terminal regions which are processed to become the mature, secreted form of BMP-7. Together, BMP5–8 constitute what is known as the *gbb-60A* subgroup of the BMP family based on their homology to this protein in *D. melanogaster* (Wharton, Thomsen, & Gelbart, 1991). Nonetheless, the protein domains, modular organization, and overall structure of BMP-7 are shared by other BMP proteins and, moreover, the entire TGF-β/BMP protein superfamily. This common architecture includes several prominent features that are critical for the signaling functions of BMP-7 (Fig. 1):

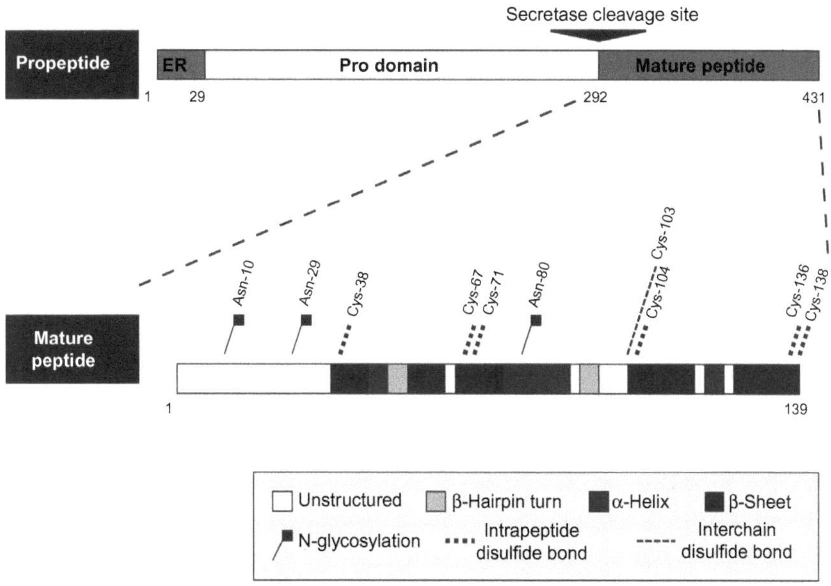

Figure 1 *Structural organization of the BMP-7 protein.* BMP-7 is synthesized as a 431-amino-acid propeptide containing a leader sequence that targets BMP-7 to secretory pathways in the ER, a large prodomain that facilitates protein folding, and a C-terminal mature peptide that is released by cleavage of the prodomain to form the cell-to-cell signaling component of BMP-7. The 139-amino-acid mature peptide adopts its characteristic cysteine-knot fold tertiary structure through structural elements including α-helicies, β-sheets, and β-turns as well as the formation of intrapeptide disulfide bonds and N-linked glycosylation, as indicated. BMP-7 monomers are then linked to create an antiparallel dimer that is stabilized by an interchain disulfide bond.

1. *Amino-Terminal Signal Sequence*—A small, hydrophobic leader sequence (Met_1-Ala_{29}) that targets BMP-7 to secretory pathways in the endoplasmic reticulum.
2. *Prodomain/Latency-Associated Peptide*—A large prodomain (Asp_{30}-Arg_{292}) that facilitates protein folding, improves solubility, and promotes the quaternary interactions of BMP-7. The precursor protein is formed in the cytoplasm and cleaved at Arg_{289}-Ser_{290}-Ile_{291}-Arg_{292} by subtilisin-like secretases (Constam & Robertson, 1999). The released prodomain then remains noncovalently associated with the mature peptide and targets it to the extracellular matrix after secretion (Gregory et al., 2005).
3. *Mature Signaling Peptide*—A mature 15.7 kDa peptide that mediates the cell-to-cell signaling functions of BMP-7. This is generated from the 139-amino-acid sequence contained between Ser_{293}–His_{431}.

4. *Glycosylation Sites*—Four N-linked glycosylation sites that facilitate protein folding, target BMP-7 for secretion following sorting in the Golgi apparatus, and contribute to protein–protein interactions. These sites are found at Asn_{187}, Asn_{302}, Asn_{321}, and Asn_{372}.
5. *Intrapeptide Disulfide Bonds*—Six conserved cysteine residues that form three intrapeptide disulfide bonds to stabilize the BMP-7 monomer in its characteristic cysteine-knot fold tertiary structure (Fig. 2). These bonds connect Cys_{330}-Cys_{396}, Cys_{359}-Cys_{428}, and Cys_{363}-Cys_{430} (Griffith et al., 1996).
6. *Interchain Disulfide Bonds*—An interchain disulfide bond that bridges Cys_{395} residues on adjacent BMP-7 monomers to form a functional 31.5 kDa homodimer (Griffith et al., 1996). Heterodimers between BMP-7 and other BMP proteins have also been shown to form both *in vitro* and *in vivo* and there is evidence that their signaling activity may even be greater than that of homodimers (Israel et al., 1996; Little & Mullins, 2009).

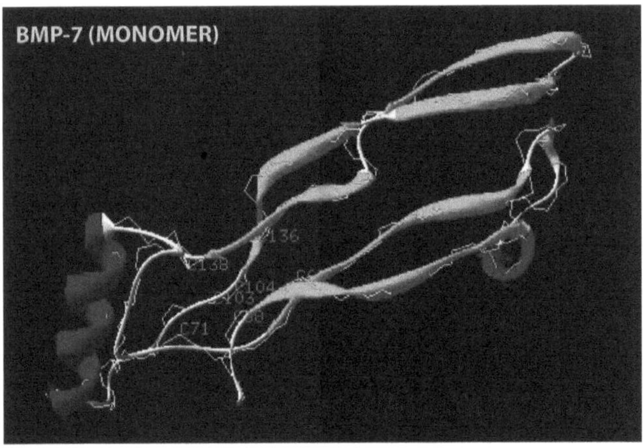

Figure 2 *The cysteine-knot fold tertiary structure of BMP-7.* The BMP-7 monomer folds into the cysteine-knot fold tertiary structure that is characteristic of TGF-β/BMP family members. As indicated, this structure is stabilized by three intrapeptide disulfide bonds that connect Cys_{38}–Cys_{104}, Cys_{67}–Cys_{136}, and Cys_{71}–Cys_{138}. An additional interchain disulfide bond links monomers by joining the Cys_{103} residues on each chain. *Structural information was obtained from entry* 1BMP *(Griffith, Keck, Sampath, Rueger, & Carlson, 1996) in the Protein Data Bank at the Research Collaboratory for Structural Bioinformatics and then a ribbon diagram was created using DeepView PDB-Viewer.* (See the color plate.)

4. TGF-β/BMP SIGNALING PATHWAYS

The functions of BMP-7 as a pleiotropic cytokine necessitate mechanisms for transducing this extracellular signal into changes in the intrinsic behavior of its target cells. Similar to other cytokines, these signaling mechanisms involve extracellular binding of the ligand to activate a transmembrane receptor, transmission to intracellular signaling cascades, and stimulation of transcriptional regulatory networks in the nucleus. The TGF-β/BMP proteins largely employ an identical strategy to accomplish these goals. However, the evolution of the TGF-β/BMP superfamily was paralleled by increases in the diversity of receptors, signaling effectors, and transcription factors in order to maintain the ability to elicit ligand-specific biological responses (Hinck, 2012). In addition, these signaling pathways have incorporated agonists, antagonists, and other complex regulatory mechanisms to allow these responses to be fine-tuned through cross talk with other signaling pathways. Here, we will review TGF-β/BMP-dependent signaling pathways and the key steps for regulating these responses. However, these mechanisms are incredibly complex and we refer the reader to several excellent reviews for a more detailed discussion (Massague, Seoane, & Wotton, 2005; Sieber, Kopf, Hiepen, & Knaus, 2009; Walsh, Godson, Brazil, & Martin, 2010; Weiskirchen & Meurer, 2013; Zhang, 2009).

4.1 SMAD-Dependent Mechanisms for TGF-β/BMP Signaling

4.1.1 Ligand Binding and Receptor Activation

The first step in the canonical TGF-β/BMP pathway (Fig. 3) is the binding of the extracellular ligand to a transmembrane receptor complex which propagates this signal across the plasma membrane (Fig. 4). Signal transduction requires the formation of a heteromeric complex consisting of a TGF-β/BMP ligand dimer, two Type I receptors, and two Type II receptors (Liu, Ventura, Doody, & Massague, 1995; Ruberte, Marty, Nellen, Affolter, & Basler, 1995). The Type I and Type II receptors are similar in that they both have an extracellular ligand-binding domain and a short, cytoplasmic tail containing a serine/threonine kinase domain. However, while Type II receptors have constitutive kinase activity, activation of the kinase activity in Type I receptors requires ligand binding and phosphorylation of a glycine-serine rich region (the GS box) in its cytoplasmic tail (Mueller & Nickel, 2012; Sieber et al., 2009). Thus, ligand binding results in

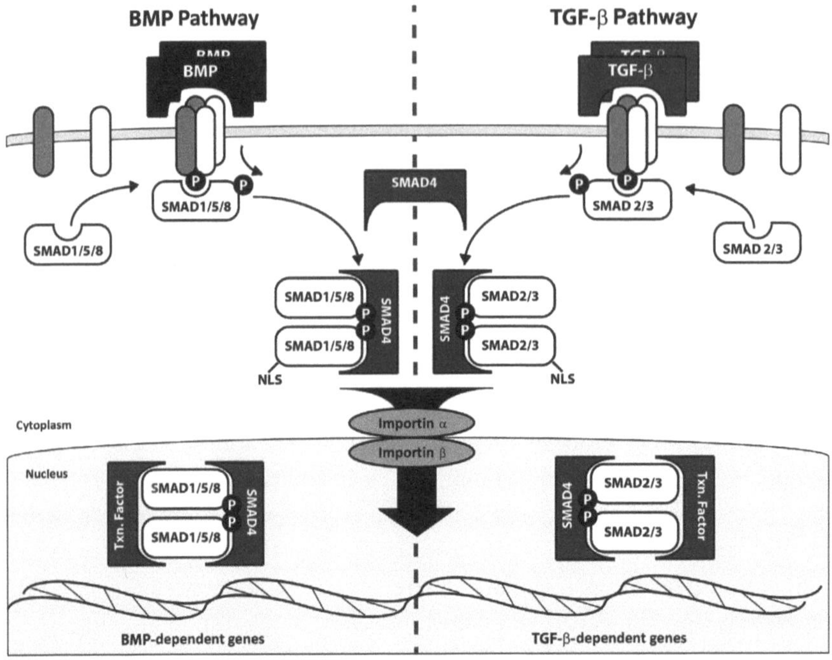

Figure 3 *The canonical, SMAD-dependent pathway for TGF-β/BMP signaling.* A ligand dimer binds and stabilizes a heteromeric receptor complex containing two Type I receptors and two Type II receptors. Subsequently, the Type II receptor phosphorylates the Type I receptor to create a docking site for binding to R-SMAD proteins. The R-SMAD protein is phosphorylated and then dissociates from the receptor complex and binds the C-SMAD protein. This heterodimer translocates to the nucleus and interacts with additional transcription factors to facilitate promoter binding and the regulation of gene transcription.

stabilization of this heteromeric receptor complex and conformational changes that bring the Type II receptor within sufficient proximity to phosphorylate and activate the Type I receptor (Greenwald et al., 2003; Kirsch, Sebald, & Dreyer, 2000).

The specificity of TGF-β-dependent and BMP-dependent responses is mediated by receptors that are exclusively devoted to each of these pathways. TGF-β ligands bind to one of four different Type I receptors (ALK1, ALK4, ALK5, and ALK7) and the Type II receptor TGFβRII. However, the mechanisms are more complicated in BMP signaling where there are three different Type I receptors (ALK2, ALK3, and ALK6) and also three different Type II receptors (BmpRII, ActRIIa, and ActRIIb) that can exist in numerous combinations (Mueller & Nickel, 2012; Sieber et al.,

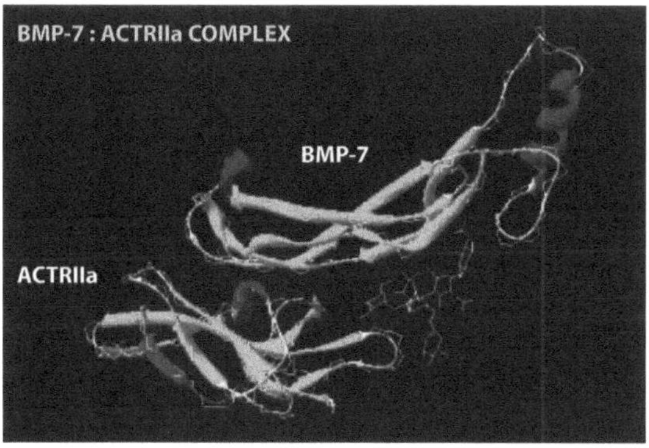

Figure 4 *Binding of BMP-7 to the Type II BMP receptor ActRIIa.* The crystal structure of the complex formed between BMP-7 and the extracellular domain of the Type II receptor ActRIIa demonstrates that specificity is mediated by interactions between the side chains on the ligand and its receptor. *Structural information was obtained from entry 1LX5 (Greenwald et al., 2003) in the Protein Data Bank at the Research Collaboratory for Structural Bioinformatics and then a ribbon diagram was created using DeepView PDB-Viewer.* (See the color plate.)

2009). Furthermore, individual BMP ligands have differing specificities for these receptors. While BMP-2 and BMP-4 signal exclusively through ALK3 and ALK6, BMP-7 binds to ALK2 and ALK6 with high affinity and ALK3 with only low affinity (Greenwald et al., 2003; Knaus & Sebald, 2001; Koenig et al., 1994; Mueller & Nickel, 2012; ten Dijke et al., 1994).

4.1.2 Activation of SMAD Transcription Factors

Following receptor activation, the phosphorylated cytoplasmic tail of the Type I receptor serves as a binding site for effector proteins which it can then phosphorylate and activate. In the canonical TGF-β/BMP pathway, these effectors are members of the SMAD protein family, which bind to the phosphorylated residues in the Type I receptor via a region rich in basic amino acids in the conserved MH2 domain at the carboxyl terminus of the protein. The SMAD protein is subsequently phosphorylated at a Ser-Ser-X-Ser motif which is also located in the MH2 domain (Derynck & Zhang, 2003; Massague et al., 2005).

The activation of SMAD proteins provides a second opportunity for establishing the specificity of TGF-β and BMP responses. While the TGF-β pathway activates Smad2 and Smad3, the BMP pathway activates Smad1, Smad5, and Smad8. Together, these proteins constitute what are known as the receptor-regulated SMAD proteins (R-SMADs)

(Derynck & Zhang, 2003; Massague et al., 2005). A region within the kinase domain of the Type I receptor known as the L45 loop determines its specificity for the appropriate R-SMAD proteins, while a complementary region within the conserved, N-terminal MH1 domain of the R-SMAD known as the L3 loop determines its specificity for Type I receptors (Feng & Derynck, 1997; Lo, Chen, Shi, Pavletich, & Massague, 1998).

4.1.3 Heterodimerization and Nuclear Translocation of SMAD Transcription Factors

Following the activating phosphorylation of R-SMAD proteins by the Type I receptor, there is a conformational shift in the R-SMAD that promotes its dissociation from the receptor complex. Subsequently, the phosphorylated MH2 domain in the R-SMAD protein serves as a docking site for binding a basic amino acid-rich region in the MH2 domain of the Smad4 protein (Chacko et al., 2001). This protein is known as common SMAD (C-SMAD) as it participates in both the TGF-β and BMP signaling pathways. Typically, the R-SMAD/C-SMAD complex is formed as a heterotrimer containing two R-SMAD proteins and one C-SMAD protein (Chacko et al., 2001). Phosphorylation of the R-SMAD protein has an additional consequence in that it triggers conformational changes that expose a nuclear localization signal in its MH1 domain. As a result, it is able to translocate to the nucleus via importin α/β-dependent mechanisms either in complex with C-SMAD or as a monomer which then binds to C-SMAD in the nucleus (Xu, Alarcon, Col, & Massague, 2003; Xu, Chen, & Massague, 2000).

4.1.4 SMAD-Dependent Regulation of Transcription

This signaling cascade culminates with the induction of transcriptional responses in the nucleus. The MH1 domains of R-SMAD and C-SMAD both contain β-hairpin structures that confer DNA-binding activity. The SMAD proteins have most commonly been shown to bind to the *CAGATCTG* sequence that is known as a SMAD Response Element or SRE (Shi et al., 1998). However, SMAD proteins have also been shown to bind to the *GCCGnCGC* sequence known as a GC-rich box (Kim, Johnson, Chen, Carroll, & Laughon, 1997; Kusanagi et al., 2000). It is important to note that, irrespective of the response element, SMAD proteins bind to DNA with low affinity and interactions with additional factors are required for high affinity binding to DNA. Indeed, a wide range of transcription factors have been implicated in the SMAD-dependent transcriptional responses triggered by BMP proteins, including Runx1-3, Hoxc8,

MyoD, Msx1, OAZ, and YY1 (Massague et al., 2005; Miyazono, Maeda, & Imamura, 2005). In contrast, little is known about the genes regulated by BMP signaling. The most established transcriptional targets are genes encoding the Id (*Id1-4*) family of transcription factors which are critical for regulating cellular proliferation and differentiation (Miyazono & Miyazawa, 2002; Ying, Nichols, Chambers, & Smith, 2003). Nonetheless, given the disparate biological functions of BMP proteins, it is clear that we are only just beginning to understand the extent of transcriptional regulation by BMP signaling and further studies directed toward identifying its target genes are greatly needed.

4.2 SMAD-Independent Mechanisms for TGF-β/BMP Signaling

The canonical, SMAD-dependent pathway for TGF-β/BMP signaling undoubtedly plays an important role and is responsible for many, if not the vast majority, of the biological responses to these signaling proteins. However, it is important to note that a number of SMAD-independent signaling pathways have also been described. This includes pathways regulated by the MAP kinases (p38, JNK, and ERK), Rho GTPases (RhoA, Rac, Cdc42), PKC, and PI3K (Moustakas & Heldin, 2005; Zhang, 2009). The significance of these and other SMAD-independent pathways is highlighted by the observation that Smad4-deficient cells treated with TGF-β or BMP proteins still exhibit many transcriptional responses in the absence of SMAD signaling (Engel, McDonnell, Law, & Moses, 1999; Hocevar, Brown, & Howe, 1999; Sirard et al., 2000).

SMAD-independent signaling pathways have been most commonly studied in the response to TGF-β. There are several conduits for the activation of SMAD-independent signaling pathways by activated TGF-β receptors. While TGF-β receptors are serine/threonine kinases, low levels of tyrosine phosphorylation also occur and create a docking site for ShcA/Grb2/Sos complexes that activate the Ras/Raf/MEK/ERK cascade (Lee et al., 2007; Mulder & Morris, 1992). In a second pathway, the activation of TGF-β receptors leads to binding and activation of the TRAF6 protein which typically functions in signaling by Toll-like receptors. TRAF6 activates the TAK1 kinase which, in turn, stimulates JNK activation via MKK4 and p38 activation via MKK3/6 (Sorrentino et al., 2008; Yamashita et al., 2008). Similar mechanisms involving the interaction of key signaling components with activated TGF-β receptors lead to the activation of Rho GTPases, PKC, and PI3K (Moustakas & Heldin, 2005; Zhang, 2009). While each of these pathways are likely to play important roles, the widespread effects of TAK1 inactivation on TGF-β-dependent responses suggests that the TAK1-MKK4-JNK and TAK1-MKK3/6-p38 pathways are of particular importance in SMAD-independent signaling by TGF-β (Shim et al., 2005).

The importance of TAK1 extends to SMAD-independent signaling by BMP proteins. Indeed, TAK1 signaling is also required for the biological responses to BMP proteins in many contexts (Shibuya et al., 1998; Shim et al., 2009). However, there are distinct mechanisms for the activation of TAK1 signaling in BMP-dependent pathways (Fig. 5). Activated BMP receptors instead associate with XIAP to bridge an interaction with TAK1-binding protein and TAK1 (Yamaguchi et al., 1999). The activation of TAK1 then stimulates the p38 and JNK pathways in a manner similar to

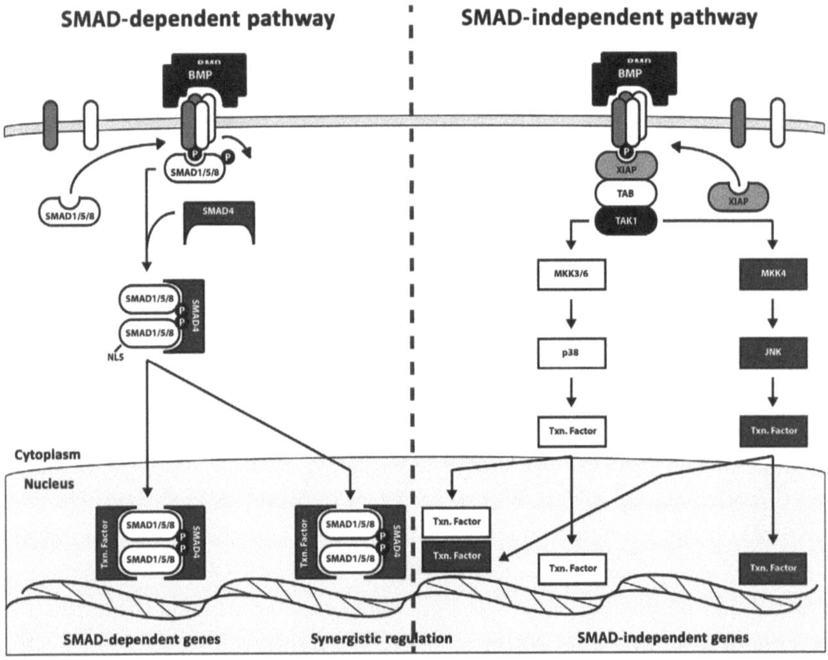

Figure 5 *The noncanonical, SMAD-independent pathway for BMP signaling.* The activation of SMAD-independent signaling pathways is mediated by the binding of XIAP to the activated receptor which bridges interactions with TAB and TAK1. The TAK1 kinase then activates the p38 pathway via MKK3/6 and the JNK pathway via MKK4. These mechanisms allow BMP proteins to trigger biological responses that are mediated by SMAD-dependent pathways, SMAD-independent pathways, or the synergistic cooperation of both pathways.

that observed in the SMAD-independent pathways triggered by TGF-β (Blank, Brown, Adams, Karolak, & Oxburgh, 2009; Hu, Wasserman, Hartwig, & Rosenblum, 2004; Leung-Hagesteijn et al., 2005).

A significant challenge remains to integrate our knowledge of SMAD-dependent and SMAD-independent pathways to develop a better understanding of the biological responses triggered by TGF-β/BMP signaling. While a number of responses have been shown to be either strictly SMAD-dependent or SMAD-independent, many responses cannot be so definitively characterized. Indeed, there are scenarios where TGF-β/BMP signaling has been shown to activate SMAD-dependent and SMAD-independent responses that operate in a redundant manner (Xu et al., 2008), combine to produce synergistic effects (Abecassis, Rogier, Vazquez, Atfi, & Bourgeade, 2004), and even function interdependently by switching between pathways in a biphasic manner (Broege et al., 2013). An additional level of complexity relates to whether there are mechanisms that allow the activation of TGF-β/BMP receptors to preferentially stimulate either SMAD-dependent or SMAD-independent pathways. While TGF-β/BMP proteins are likely to activate both of these pathways in many cases, there is evidence that differing mechanisms of receptor assembly allow SMAD-dependent or SMAD-independent responses to be activated independently (Nohe et al., 2002). Although there are still many unanswered questions, the importance of SMAD-independent mechanisms has been established in numerous processes and the contributions of this noncanonical pathway for TGF-β/BMP signaling cannot be overlooked.

4.3 Key Regulatory Steps in TGF-β/BMP Signaling

To this point, we have presented TGF-β/BMP signaling as a linear pathway in which responses are strictly binary and dependent only on the presence of the ligand. However, this is an over simplification. The TGF-β and BMP signaling pathways have incorporated numerous regulatory steps that allow biological responses to be fine-tuned according to a wide range of contextual cues. This is largely mediated by several key signaling nodes in TGF-β/BMP signaling pathways that provide opportunities for cross talk with other signaling pathways.

4.3.1 Ligand Expression

Differences in the spatiotemporal expression patterns of TGF-β/BMP proteins play an integral role in the distinct biological functions of individual members of the TGF-β and BMP families. Indeed, separate mechanisms for regulating ligand expression are absolutely necessary since many of the downstream signaling pathways are shared between family members. The expression of TGF-β/BMP proteins is most commonly regulated at the mRNA level. As an example, this is particularly evident in BMP-7 where changes in *Bmp7* mRNA expression occur following renal injury (Simon et al., 1999) and during tumorigenesis (Basic-Jukic et al., 2011). While the mechanisms regulating *Bmp7* transcription are poorly understood, the *Bmp7* promoter has been cloned and shown to contain numerous putative transcription factor binding sites (Dorai et al., 2001; Simon, Feliers, Arar, Bhandari, & Abboud, 2002). Another level of complexity has recently emerged with the demonstration that *Bmp7* mRNA can be silenced by the microRNA *miR-22* to regulate BMP-7 expression (Long et al., 2013). Similar strategies are employed for regulating the expression of other TGF-β/BMP ligands.

4.3.2 Ligand Availability

An additional level of regulation occurs in the form of soluble antagonists that sequester ligands in the extracellular space. The importance of this regulatory mechanism is suggested by the large number of BMP antagonists that have been identified including Noggin, Chordin, USAG-1, Gremlin-1, CTGF, DAN, Cerebus-1, Coco, PRDC, Follistatin, Sclerostin, and Twisted Gastrulation (Walsh et al., 2010; Yanagita, 2005). The structure of the Noggin: BMP-7 complex has provided insights into the inhibitory mechanisms of BMP antagonists (Fig. 6). Similar to TGF-β/BMP proteins, Noggin adopts a cysteine-knot fold tertiary structure and forms homodimers. This Noggin homodimer binds to a BMP-7 homodimer to form a ring-like structure where the finger/clip regions extend from Noggin and binds to BMP-7 through its cysteine-rich domains (*CXXCXC* and *CCXXC*). As a result, this interaction blocks key residues on BMP-7 from interacting with both Type I and Type II BMP receptors in a reversible manner (Groppe et al., 2002). Most of the other known BMP antagonists are structurally related to Noggin and thought to operate by similar mechanisms. However, a second mode of BMP antagonism is mediated by the membrane-bound antagonist BAMBI, which acts as a pseudoreceptor to sequester ligands and inactivate heteromeric receptor complexes in a dominant negative manner (Onichtchouk et al., 1999). Conversely, KCP and Crossveinless-2 act as agonists that enhance BMP signaling by mechanisms that are not yet clear. While regulation by extracellular antagonists/agonists

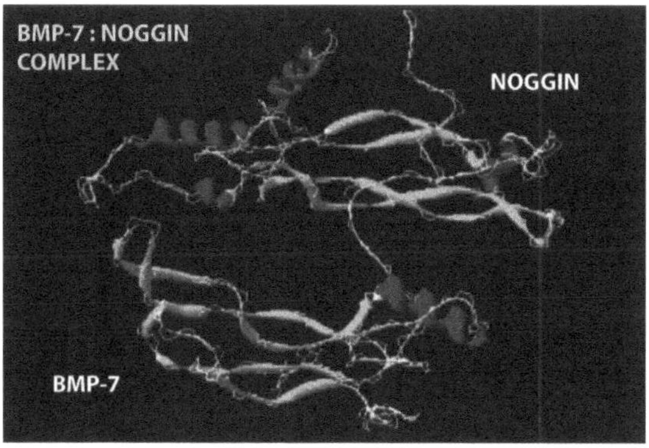

Figure 6 *Binding of the BMP Antagonist Noggin to BMP-7.* Noggin folds into a cysteine-knot fold structure that resembles the structure of TGF-β/BMP family members. A Noggin homodimer then binds to a BMP-7 homodimer to form a ring-like structure where the finger/clip regions extend from Noggin and bind to BMP-7 to block key residues on BMP-7 from interacting with its receptors. *Structural information was obtained from entry* 1M4U *(Groppe et al., 2002) in the Protein Data Bank at the Research Collaboratory for Structural Bioinformatics and then a ribbon diagram was created using DeepView PDB-Viewer.* (See the color plate.)

does not appear to occur as frequently in the TGF-β pathway, it is clear that this is an important regulatory step in the BMP pathway.

4.3.3 Receptor Activation

There are also several mechanisms for modulating TGF-β/BMP signaling at the receptor level: (1) First, membrane-bound proteins known as coreceptors (Dragon, RGM proteins, betaglycan, and endoglin) facilitate binding of the ligand to the receptor and enhance receptor signaling (Corradini, Babitt, & Lin, 2009). (2) In contrast, phosphatases act on the receptor to negatively regulate signaling. Dullard is a membrane-bound phosphatase that dephosphorylates activated Type I receptors and triggers the ubiquitin-mediated degradation of Type II receptors (Satow, Kurisaki, Chan, Hamazaki, & Asashima, 2006). PP2a is a cytoplasmic phosphatase that inhibits signaling through similar mechanisms (Bengtsson et al., 2009). (3) Finally, receptor signaling is autoregulated by negative feedback loops. BMP proteins and TGF-β proteins induce expression of the inhibitory SMAD proteins Smad6 and Smad7, respectively. These proteins bind to activated Type I receptors in a manner similar to R-SMAD proteins. However, the inhibitory SMAD proteins cannot be phosphorylated and instead impede the binding sites on the receptor. Furthermore, Smad6 and Smad7 recruit the ubiquitin ligases Smurf1 and Smurf2, respectively, which then trigger the degradation of activated receptors (Ebisawa et al., 2001; Hayashi et al., 1997; Imamura et al., 1997; Kavsak et al., 2000).

4.3.4 SMAD Activation

TGF-β/BMP signaling is also regulated downstream of the receptors at the level of SMAD activation. The previously mentioned inhibitory functions of Smad6 and Smad7 proteins extend to binding to activated R-SMAD proteins, preventing their interaction with C-SMAD, and recruiting Smurf1 and Smurf2 to promote their ubiquitin-mediated degradation (Lin, Liang, & Feng, 2000; Murakami, Watabe, Takaoka, Miyazono, & Imamura, 2003). Thus, the negative feedback loop provided by Smad6 and Smad7 operates at multiple levels to attenuate TGF-β/BMP signaling. Additional regulatory mechanisms are mediated by the posttranslational modification of SMAD proteins. The cytoplasmic phosphatase PP2a inhibits signaling by dephosphorylating activated R-SMAD proteins. In addition, kinases in other signaling pathways (ERK, JNK, PKC, CamKII) are capable of phosphorylating R-SMAD proteins in the linker region joining the conserved MH1 and MH2 domains (Derynck & Zhang, 2003; Massague et al., 2005). This subsequently impacts their ability to translocate to the nucleus, bind to DNA, and interact with other transcription factors. Cumulatively, these mechanisms provide a wide range of opportunities for molecular cross talk with other signaling pathways.

4.3.5 Formation of Transcription Factor Complexes

After SMAD proteins move into the nucleus, there is an even greater potential for cross talk with other signaling pathways. Transcriptional regulation by SMAD proteins, as mentioned, first necessitates the formation of a multimeric complex with additional DNA-binding proteins. The major consequence of this requirement is that input from other signaling pathways has a significant impact on not only the activity of SMAD transcription factor complexes, but also the selection of target genes (Derynck & Zhang, 2003; Massague et al., 2005). Furthermore, SMAD proteins have been shown to interact with coactivators such as the p300 and CBP histone acetylases (Pouponnot, Jayaraman, & Massague, 1998) and corepressors such as c-Ski and SnoN (Takeda et al., 2004; Wang, Mariani, Harland, & Luo, 2000). These interactions allow SMAD proteins to dynamically switch between the transcriptional activation and transcriptional repression of target genes. It is precisely this tractability in regulating transcriptional responses which has allowed TGF-β/BMP-dependent pathways to serve as critical signaling nodes in a multitude of diverse biological processes.

5. THE ROLE OF BMP-7 IN KIDNEY DEVELOPMENT

The importance of BMP-7 and its downstream signaling pathways in the kidneys was first recognized in the setting of kidney development (Dudley et al., 1995; Luo et al., 1995). Renal organogenesis is an incredibly intricate process that relies upon the coordination of complex cell-to-cell signaling networks to direct the morphogenesis of a mature kidney, which will eventually contain more than 30 distinct cell types organized into about 10,000 and 300,000–1,000,000 nephrons in the mouse and human, respectively (Bertram, Douglas-Denton, Diouf, Hughson, & Hoy, 2011). Nonetheless, congenital abnormalities of the kidneys and urinary tract (CAKUT) occur frequently in children and are the leading causes of pediatric kidney disease (Collins et al., 2014; Schedl, 2007). Thus, important research areas include determining the specific functions of BMP-7 in the developing kidney, the consequences of genetic defects affecting BMP-7 signaling, and their role in the congenital abnormalities responsible for pediatric kidney disease.

The tissues of the mammalian kidney develop from both mesenchymal and epithelial cell types derived from the intermediate mesoderm. Formation of the kidneys begins with the budding of the ureteric bud (UB) from the nephric duct and its subsequent outgrowth and invasion into the surrounding metanephric mesenchyme (MM). Signals from the MM stimulate the epithelial UB structure to undergo dichotomous branching and form a network that will eventually mature into the collecting duct system (Dressler, 2006; Michos, 2009). Simultaneously, the UB induces cells of the MM to condense around its tips and undergo mesenchymal–epithelial transition (MET) to generate renal vesicles, each of which represents the precursor of a single nephron. Other cells of the MM differentiate further into the stromal cell populations that constitute the renal interstitium. This cycle of UB branching, MM condensation, and MET to generate renal vesicles repeats for many iterations until the population of nephrogenic progenitor cells in the MM becomes exhausted. This marks the cessation of nephrogenesis and occurs at about P4 in mice and shortly before birth in humans (Dressler, 2006).

Over many years, genetic and fate mapping studies have begun to reveal the cellular and molecular mechanisms underlying nephrogenesis. These studies have shown that there are many reciprocal inductive signaling interactions between the UB, MM, and the stroma, which are required for nephrogenesis to proceed normally. It is in this context that the most well-defined roles for BMP-7 in kidney development have been established, as we will detail below.

The subsequent developmental processes that result in maturation of the renal vesicle into a fully differentiated and segmented nephron have been described morphologically but the roles of many important renal morphogens, including BMP-7, in these steps remain relatively unclear. The nascent renal vesicle becomes polarized and then invaginates to form a proximal cleft and generate a structure known as the comma-shaped body. The formation of a second, more distal cleft defines the structure known as an S-shaped body. The glomerulus forms as endothelial cells and mesangial cells invade the proximal cleft and separate the epithelial cells that will become the visceral and parietal epitheliums. The distal end of the S-shaped body fuses to the UB to form a continuous, epithelial tubule and differentiation and patterning proceeds in a proximal-to-distal manner to generate a mature nephron. In parallel, stromal cell populations form the vascular and neural support networks that are required for proper kidney patterning and functioning (Dressler, 2006).

5.1 Effects of BMP-7 Genetic Ablation on Development

Given the important signaling functions of BMP-7 and its widespread expression during mammalian development, two research groups generated mouse lines with germline deletions of *Bmp7* and reported their phenotypes in 1995 (Dudley et al., 1995; Luo et al., 1995). Although *Bmp7*$^{+/-}$ are viable and appear to be phenotypically normal, *Bmp7*$^{-/-}$ mice exhibit significant abnormalities. These mice are born at the expected Mendalian ratio but are moderately reduced in size as neonates and die within 24–48 h of birth. Upon gross inspection of *Bmp7*$^{-/-}$ embryos, the vast majority of *Bmp7*$^{-/-}$ embryos can be identified by their profound eye abnormalities that are characterized by microphthalmia or anophthalmia. Additionally, there are numerous skeletal abnormalities in *Bmp7*$^{-/-}$ embryos with the most prevalent defect being unilateral hindlimb polydactyly (Dudley et al., 1995; Luo et al., 1995).

While germline deletion of *Bmp7* has significant effects in the optic and skeletal systems, the perinatal lethality in *Bmp7*$^{-/-}$ mice results from severe defects in kidney development. *Bmp7*$^{-/-}$ mice have small, dysgenic kidneys with significantly reduced numbers of glomeruli and nephrons. The overall architecture of these maldeveloped kidneys is highly disorganized with an expanded renal interstitium. The vast majority of *Bmp7*$^{-/-}$ mice also develop either unilateral or bilateral hydroureter accompanied by a dilated collecting duct system. Despite these severe renal defects, the rudimentary kidneys in *Bmp7*$^{-/-}$ mice have limited function, as evidenced by the accumulation of urinary filtrate in the bladders of neonates, and the remainder of the urinary tract system appears to be phenotypically normal (Dudley et al., 1995; Luo et al., 1995).

Although mice with germline deletions of other BMP family members have been generated, thus far only BMP-7 has been demonstrated to be absolutely required for kidney development. Nonetheless, other BMP family members contribute to normal kidney development. *Bmp4*$^{+/-}$ mice have renal hypoplasia/dysplasia, hydroureter, a duplicated collecting duct system, and other congenital abnormalities (Miyazaki, Oshima, Fogo, Hogan, & Ichikawa, 2000). In addition, more than 50% of mice with a null mutation in the *Bmp5* locus develop hydronephrosis and/or hydroureter (Green, 1968).

These findings raise the possibility that many important roles for BMP-7 may be masked in knockout mice as a result of partial compensation by other BMP family members. Indeed, the defects in kidney development observed in *Bmp7*$^{-/-}$ mice are rescued in mice engineered to instead express either *Bmp4* or *Bmp6* from the *Bmp7* locus (Oxburgh et al., 2005). This strongly suggests that in the context of kidney development, the requirement for BMP-7 function is due to its distinct spatiotemporal patterns of expression (Dudley & Robertson, 1997) rather than any signaling properties that are intrinsically different from other BMP family members. Furthermore, since radiolabeled BMP-7 that is injected into pregnant females is capable of crossing the placental barrier and localizing to the fetal kidney, maternally produced BMP-7 may also partially compensate for the effects of germline *Bmp7* deletion (Borovecki et al., 2004). Despite the limitations of *Bmp7* knockout models, these studies have clearly demonstrated a requirement for BMP-7 in kidney development. More detailed analyses of null tissues and *in vitro* studies using reductionist approaches have begun to provide additional insights into the specific developmental functions of BMP-7 in the kidneys (Fig. 7).

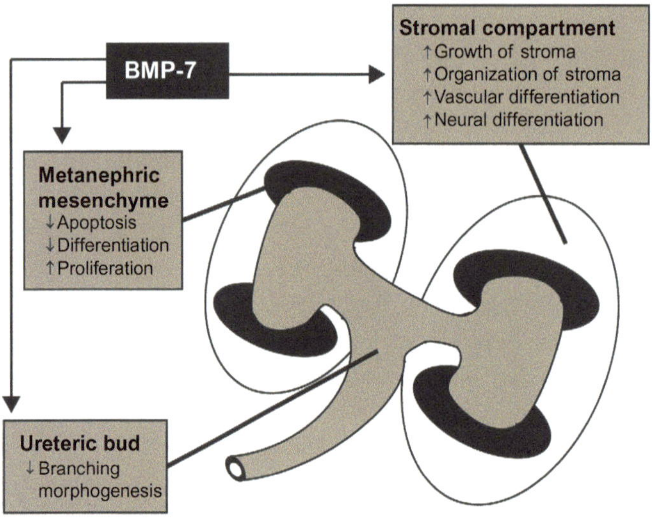

Figure 7 *The key functions of BMP-7 during kidney development.* The requirement for BMP-7 during renal organogenesis can be attributed to its important functions in each of the major tissue compartments in the developing kidney. In the MM, BMP-7 maintains the nephrogenic progenitor cell populations by inhibiting apoptosis and differentiation while stimulating proliferation. In the UB, BMP-7 controls the timing and localization of branching morphogenesis. In the stroma, BMP-7 regulates the growth, organization, and differentiation of stromal progenitor cell populations.

5.2 Functions of BMP-7 in the Metanephric Mesenchyme

The most important roles for BMP-7 signaling in kidney development have been attributed to its functions in the MM during nephrogenesis. The UB invades the MM at about E10.5 in mice and condensation of the MM and nephrogenesis begin shortly thereafter. These processes initially occur properly in *Bmp7*$^{-/-}$ mice and kidney development proceeds normally through E14.5 with branching of the UB, induction and condensation of the MM, and formation of comma- and S-shaped bodies. However, the kidneys in *Bmp7*$^{-/-}$ mice are reduced in size at E14.5 and by E16.5 there is aberrant organization of the nephrogenic zone characterized by the lack of condensed MM in the periphery, loss of markers of MM induction, and no evidence of active formation of renal vesicles (Dudley et al., 1995; Luo et al., 1995). Together, these findings demonstrate that the renal hypoplasia in *Bmp7*$^{-/-}$ mice results primarily from the premature cessation of nephrogenesis.

An early study suggested that BMP-7 expression may be required for the induction of nephrogenesis (Vukicevic, Kopp, Luyten, & Sampath, 1996).

However, this is not supported by the phenotype of *Bmp7*$^{-/-}$ mice since induction of the MM and the formation of nephrons initially proceed normally despite the absence of BMP-7. This has been further corroborated by *in vitro* renal organogenesis models in which notochord is used to induce MM isolated from rudimentary kidneys. In this model, *Bmp7*$^{-/-}$ notochord is able to stimulate the expression of markers of induction and tubulogenesis in MM derived from *Bmp7*$^{-/-}$ mice. Additionally, treatment with recombinant BMP-7 is unable to induce MM *in vitro* (Godin, Robertson, & Dudley, 1999). These findings clearly demonstrate that BMP-7 is not an inductive signal for the MM in the developing kidney.

Nonetheless, BMP-7 has several other important functions that are required for supporting nephrogenesis:

1. First, BMP-7 acts as a survival factor for progenitor cell populations in the MM, as demonstrated by the excessive levels of apoptosis found in the nephrogenic zone of the rudimentary kidneys in *Bmp7*$^{-/-}$ mice. The prosurvival functions of BMP-7 have also been shown *in vitro* using explants of MM. While these cells normally undergo apoptosis in culture in the absence of exogenously added survival factors, treatment with BMP-7 allows the long-term survival and *in vitro* culture of MM explants (Godin, Takaesu, Robertson, & Dudley, 1998).
2. This finding also revealed a second important role for BMP-7 in the MM, the inhibition of differentiation. While other growth factors have been shown to allow the survival of cells cultured from MM explants, these surviving cells typically undergo differentiation. However, treatment with BMP-7 allows surviving MM cells to remain competent to respond to inductive signals. In addition, BMP-7 is able to suppress the induction of tubulogenesis in MM explants cocultured with notochord (Godin et al., 1999).
3. Third, BMP-7 has been shown to promote the proliferation of nephrogenic progenitor cells *in vitro* by the activation of SMAD-independent and JNK-dependent signaling mechanisms. The *in vivo* relevance of these pathways is demonstrated by decreased levels of JNK activation and cellular proliferation in the nephrogenic zone of kidneys in *Bmp7*$^{-/-}$ mice (Blank et al., 2009).

Together, these findings establish a model where BMP-7 expands the progenitor cell population in the MM by promoting the proliferation and survival of these cells. The inhibition of differentiation by BMP-7 also maintains these progenitor cells in an undifferentiated, stem cell state and retains their capacity for nephrogenesis. Conversely, disruptions in

BMP-7 signaling (such as in *Bmp7*$^{-/-}$ mice) lead to accelerated depletion of progenitor cell populations in the MM and a premature cessation of nephrogenesis.

5.3 Functions of BMP-7 in the Ureteric Bud

The importance of BMP signaling during renal organogenesis also extends to the regulation of cell populations derived from the UB and the development of the collecting duct system. This is notable as the UB plays a fundamental role in patterning the kidney and reciprocal interactions between the UB and MM significantly impact nephrogenesis. The morphogenesis of the collecting duct system is largely driven by three processes: budding of the UB from the nephric duct, elongation/growth of the UB stalk, and branching at the UB tips. While the underlying mechanisms of regulation are extremely complex, the prevailing model is that the opposing actions of positive and negative regulatory factors control the timing and localization of these morphogenic processes (Michos, 2009).

BMP signaling has been implicated in the budding, elongation, and branching of the UB, primarily as a negative regulatory factor. This is evident in the phenotypes of mice with genetic alterations that increase BMP signaling. Mice with a germline deletion of the BMP antagonist *Gremlin1* exhibit bilateral renal agenesis due to a complete block in UB budding and outgrowth (Michos et al., 2004). This phenotype is due to excessive BMP signaling, as demonstrated by the finding that UB budding and outgrowth and normal kidney development are restored in *Grem1*$^{-/-}$; *Bmp4*$^{+/-}$ mice (Michos et al., 2007). Similarly, mice engineered to express a constitutively active version of the BMP receptor ALK3 under control of the UB-specific *HoxB7* promoter develop renal dysplasia characterized by decreased branching of the UB and maldevelopment of the collecting duct system (Hu, Piscione, & Rosenblum, 2003).

While elevated BMP signaling has adverse consequences, BMP signaling is still required for the normal development of the collecting duct system. Conditional knockout of ALK3 in lineages derived from the UB results in renal dysplasia. At early developmental time points, there is an increase in the number of primary and secondary UB branches including abnormal structures with three or more branches. This, in turn, leads to impairment of higher order UB branching and maldevelopment of the collecting duct system (Hartwig et al., 2008). Similarly, *Bmp4*$^{+/-}$ have a wide spectrum of defects related to abnormal UB budding and branching. Most prominently, ectopic ureteral budding results in duplication of the collecting duct system and ureterovesical junctions (Miyazaki et al., 2000). Together, these findings suggest a model where BMP signaling ensures the proper timing of morphogenic events and localization of developing structures in the collecting duct system by inhibiting UB budding and branching.

While these inhibitory functions likely extend to BMP-7, examining the role of BMP-7 in the collecting duct system presents technical challenges. *Bmp7*$^{-/-}$ mice exhibit reduced branching of the ureter and collecting duct system, but it is difficult to discern whether this phenotype is secondary to the premature cessation of nephrogenesis in *Bmp7*$^{-/-}$ mice given the numerous reciprocal interactions between the UB and MM. However, the importance of BMP-7 in the UB is strongly suggested by findings that the defects in UB budding and outgrowth in *Grem1*$^{-/-}$ mice are reversed in a *Bmp7*$^{-/-}$ background (Goncalves & Zeller, 2011).

A role for BMP-7 in the collecting duct system is further supported by *in vitro* studies. Treatment with BMP-7 inhibits UB branching morphogenesis in whole kidney explants *in vitro* (Piscione et al., 1997). Although these explants also contain MM, this phenomenon can also be reproduced in inner medullary collecting duct cells where BMP-7 inhibits branching morphogenesis and tubule formation in a Smad-dependent manner (Piscione, Phan, & Rosenblum, 2001). Interestingly, in both of these *in vitro* models, low doses of BMP-7 can instead stimulate branching morphogenesis, suggesting a complex system that depends upon the context and magnitude of the BMP-7 signal (Piscione et al., 1997, 2001).

While these findings demonstrate the importance of BMP-7 in the developing collecting duct system, its specific roles remain poorly defined. The high levels of BMP-7 expression in UB lineages and broad importance of BMP signaling in collecting duct morphogenesis, however, advocate the need for additional studies in this area. Our understanding of the roles of BMP-7 in the collecting duct system will likely continue to evolve as improved methods for examining UB morphogenesis are developed and applied to the study of BMP-7.

5.4 Functions of BMP-7 in Stromal Cell Populations

Of the three major tissue compartments found in the developing kidney, the least amount of research has been devoted to understanding the stromal compartment since these cell populations do not contribute directly to forming the epithelial structures of the mature nephron. A recent study, however, demonstrated a critical role for stromal cell populations by using a genetic ablation strategy that expressed diphtheria toxin under control of the stroma-specific *FoxD1* promoter. Ablation of the renal stroma leads to expansion of the cap mesenchyme, inhibition of nephrogenesis, and perinatal lethality. These findings suggest a model whereby signals from the stromal compartment inhibit self-renewal and promote differentiation in nephrogenic progenitor cells (Das et al., 2013). Furthermore, this study establishes the importance of cross talk between the stroma and both the MM and UB in kidney development.

While the significance of stromal cell populations has only recently been recognized, several studies have begun to reveal roles for BMP-7 in this compartment. First, combined treatment with BMP-7 and FGF-2 has been shown to promote the growth of whole kidney explants *in vitro* through the expansion of stromal progenitor cell populations (Dudley, Godin, & Robertson, 1999). This is likely to, in turn, have downstream effects on nephrogenesis, branching morphogenesis, and other developmental processes.

Second, BMP signaling has been implicated in the patterning of the mesenchymal and stromal progenitor cell compartments. While both of these cell populations originate from the intermediate mesoderm, they differentiate into separate lineages and typically do not mix following the onset of nephrogenesis. However, alterations in BMP signaling have been shown to cause complex phenotypes involving the mixing of cap mesenchyme and cortical interstitial cells. This has been observed in both mice with a conditional knockout of *Smad4* in the BMP-7 expression domain (*Bmp7-Cre*) and mice with a germline deletion of the BMP signaling amplifier *Crossveinless-2* (*Cv2*) (Oxburgh, Brown, Fetting, & Hill, 2011). These defects are likely not trivial since they have the potential for widespread disruption of signaling interactions.

Finally, BMP signaling regulates the differentiation of vascular and neural cells, both of which are in close apposition to the epithelial sources of BMP-7 and derive from the stromal compartment in the developing kidneys. Expression of a constitutively active version of the BMP receptor ALK-3 or treatment with BMP-7 stimulates the differentiation of cells cultured from the neural crest into adrenergic neurons (Varley, McPherson, Zou, Niswander, & Maxwell, 1998; Varley, Wehby, Rueger, & Maxwell, 1995). Additionally, mice with a germline deletion of the BMP effector *Smad5* are embryonic lethal due to defects in angiogenesis and mice with transgenic overexpression of either *Bmp4* or the BMP antagonist

Noggin in podocytes results in severe defects in glomerular capillary formation (Ueda et al., 2008). Together, these findings demonstrate that BMP signaling contributes to the generation of the neural and vascular structures that are essential for the proper patterning and function of the kidneys.

6. BMP-7, CONGENITAL RENAL ABNORMALITIES, AND PEDIATRIC KIDNEY DISEASE

Basic science studies have greatly improved our knowledge of the morphogenic processes and molecular mechanisms underlying normal kidney development. An important goal remains to translate these findings into a better understanding of pediatric kidney disease by defining the contributions of specific genetic defects to the development of CAKUT. However, this has proven to be challenging given that the vast majority of cases of CAKUT occur sporadically rather than through familial inheritance, thus precluding the use of many standard methods for genetic analysis. Additionally, mutations that have been identified often associate with a varying range of congenital abnormalities, suggesting complex interactions with other genetic and/or environmental factors. Although the potential for monogenic mutations to cause CAKUT has been demonstrated in numerous mouse models, studies typically fail to detect mutations in any key developmental genes in most patients with CAKUT (~80–98%) (Schedl, 2007; Yosypiv, 2012).

Despite these limitations, a recent study was able to identify a family containing two individuals with CAKUT linked to mutations in BMP-7. These individuals had a G661A nucleotide substitution in the *Bmp7* genetic locus that resulted in an E221K missense mutation in the BMP-7 protein. Both of these individuals exhibited defects in the ureteral and collecting duct system, urinary tract obstruction, and hydronephrosis (Hwang et al., 2014). The similarities between these congenital abnormalities and the phenotypes of *Bmp7*$^{-/-}$ mice strongly suggest a causal relationship with the BMP-7 mutation detected in these individuals (Dudley et al., 1995; Luo et al., 1995). While mutations in BMP-7 do not appear to be prevalent in patients with CAKUT, only a very limited number of studies have been performed thus far. Additionally, the possibility of mutations in other important components of the BMP signaling pathway (e.g., receptors, effector proteins, antagonists) has yet to be explored.

The strongest evidence for the contribution of BMP-7 to the development of CAKUT relates not to its mutational inactivation during the initiation of disease progression, but instead to its dysregulation as a secondary defect that amplifies disease progression. For example, mutations in *Sall1* cause Townes–Brocks syndrome, a multiorgan syndrome that includes renal hypoplasia and dysplasia (Kohlhase, Wischermann, Reichenbach, Froster, & Engel, 1998). This phenotype is reproduced in *Sall1*$^{-/-}$ mice and accompanied by a dramatic decrease in BMP-7 expression. Furthermore, in a manner similar to *Bmp7*$^{-/-}$ mice, the renal hypoplasia and dysplasia in *Sall1*$^{-/-}$ mice are characterized by the accelerated depletion of progenitor cell populations and premature cessation of nephrogenesis (Basta, Robbins, Kiefer, Dorsett, & Rauchman, 2014; Nishinakamura et al., 2001). This strongly suggests that the secondary loss of BMP-7 expression contributes to the deficits in renal development observed in *Sall1*$^{-/-}$ mice and Townes–Brocks syndrome. Similarly, the loss of BMP-7 expression occurs in *Six2*$^{-/-}$ mice during the premature cessation of nephrogenesis and the development of renal hypoplasia/dysplasia (Self et al., 2006) and mutations in *Six2* have recently been identified in a number of patients with CAKUT (Hwang et al., 2014; Weber et al., 2008).

When considered together, studies investigating the underlying genetic causes of CAKUT have demonstrated that a wide range of mutations have the potential to lead to impaired renal organogenesis and are observed in patients. Nonetheless, the renal hypoplasia/dysplasia occurring in patients often exhibits common histopathologic features irrespective of the underlying genetic cause. These observations suggest that there are shared mechanisms that contribute to the development of CAKUT. Jain et al. hypothesized that given the wide array of reciprocal interactions required for renal organogenesis, the mutation of any of a number of key genes may result in the disruption of core transcriptional programs involved in kidney development as well as the shared histopathologic features of renal hypoplasia/dysplasia. They subsequently performed microarray expression analysis on tissue isolated from normal and dysplastic kidneys at midgestation (17–22 weeks) and found that this was indeed the case. Interestingly, one of the most significant changes that they observed was a 90.9% decrease in *Bmp7* mRNA expression, which was also confirmed at the protein level (Jain, Suarez, McGuire, & Liapis, 2007). Thus, this elegant study emphasizes the potential importance of impaired BMP-7 signaling as a secondary defect that contributes to the development of CAKUT.

While genetic defects can lead to the loss of BMP-7 expression early in kidney development, this potential is only exacerbated later in kidney development. Congenital defects that impair kidney function subsequently result in urinary obstruction, ischemia, the accumulation of toxic metabolites, and/or other injurious conditions. As will be discussed in the following sections, renal injury caused by a wide range of stimuli results in the loss of BMP-7 expression in the mature kidney. Our laboratory has recently found that these mechanisms extend to the developing kidney since surgical ligation of the ureter at P1 results in urinary obstruction, the loss of BMP-7 expression, and apoptosis in the nephrogenic zone that mimics the phenotype of *Bmp7*$^{-/-}$ mice (Manson SR et al., *unpublished studies*). When considered alongside the study by Jain et al., these findings strongly suggest that alterations in BMP-7 signaling are much more widespread in CAKUT than would otherwise be indicated by the results of mutational screens conducted in these patients.

The dysregulation of BMP-7 signaling is likely to have significant implications during disease progression in patients with CAKUT regardless of whether the cause is a primary mutation, a secondary effect of other mutations, or a secondary effect of renal injury. As mentioned previously, *Bmp7*$^{-/-}$ mice die shortly after birth due to a premature cessation of nephrogenesis (Dudley et al., 1995; Luo et al., 1995). More recently, it was shown in a conditional *Bmp-7* knockout model that even loss of BMP-7 expression at time points late in gestation lead to deficits in nephron endowment (Tomita et al., 2013).

These findings lead to a potential explanation of why the role of BMP-7 in CAKUT has not yet been better established. Only cases with relatively severe defects in the development of the kidneys and urinary tract are classified as CAKUT and, therefore, included in study populations. However, the number of nephrons found in a human kidney is subject to significant variability, ranging from approximately 200,000 to more than 2.5 million, and the clinical definition of "normal" is widely debated (Bertram et al., 2011). This raises the possibility that, while severe disruptions in BMP signaling are likely lethal during the gestational period (as in *Bmp7*$^{-/-}$ mice), moderate alterations in BMP signaling may allow kidney development to proceed normally (as in *Bmp7*$^{+/-}$ mice) but instead manifest as a decrease in nephron number that does not fall within the clinical definitions of renal hypoplasia or CAKUT. Such a possibility would still represent a clinically important outcome since there is a significant correlation between nephron endowment and susceptibility to CKD later in life (Hershkovitz, Burbea, Skorecki, & Brenner, 2007).

7. THE ROLE OF BMP-7 IN THE PATHOGENESIS OF CKD

While the importance of BMP-7 in the kidneys was first recognized in the setting of kidney development, critical roles for BMP-7 have also been established in the mature kidney where it modulates the molecular and cellular responses to renal injury. These functions have significant implications for the pathogenesis and treatment of CKD in the adult population. Although a wide range of acute (e.g., ischemia) and chronic (e.g., diabetes, hypertension) conditions lead to CKD in adults, these conditions are linked by their propensity to cause renal injury (Collins et al., 2014). Accordingly, there has been an intense research focus on elucidating the mechanisms underlying the progression of renal injuries to CKD.

The kidneys ordinarily possess a remarkably capacity for structural repair and functional recovery following AKI (Benigni, Morigi, & Remuzzi, 2010; Guo & Cantley, 2010). The regeneration of renal architecture parallels cutaneous wound healing (Hirschberg, 2005), recapitulates developmental paradigms (Little, 2006), and results from the careful orchestration of a variety of biologic processes including the removal of injured cells through apoptosis (Havasi & Borkan, 2011), recruitment of repair-promoting inflammatory and stem cell populations (Cantley, 2005; Ferenbach, Kluth, & Hughes, 2007; Humphreys & Bonventre, 2007, 2008; Ricardo, van Goor, & Eddy, 2008), deposition and remodeling of extracellular matrix (Lelongt & Ronco, 2003; Ronco & Chatziantoniou, 2008), along with cycles of cellular dedifferentiation, proliferation, and redifferentiation that restore the tubular epithelium (Bonventre, 2003; Humphreys et al., 2008; Ishibe & Cantley, 2008).

In contrast, chronic renal injury leads to the erosion of the innate repair mechanisms of the kidney and, in turn, chronic inflammation, excessive apoptosis, renal fibrosis, and an irreversible loss of kidney function. For these reasons, CKD has been characterized as a syndrome resulting from maladaptive kidney repair (Chawla & Kimmel, 2012; Yang, Humphreys, & Bonventre, 2011). Critical areas of research include defining the mechanisms that mediate kidney repair, identifying the molecular steps that become impaired or dysregulated following chronic renal injury, and developing therapeutic strategies to stimulate kidney repair and preserve renal function in patients. Here, we will summarize the wide body of literature that has described the important roles for BMP-7 in these processes and the therapeutic effects of recombinant BMP-7 in models of CKD.

7.1 Therapeutic Effects of Recombinant BMP-7 in Models of CKD

Soon after studies demonstrated the requirement for BMP-7 in kidney development (Dudley et al., 1995; Luo et al., 1995), several groups sought to examine the role of BMP-7 in the context of CKD. These studies were largely motivated by the observation that BMP-7 remains highly expressed in the kidneys after birth. Accordingly, it was hypothesized that functions of BMP-7 analogous to those during development may serve to maintain tissue homeostasis in the mature kidney. In testing this hypothesis, Vukicevic et al. (1998) found that recombinant BMP-7 inhibits the pathogenesis of renal injury and preserves kidney function in a model of acute ischemic injury. Subsequently, Hruska et al. (2000) found that the renal protective effects of BMP-7 extend to models of CKD when they obtained similar results in a model of obstructive uropathy. Since these initial studies, the therapeutic efficacy of recombinant BMP-7 has been demonstrated by numerous research groups and in a wide range of models of CKD. These studies have largely reported that administering BMP-7 has similar renal protective effects during disease progression, irrespective of the model employed (summarized in Table 1).

The early stages of disease progression in models of renal injury are characterized by marked increases in apoptosis and inflammation. Treatment with recombinant BMP-7 suppresses apoptosis following acute ischemic injury (Celic et al., 2011; Vukicevic et al., 1998). The antiapoptotic effects of BMP-7 are even more pronounced following chronic renal injury in models of obstructive uropathy and diabetic nephropathy (Hruska et al., 2000; Wang et al., 2006). It is likely that the inhibition of apoptosis by recombinant BMP-7 in models of chronic renal injury is, at least in part, due to reduced levels of inflammatory injury. Indeed, treatment with recombinant BMP-7 reduces the infiltration of neutrophils following ischemic injury and macrophages in models of obstructive uropathy and diabetic nephropathy (Hruska et al., 2000; Li et al., 2015; Vukicevic et al., 1998).

The later stages of disease progression in models of CKD are characterized by the development of renal fibrosis. While treatment with recombinant BMP-7 has been shown to have a multitude of renal protective effects, the most commonly reported finding is a dramatic reduction in renal fibrosis. The suppression of myofibroblast activation by BMP-7 has been demonstrated in models of obstructive uropathy, nephrotoxic nephritis, lupus, and Alport syndrome (Hruska et al., 2000; Zeisberg, Bottiglio,

Table 1 Summary of Studies Demonstrating the Renal Protective Effects of BMP-7

		Measures of Injury and Disease Progression						
	Model / References	*Renal Architecture*	*Apoptosis*	*Inflammation*	*Fibrosis*	*Proliferation*	*Fibroblast Activation*	*Renal Function*
Acute	**Ischemia** Vukicevic et al. (1998) Celic et al. (2011) Zhen-Qiang et al. (2012)	↑↑↑	↓↓↓	↓		↑↑		↑↑
Chronic	**Obstructive uropathy** Hruska et al. (2012) Morissey et al. (2002) Manson et al. (2011a) Manson et al. (2011b)	↑↑↑↑	↓↓	↓	↓↓↓↓		↓	↑↑
Chronic	**Diabetic nephropathy** Wang et al. (2003) Wang et al. (2006) Sugimoto, Grahovac, Zeisberg and Kalluri (2007) Li et al. (2015)	↑↑↑↑	↓	↓	↓↓↓			↑↑↑↑
Chronic	**Nephrotoxic nephritis** Zeisberg, Bottiglio et al. (2003) Zeisberg, Hanai et al. (2003) Zeisberg, Shah, and Kalluri (2005)	↑↑			↓↓		↓↓	↑
Chronic	**Alport syndrome** Zeisberg, Bottiglio et al. (2003) Zeisberg, Hanai et al. (2003)	↑			↓		↓	↑
Chronic	**Lupus** Zeisberg, Bottiglio et al. (2003) Zeisberg, Hanai et al. (2003)	↑			↓		↓	↑
ESRD	**Remnant kidney model** Dube et al. (2004)	↑				↑		n.c.
ESRD	**Protein overload** Ikeda et al. (2004)	↑			n.c.			

The therapeutic effects of administering recombinant BMP-7 have been demonstrated in numerous *in vivo* models of AKI and CKD. These renal protective effects include the stimulation of proliferation and inhibition of apoptosis, inflammation, myofibroblast activation, and renal fibrosis as well as the preservation of renal architecture and kidney function.

et al., 2003; Zeisberg, Hanai, et al., 2003; Zeisberg, Shah, & Kalluri, 2005). In each of these studies, reduced myofibroblast activation was also found to be accompanied by decreased levels of renal fibrosis. Additionally, studies have shown that the antifibrotic effects of recombinant BMP-7 extend to models of diabetic nephropathy (Sugimoto, Grahovac, Zeisberg, & Kalluri, 2007; Wang et al., 2003).

In accordance with the numerous renal protective effects of recombinant BMP-7, each of the aforementioned studies has demonstrated the preservation of tubular architecture in the kidney. This has even been observed in models of ESRD including the remnant kidney model (5/6 nephrectomy) and a model of CKD-induced protein overload (Dube et al., 2004; Ikeda et al., 2004). Most importantly, recombinant BMP-7 has been shown to preserve kidney function following acute ischemic injury as well as chronic renal injury in models of obstructive uropathy, diabetic nephropathy, nephrotoxic nephritis, lupus, and Alport syndrome (Hruska et al., 2000; Vukicevic et al., 1998; Wang et al., 2003; Zeisberg, Bottiglio, et al., 2003; Zeisberg, Hanai, et al., 2003). Together, these findings illustrate the potential clinical utility of BMP-7 in the development of novel therapies for CKD.

7.2 Renal Protective Mechanisms of Recombinant BMP-7

Given the therapeutic effects of treatment with recombinant BMP-7 in models of CKD, there has been significant interest in elucidating the renal protective mechanisms of BMP-7. Most studies have examined the antifibrotic effects of BMP-7 since renal fibrosis is a common pathologic endpoint in the progression of CKD and an effective predictor of the likelihood for progression to ESRD (Nangaku, 2004). However, important roles for BMP-7 have also been revealed in a wide range of other processes that contribute to CKD.

7.2.1 Inhibition of Renal Fibrosis

Studies of the antifibrotic effects of BMP-7 have largely focused on its ability to modulate responses to the profibrotic cytokine TGF-β. While many pathways have been implicated in the pathogenesis of renal fibrosis, the most important roles have been widely attributed to TGF-β. Indeed, TGF-β is upregulated in response to a wide range of injurious stimuli and secreted from activated tubular epithelial cells and infiltrating inflammatory cells. This results in a number of profibrotic responses including myofibroblast activation and extracellular matrix synthesis. The central role of TGF-β signaling is demonstrated by findings that transgenic overexpression of TGF-β is sufficient to cause renal fibrosis (Kopp et al., 1996; Sanderson et al., 1995) and that inhibition of TGF-β with neutralizing antibodies or antisense RNA ameliorates renal fibrosis in numerous models of CKD (Border, Okuda, Languino, Sporn, & Ruoslahti, 1990; Gagliardini & Benigni, 2006; Isaka et al., 2000).

In examining the effects of BMP-7 on responses to TGF-β, early studies showed that BMP-7 inhibits TGF-β-induced epithelial–mesenchymal transition (EMT), a process whereby epithelial cells transdifferentiate into mesenchymal cells resembling myofibroblasts. BMP-7 reverses TGF-β-induced EMT *in vitro* by suppressing the expression of myofibroblast markers and restoring the expression of epithelial markers. Similarly, treatment with BMP-7 restores the expression of epithelial markers, suppresses myofibroblast activation, and inhibits renal fibrosis *in vivo* in models of CKD (Zeisberg, Hanai, et al., 2003; Zeisberg et al., 2005).

While EMT was accepted as a major source of renal fibroblasts for a number of years (Iwano et al., 2002), studies using lineage tracing techniques recently demonstrated that cells of epithelial origin, and therefore EMT, do not directly contribute to the generation of renal fibroblasts during the pathogenesis of fibrosis *in vivo* (Humphreys et al., 2010; Lin, Kisseleva, Brenner, & Duffield, 2008). Although the cellular origins of renal fibroblasts remain widely debated (Falke, Gholizadeh, Goldschmeding, Kok, & Nguyen, 2015), there is convincing *in vivo* evidence that perivascular cells known as pericytes are the primary source of renal fibroblasts (Humphreys et al., 2010; Lin et al., 2008).

These recent advances suggest a need to reevaluate the antifibrotic mechanisms of BMP-7. Irrespective of the controversies related to the *in vivo* role of EMT, these studies clearly demonstrated that BMP-7 inhibits TGF-β signaling and the expression of TGF-β-dependent, profibrotic genes in renal epithelial cells *in vitro* (Xu, Wan, Jiang, & Wu, 2009; Zeisberg, Hanai, et al., 2003). The inhibitory functions of BMP-7 also extend to other cell types. In cultured renal fibroblasts, BMP-7 inhibits collagen and fibronectin synthesis (Zeisberg, Bottiglio, et al., 2003). Similarly, BMP-7 inhibits TGF-β-stimulated signaling pathways and matrix synthesis in mesangial cells, which have been characterized as specialized glomerular fibroblasts (Wang & Hirschberg, 2003, 2004). Together, these findings establish an important role for BMP-7 in suppressing the profibrotic responses to TGF-β in fibroblasts.

Studies have begun to reveal the mechanisms by which TGF-β signaling is inhibited by BMP-7. The profibrotic responses to TGF-β occur primarily through Smad-dependent signaling mechanisms, as evidenced by the finding that renal fibrosis is markedly reduced in *Smad3*$^{-/-}$ mice (Inazaki et al., 2004). As mentioned, Smad4 is a shared signaling partner in the TGF-β and BMP pathways. Our laboratory has found that treatment with BMP-7 promotes the formation of Smad1/5/8–Smad4 protein complexes, but decreases the formation of Smad2/3–Smad4 protein complexes as well as the expression of TGF-β-dependent, profibrotic genes. Since BMP-7 does not affect TGF-β levels or Smad2/3 activation, this suggests a mechanism where TGF-β and BMP signals are mutually antagonistic by competing for a limited pool of Smad4 to induce transcriptional responses (Inazaki et al., 2004; Manson, Niederhoff, Hruska, & Austin, 2011a,2011b).

The inhibition of TGF-β signaling by BMP-7 likely has further implications for the regulation of pericyte-derived myofibroblasts. It was recently shown that TGF-β triggers the differentiation of pericytes into myofibroblasts *in vitro* and that TGF-β blockade prevents pericyte–myofibroblast transition and the development of renal fibrosis *in vivo* (Wu et al., 2013). Several findings point toward a potential role for BMP-7 in preventing myofibroblast activation. First, BMP-7 expression is decreased in pericytes as they differentiate into renal myofibroblasts, suggesting the possibility that BMP-7 inhibits pericyte differentiation and/or myofibroblast function in an autocrine manner (Lin et al., 2008). Furthermore, treatment with BMP-7 following chronic renal injury suppresses the TGF-β-dependent expression of *α-SMA* (a cytoskeletal protein that is necessary for myofibroblast contractility and upregulated during myofibroblast differentiation) and *Collagen Iα1* (the primary matrix protein secreted by myofibroblasts in fibrosis) (Manson et al., 2011a,2011b). Finally, treatment with BMP-7 prevents the accumulation of myofibroblasts and renal fibrosis in models of CKD (see Table 1). These findings clearly establish that BMP-7 suppresses TGF-β-dependent signals that are required for myofibroblast activation during the pathogenesis of renal fibrosis. However, it remains to be seen whether BMP-7 directly regulates pericyte–myofibroblast transition and further studies addressing this possibility are needed.

7.2.2 Other Important Renal Protective Mechanisms

While most studies of BMP-7 have focused on is ability to inhibit the pathogenesis of renal fibrosis, the therapeutic effects of recombinant BMP-7 in models of CKD suggested that BMP-7 has additional renal protective mechanisms. A universal finding in these studies was that BMP-7 preserves renal architecture, and in some cases even renal function, in the injured kidney (see Table 1). This phenomenon is likely due to the concerted actions of BMP-7 in regulating a number of important biological processes:

1. *Inhibition of Inflammation*—A significant component of disease progression in models of CKD is the onset of chronic inflammation, as demonstrated by findings that macrophage ablation ameliorates renal injury and fibrosis (Meng, Nikolic-Paterson, & Lan, 2014). Several studies have shown that treatment with BMP-7 inhibits inflammation in models of CKD (Table 1). While the mechanisms underlying the anti-inflammatory effects of BMP-7 are unclear, *in vitro* studies have shown that BMP-7 prevents the upregulation of proinflammatory cytokines (IL-6, IL-1β) and chemoattractants (MCP-1, IL-8) in renal epithelial cells stimulated with TNF-α (Gould, Day, Jones, & Dorai, 2002). Treatment of renal epithelial cells with BMP-7 also reduces the expression of ICAM-1, a cell surface protein that mediates interactions with monocytes. These findings are significant since renal fibrosis is abrogated in both *ICAM-1*$^{-/-}$ and *MCP-1*$^{-/-}$ mice (Chow, Nikolic-Paterson, Ozols, Atkins, & Tesch, 2005; Chow et al., 2006).
2. *Inhibition of Apoptosis*—While treatment with BMP-7 decreases apoptosis in models of CKD, it is unclear whether this is a direct effect or secondary to the suppression of inflammatory injury. However, *in vitro* studies have shown that BMP-7 acts as a survival factor in podocyte epithelial cells by suppressing apoptosis in response to TGF-β and other stimuli in a SMAD-dependent manner (Mitu, Wang, & Hirschberg, 2007). This finding likely has *in vivo* relevance since a significant extent of apoptosis following chronic renal injury is triggered by TGF-β-dependent mechanisms (Miyajima et al., 2000).
3. *Maintenance of Epithelial Differentiation*—Renal injury is characterized by progressive dedifferentiation of the tubular epithelium (Bonventre, 2003). As mentioned, BMP-7 inhibits TGF-β-induced EMT *in vitro*. While EMT is no longer widely accepted as a major source of fibroblasts, epithelial dedifferentiation *in vivo* resembles a partial-EMT in which cells do not cross the basement membrane. This is evidenced by decreased expression of epithelial markers (e.g., E-cadherin) and increased expression of mesenchymal markers (e.g., vimentin) in surviving epithelial cells (Witzgall, Brown, Schwarz, & Bonventre, 1994). Since treatment with BMP-7 reverses these changes in models of CKD, these findings still

indicate a role for BMP-7 in maintaining epithelial differentiation (Zeisberg, Bottiglio, et al., 2003; Zeisberg, Hanai, et al., 2003).

4. *Stimulation of Proliferation*—Several studies have shown that treatment with BMP-7 stimulates proliferation in models of CKD (Table 1). While the effects of BMP-7 on cellular proliferation have not been studied in detail, BMP-7 has most commonly been reported to inhibit proliferation *in vitro*. However, these effects are context-dependent and stimulatory effects have also been observed (Otani et al., 2007; Piscione et al., 2001). Since TGF-β causes cell cycle arrest in tubular epithelial cells (Wu et al., 2013), it is possible that the suppression of TGF-β signaling by BMP-7 could lead to the increases in proliferation observed in models of CKD.
5. *Maintenance of the Microvasculature*—Recently, there has been increased focus on the roles of angiogenesis and microvascular injury in the progression of CKD. Since pericytes normally function as vascular support cells, their differentiation into interstitial myofibroblasts eliminates an important source of vascular growth factors and leads to capillary rarefaction (Schrimpf, Teebken, Wilhelmi, & Duffield, 2014). The ability of BMP-7 to inhibit TGF-β signaling may suppress pericyte–myofibroblast transition and restore vascular stability. While few studies have examined the effects of BMP-7 on angiogenesis, it has been reported to stimulate proliferation and inhibit apoptosis in endothelial cells, increase expression of the vascular growth factor VEGF, as well as stimulate angiogenesis *in vitro* (Dai et al., 2004; Ramoshebi & Ripamonti, 2000; Teichert-Kuliszewska et al., 2006). These findings suggest a need for studies examining the effects of BMP-7 on microvascular injury and angiogenesis in models of CKD.

8. A MODEL FOR THE ROLE OF ENDOGENOUS BMP-7 IN REGULATING THE RESPONSES TO RENAL INJURY

The therapeutic effects of treatment with recombinant BMP-7 in models of CKD are well established. However, a key research objective remains to define the physiologic roles of *endogenous* BMP-7 and its renal protective functions, a task which is complicated by the lethality of *Bmp7*$^{-/-}$ mice. The vast majority of studies to date have utilized recombinant BMP-7 and, thus, examined the effects of BMP-7 independent of the spatiotemporal constraints that regulate endogenous BMP-7 expression.

Nonetheless, studies have recently begun to establish an equally important role for endogenous BMP-7 by assessing the impact of genetic alterations affecting its activity. Mice deficient for the BMP antagonists USAG-1 or Twisted Gastrulation have decreased susceptibility to renal injury (Yamada et al., 2014; Yanagita et al., 2006). Conversely, mice deficient for the BMP agonist KCP have increased susceptibility to renal injury (Lin et al., 2005). While these genetic manipulations alter the activity of other BMP family members, the importance of endogenous BMP-7 is further supported by the finding that mice deficient for *miR-22*, a microRNA that silences *Bmp-7*, have decreased susceptibility to renal injury (Long et al., 2013). These studies, together, establish an important role for endogenous BMP-7 in regulating the responses to renal injury.

Based on the previously mentioned studies which demonstrated that BMP-7 suppresses TGF-β signaling, our laboratory has hypothesized that BMP-7 functions to counterregulate the profibrotic responses to TGF-β during injury repair. It is important to note that, while fibrosis is commonly viewed as a pathogenic process, it likely also has important physiologic roles during injury repair such as augmenting the mechanical strength of injured tissue to limit further damage and providing a scaffold for subsequent regenerative processes. Indeed, fibrosis is a conserved response to chronic injury in the kidneys, lungs, liver, and heart (Weiskirchen & Meurer, 2013). The requirement for TGF-β in the development of fibrosis in each of these organ systems suggests that the TGF-β pathway must be strictly regulated to ensure that the physiologic, repair-promoting functions of TGF-β do not become pathogenic.

To address these hypotheses, our laboratory has utilized a reversible model of unilateral ureteral obstruction in which renal recovery can be examined after varying durations of obstruction-induced renal injury. As such, we have found that, while short-term obstruction (2 days) results in injuries that are effectively repaired by the kidney, prolonged obstruction (7 days) leads to irreversible renal injury. This model allowed our laboratory to compare profibrotic responses during injury repair and progression to CKD (Manson et al., 2011a,2011b).

Using this approach, we found that the TGF-β-dependent, profibrotic genes *α-SMA* and *Collagen Iα1* are transiently upregulated following short-term obstruction, but efficiently turned off when repair is completed. This contrasts the response following prolonged obstruction where these genes remain persistently activated. In investigating the cessation of the TGF-β response during repair, we surprisingly found that *α-SMA* and

Collagen Iα1 are downregulated even as the levels of TGF-β and activated Smad3 remain highly elevated (Manson et al., 2011a,2011b). This suggested to us that the downstream steps in the TGF-β pathway are suppressed during the cessation of the TGF-β response and led us to examine the role of BMP-7.

Notably, when BMP-7 function is blocked by treatment with BMP-7-neutralizing antibodies during renal recovery, the downregulation of TGF-β-dependent profibrotic genes at the completion of repair is markedly inhibited. Furthermore, while renal injuries resulting from short-term obstruction are effectively repaired by the kidney, treatment with BMP-7-neutralizing antibodies inhibits the restoration of tubular architecture and the resolution of fibrotic changes in the kidney (Manson et al., 2011a,2011b). Together, these findings demonstrate that BMP-7 plays a critical role in the innate repair mechanisms of the kidney that respond to renal injury by suppressing TGF-β-dependent profibrotic pathways at the completion of injury repair.

Given that prolonged obstruction results in irreversible renal injury, we next examined whether the BMP-7 pathway is impaired following chronic injury. Indeed, we found that prolonged obstruction leads to a marked decrease in BMP-7 expression. In examining the importance of this event, we found that the restoration of BMP-7 function through treatment with recombinant BMP-7 again enabled the suppression of TGF-β-dependent profibrotic genes. Furthermore, while there is minimal kidney repair following prolonged obstruction, treatment with recombinant BMP-7 significantly enhances repair by promoting the restoration of renal architecture and the resolution fibrotic changes in the kidney (Manson et al., 2011a, 2011b). Together, these results demonstrate that the loss of BMP-7 expression contributes to the dysregulation of kidney repair following chronic renal injury.

When considered together, these findings suggest an intriguing model (summarized in Fig. 8) to explain the role of endogenous BMP-7 in regulating the responses to renal injury. The upregulation of TGF-β expression in injured cells triggers localized wound-healing responses including apoptosis, cellular dedifferentiation, activation of myofibroblasts, recruitment of inflammatory cells, and matrix deposition. While these processes are facilitated by the simultaneous loss of BMP-7 expression in injured cells, the restoration of a differentiated tubular epithelium restores BMP-7 expression and leads to the cessation of the TGF-β response. The opposing actions of TGF-β and BMP-7 not only provide temporal control, but also spatial

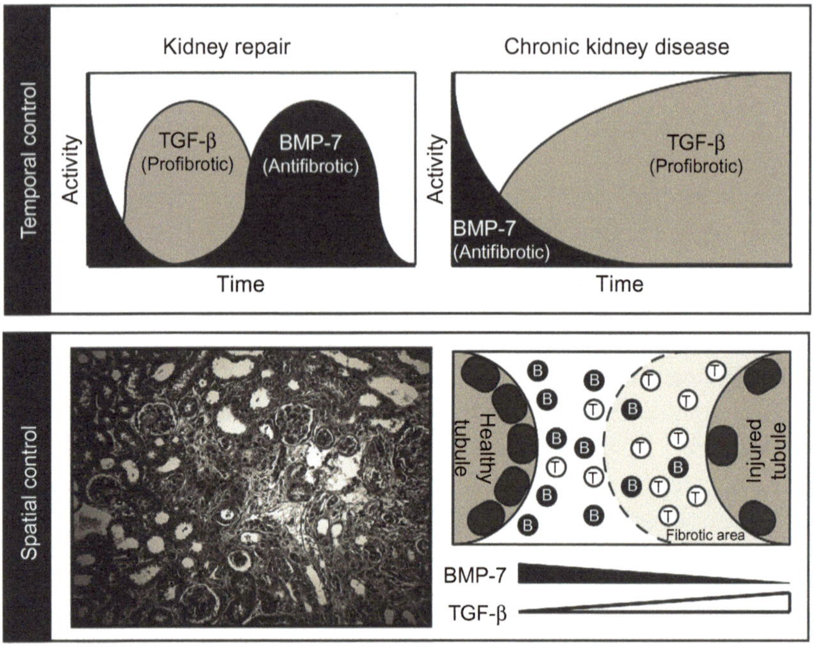

Figure 8 *A model for the regulation of TGF-β-dependent, profibrotic responses by spatiotemporal gradients of BMP-7.* During kidney repair, the physiologic functions of TGF-β are opposed by BMP-7 to provide mechanisms for the spatiotemporal regulation of wound-healing responses. The restoration of epithelial differentiation is paralleled by an increase in BMP-7 expression that leads to the cessation of the TGF-β response. Gradients of BMP-7 may also restrict wound-healing responses to sites of injury, as illustrated by the sporadic scarring of the kidney following obstruction-induced renal injury. During the progression of CKD, persistent activation of TGF-β and suppression of BMP-7 leads to the erosion of the innate repair mechanisms of the kidneys, renal fibrosis, and ESRD. (See the color plate.)

control of wound-healing responses. Indeed, the early stages of the pathogenesis of renal fibrosis are characterized by sporadic scarring of the kidney. BMP-7 signaling may serve as a key mechanism for restricting TGF-β-dependent wound-healing responses to sites of injury.

While little is known about the spatiotemporal control of wound-healing responses due to the numerous technical challenges, there is precedence for these signaling mechanisms since gradients of BMP activity are essential for controlling morphogenic processes during dorsal–ventral patterning in *Drosophila* and *Xenopus* (Ramel & Hill, 2012). In addition, the properties of BMP-7 and its signaling modifiers suggest that the maintenance of spatial gradients is a critical component of the BMP-7 pathway. The diffusion of BMP-7 in the extracellular space is restricted by its binding to matrix proteins and proteoglycans (Gregory et al., 2005; Ruppert, Hoffmann, & Sebald, 1996; Vukicevic et al., 1994). Similarly, the BMP agonist Crossveinless-2 binds BMP-7 and associates with heparin sulfate-containing membrane proteoglycans to maintain a localized pool of BMP-7 (Ambrosio et al., 2008; Rentzsch, Zhang, Kramer, Sebald, & Hammerschmidt, 2006). In contrast, soluble BMP antagonists bind to BMP-7 and allow its signals to be transmitted over longer distances by permitting diffusion within the extracellular space (Larrain et al., 2001). The exploration of these additional levels of signaling complexity may provide further insights into the potential roles of endogenous BMP-7 in regulating the responses to renal injury.

9. CLINICAL IMPLICATIONS FOR BMP-7 IN PATIENTS WITH CKD

Although the physiologic roles of endogenous BMP-7 remain an active area of investigation, it is has become increasingly evident that BMP-7 is broadly important in the progression of CKD. The loss of BMP-7 expression and its renal protective functions occurs experimentally following severe acute injury due to ischemia (Almanzar et al., 1998; Simon et al., 1999; Vukicevic et al., 1998) and nephrotoxicity (Tuglular et al., 2004) as well as chronic renal injury in models of diabetic nephropathy (Qu et al., 2012; Wang, Lapage, & Hirschberg, 2001; Wang et al., 2003; Yeh, Chang, Cheng, Lin, & Cheng, 2009) and obstructive uropathy (Hruska et al., 2000; Manson, Song, Hruska, & Austin, 2014). These findings have recently been extended to human disease in patients with obstructive uropathy, diabetic nephropathy, hypertensive nephrosclerosis, pyelonephritis, and acute nephrotoxicity (Bramlage et al., 2010; Duranay et al., 2011; Manson, Song, Guo, Liapis, & Austin, 2015; Spanjol et al., 2010; Turk et al., 2009).

The likelihood that the loss of BMP-7 expression contributes to disease progression in patients with CKD is supported by the numerous studies, which have demonstrated the renal protective effects of recombinant BMP-7 in murine models of CKD (see Table 1). Moreover, several recent biomarker studies have confirmed the importance of BMP-7 in the setting of human disease. Wong et al. demonstrated that decreases in circulating BMP-7 are predictive of ESRD in patients with diabetes. The prognostic value of BMP-7 was increased beyond that of the best available risk markers for diabetic nephropathy when combined with the detection of increased TGF-β levels (Wong et al., 2013). Using an analogous approach, Neuwirt et al. (2014) found that a biomarker panel containing BMP-7 accurately predicts the loss of kidney function in patients with proteinuric diseases. Together, these studies serve as a "proof-of-concept" for the potential clinical utility of BMP-7 as a prognostic biomarker in patients with CKD.

Nonetheless, most studies of BMP-7 have focused on developing therapeutic approaches for exploiting the renal protective functions of BMP-7 in patients with CKD. This interest has intensified, as studies have demonstrated that BMP-7 not only inhibits the development of renal injuries but also stimulates the repair of established renal injuries. Indeed, treatment with recombinant BMP-7 is capable of promoting structural regeneration and functional recovery following acute renal injury (Zhen-Qiang et al., 2012) as well as reversing the progression of chronic renal injuries (Manson et al., 2011a, 2011b; Morrissey et al., 2002; Zeisberg, Hanai, et al., 2003; Zeisberg et al., 2005). The clinical prospects for the use of recombinant BMP-7 during the treatment of patients with CKD are enhanced by the existence of FDA-approved treatment protocols that utilize BMP-7 to promote the healing of bone fractures (White et al., 2007).

A potential caveat for the use of recombinant BMP-7 in therapies for CKD is the possibility of unintended consequences in the wide range of tissues expressing BMP signaling components. Since many extrarenal organs rely upon the ALK2 and ALK6 receptors to mediate BMP-7-dependent responses, a recent study by Sugimoto et al. (2012) reported the development of an ALK3 agonist which is capable of stimulating BMP-7 signaling pathways and inhibiting the pathogenesis of CKD. Other approaches have evaluated approaches for restoring the expression of endogenous BMP-7 in patients with CKD. This strategy has been validated with the HDAC inhibitor TSA and continued studies may identify HDAC signaling partners that can serve as targets for specifically stimulating BMP-7 expression in the kidneys (Manson et al., 2014). Similar strategies have been suggested for the modulation of BMP agonists and antagonists in the kidney (Walsh et al., 2010). Finally, therapies utilizing mesenchymal stem cells have been proposed since these cells are recruited to the injured kidney, stimulate repair by secreting cytokines including BMP-7, and able to be genetically modified to produce high levels of BMP-7 (Humphreys & Bonventre, 2008; Lv et al., 2014; Zhen-Qiang et al., 2012).

While there are numerous exciting possibilities for the application of BMP-7 in the clinical setting, the successful translation of these strategies to the improved treatment of patients with CKD will likely rely upon our ability to continually expand our knowledge of the critical roles of BMP-7 in both kidney development and the responses to renal injury. This includes further assessing the contributions of BMP-7 to the congenital abnormalities responsible for pediatric kidney disease and defining the physiologic roles of BMP-7 in the innate repair mechanisms of the kidney that respond to renal injury. Although many important research questions remain to be answered, it is clear that we are just beginning to fully realize the importance of BMP-7 signaling in normal health and disease in the kidneys.

ACKNOWLEDGMENTS

This work was supported by a Young Investigator's Award from the National Kidney Foundation to S.R.M., a research grant from the Midwest Stone Institute to S.R.M., and a research grant from the NIH/NIDDK to P.F.A. (1R01DK096177).

REFERENCES

Abecassis, L., Rogier, E., Vazquez, A., Atfi, A., & Bourgeade, M. F. (2004). Evidence for a role of MSK1 in transforming growth factor-beta-mediated responses through p38alpha and Smad signaling pathways. *The Journal of Biological Chemistry*, *279*(29), 30474–30479. http://dx.doi.org/10.1074/jbc.M403294200.

Almanzar, M. M., Frazier, K. S., Dube, P. H., Piqueras, A. I., Jones, W. K., Charette, M. F., et al. (1998). Osteogenic protein-1 mRNA expression is selectively modulated after acute ischemic renal injury. *Journal of the American Society of Nephrology*, *9*(8), 1456–1463.

Ambrosio, A. L., Taelman, V. F., Lee, H. X., Metzinger, C. A., Coffinier, C., & De Robertis, E. M. (2008). Crossveinless-2 is a BMP feedback inhibitor that binds chordin/BMP to regulate Xenopus embryonic patterning. *Developmental Cell*, *15*(2), 248–260. http://dx.doi.org/10.1016/j.devcel.2008.06.013.

Basic-Jukic, N., Hudolin, T., Radic-Antolic, M., Coric, M., Zadro, R., Kastelan, Z., et al. (2011). Bone morphogenetic protein-7 expression is down-regulated in human clear cell renal carcinoma. *Journal of Nephrology*, *24*(1), 91–97.

Basta, J. M., Robbins, L., Kiefer, S. M., Dorsett, D., & Rauchman, M. (2014). Sall1 balances self-renewal and differentiation of renal progenitor cells. *Development*, *141*(5), 1047–1058. http://dx.doi.org/10.1242/dev.095851.

Bengtsson, L., Schwappacher, R., Roth, M., Boergermann, J. H., Hassel, S., & Knaus, P. (2009). PP2A regulates BMP signalling by interacting with BMP receptor complexes and by dephosphorylating both the C-terminus and the linker region of Smad1. *Journal of Cell Science*, *122*(Pt. 8), 1248–1257. http://dx.doi.org/10.1242/jcs.039552.

Benigni, A., Morigi, M., & Remuzzi, G. (2010). Kidney regeneration. *Lancet*, *375*(9722), 1310–1317. http://dx.doi.org/10.1016/S0140-6736(10)60237-1.

Bertram, J. F., Douglas-Denton, R. N., Diouf, B., Hughson, M. D., & Hoy, W. E. (2011). Human nephron number: Implications for health and disease. *Pediatric Nephrology*, *26*(9), 1529–1533. http://dx.doi.org/10.1007/s00467-011-1843-8.

Blank, U., Brown, A., Adams, D. C., Karolak, M. J., & Oxburgh, L. (2009). BMP7 promotes proliferation of nephron progenitor cells via a JNK-dependent mechanism. *Development, 136*(21), 3557–3566. http://dx.doi.org/10.1242/dev.036335.

Bonventre, J. V. (2003). Dedifferentiation and proliferation of surviving epithelial cells in acute renal failure. *Journal of the American Society of Nephrology, 14*(Suppl. 1), S55–S61.

Border, W. A., Okuda, S., Languino, L. R., Sporn, M. B., & Ruoslahti, E. (1990). Suppression of experimental glomerulonephritis by antiserum against transforming growth factor beta 1. *Nature, 346*(6282), 371–374. http://dx.doi.org/10.1038/346371a0.

Borovecki, F., Jelic, M., Grgurevic, L., Sampath, K. T., Bosukonda, D., & Vukicevic, S. (2004). Bone morphogenetic protein-7 from serum of pregnant mice is available to the fetus through placental transfer during early stages of development. *Nephron. Experimental Nephrology, 97*(1), e26–e32. http://dx.doi.org/10.1159/000077595.

Bramlage, C. P., Tampe, B., Koziolek, M., Maatouk, I., Bevanda, J., Bramlage, P., et al. (2010). Bone morphogenetic protein (BMP)-7 expression is decreased in human hypertensive nephrosclerosis. *BMC Nephrology, 11*, 31. http://dx.doi.org/10.1186/1471-2369-11-31.

Broege, A., Pham, L., Jensen, E. D., Emery, A., Huang, T. H., Stemig, M., et al. (2013). Bone morphogenetic proteins signal via SMAD and mitogen-activated protein (MAP) kinase pathways at distinct times during osteoclastogenesis. *The Journal of Biological Chemistry, 288*(52), 37230–37240. http://dx.doi.org/10.1074/jbc.M113.496950.

Cantley, L. G. (2005). Adult stem cells in the repair of the injured renal tubule. *Nature Clinical Practice. Nephrology, 1*(1), 22–32. http://dx.doi.org/10.1038/ncpneph0021.

Celic, T., Spanjol, J., Grskovic, A., Markic, D., Prebilic, I., Fuckar, Z., et al. (2011). Bone morphogenetic protein-7 reduces cold ischemic injury in rat kidney. *Transplantation Proceedings, 43*(7), 2505–2509. http://dx.doi.org/10.1016/j.transproceed.2011.05.056.

Chacko, B. M., Qin, B., Correia, J. J., Lam, S. S., de Caestecker, M. P., & Lin, K. (2001). The L3 loop and C-terminal phosphorylation jointly define Smad protein trimerization. *Nature Structural Biology, 8*(3), 248–253. http://dx.doi.org/10.1038/84995.

Chawla, L. S., & Kimmel, P. L. (2012). Acute kidney injury and chronic kidney disease: An integrated clinical syndrome. *Kidney International, 82*(5), 516–524. http://dx.doi.org/10.1038/ki.2012.208.

Chow, F. Y., Nikolic-Paterson, D. J., Ozols, E., Atkins, R. C., Rollin, B. J., & Tesch, G. H. (2006). Monocyte chemoattractant protein-1 promotes the development of diabetic renal injury in streptozotocin-treated mice. *Kidney International, 69*(1), 73–80. http://dx.doi.org/10.1038/sj.ki.5000014.

Chow, F. Y., Nikolic-Paterson, D. J., Ozols, E., Atkins, R. C., & Tesch, G. H. (2005). Intercellular adhesion molecule-1 deficiency is protective against nephropathy in type 2 diabetic db/db mice. *Journal of the American Society of Nephrology, 16*(6), 1711–1722. http://dx.doi.org/10.1681/ASN.2004070612.

Collins, A. J., Foley, R. N., Chavers, B., Gilbertson, D., Herzog, C., Ishani, A., et al. (2014). US renal data system 2013 annual data report. *American Journal of Kidney Diseases, 63*(1 Suppl.), A7. http://dx.doi.org/10.1053/j.ajkd.2013.11.001.

Constam, D. B., & Robertson, E. J. (1999). Regulation of bone morphogenetic protein activity by pro domains and proprotein convertases. *The Journal of Cell Biology, 144*(1), 139–149.

Corradini, E., Babitt, J. L., & Lin, H. Y. (2009). The RGM/DRAGON family of BMP co-receptors. *Cytokine & Growth Factor Reviews, 20*(5-6), 389–398. http://dx.doi.org/10.1016/j.cytogfr.2009.10.008.

Dai, J., Kitagawa, Y., Zhang, J., Yao, Z., Mizokami, A., Cheng, S., et al. (2004). Vascular endothelial growth factor contributes to the prostate cancer-induced osteoblast differentiation mediated by bone morphogenetic protein. *Cancer Research, 64*(3), 994–999.

Das, A., Tanigawa, S., Karner, C. M., Xin, M., Lum, L., Chen, C., et al. (2013). Stromal-epithelial crosstalk regulates kidney progenitor cell differentiation. *Nature Cell Biology, 15*(9), 1035–1044. http://dx.doi.org/10.1038/ncb2828.

Derynck, R., & Zhang, Y. E. (2003). Smad-dependent and Smad-independent pathways in TGF-beta family signalling. *Nature, 425*(6958), 577–584. http://dx.doi.org/10.1038/nature02006.

Dorai, H., Shepard, A., Ozkaynak, E., Lin, K., Jones, S., Oppermann, H., et al. (2001). The 5′ flanking region of the human bone morphogenetic protein-7 gene. *Biochemical and Biophysical Research Communications, 282*(3), 823–831. http://dx.doi.org/10.1006/bbrc.2001.4645.

Dressler, G. R. (2006). The cellular basis of kidney development. *Annual Review of Cell and Developmental Biology, 22*, 509–529. http://dx.doi.org/10.1146/annurev.cellbio.22.010305.104340.

Dube, P. H., Almanzar, M. M., Frazier, K. S., Jones, W. K., Charette, M. F., & Paredes, A. (2004). Osteogenic Protein-1: Gene expression and treatment in rat remnant kidney model. *Toxicologic Pathology, 32*(4), 384–392.

Dudley, A. T., Godin, R. E., & Robertson, E. J. (1999). Interaction between FGF and BMP signaling pathways regulates development of metanephric mesenchyme. *Genes & Development, 13*(12), 1601–1613.

Dudley, A. T., Lyons, K. M., & Robertson, E. J. (1995). A requirement for bone morphogenetic protein-7 during development of the mammalian kidney and eye. *Genes & Development, 9*(22), 2795–2807.

Dudley, A. T., & Robertson, E. J. (1997). Overlapping expression domains of bone morphogenetic protein family members potentially account for limited tissue defects in BMP7 deficient embryos. *Developmental Dynamics, 208*(3), 349–362. http://dx.doi.org/10.1002/(SICI)1097-0177(199703)208:3<349::AID-AJA6>3.0.CO;2-I.

Duranay, M., Segall, L., Sen, N., Yilmaz, F. M., Cetin, M., Isleyen, A., et al. (2011). Bone morphogenic protein-7 serum level decreases significantly in patients with contrast-induced nephropathy. *International Urology and Nephrology, 43*(3), 807–812. http://dx.doi.org/10.1007/s11255-010-9871-z.

Ebisawa, T., Fukuchi, M., Murakami, G., Chiba, T., Tanaka, K., Imamura, T., et al. (2001). Smurf1 interacts with transforming growth factor-beta type I receptor through Smad7 and induces receptor degradation. *The Journal of Biological Chemistry, 276*(16), 12477–12480. http://dx.doi.org/10.1074/jbc.C100008200.

Engel, M. E., McDonnell, M. A., Law, B. K., & Moses, H. L. (1999). Interdependent SMAD and JNK signaling in transforming growth factor-beta-mediated transcription. *The Journal of Biological Chemistry, 274*(52), 37413–37420.

Falke, L. L., Gholizadeh, S., Goldschmeding, R., Kok, R. J., & Nguyen, T. Q. (2015). Diverse origins of the myofibroblast-implications for kidney fibrosis. *Nature Reviews. Nephrology, 11*(4), 233–244. http://dx.doi.org/10.1038/nrneph.2014.246.

Feng, X. H., & Derynck, R. (1997). A kinase subdomain of transforming growth factor-beta (TGF-beta) type I receptor determines the TGF-beta intracellular signaling specificity. *The EMBO Journal, 16*(13), 3912–3923. http://dx.doi.org/10.1093/emboj/16.13.3912.

Ferenbach, D., Kluth, D. C., & Hughes, J. (2007). Inflammatory cells in renal injury and repair. *Seminars in Nephrology, 27*(3), 250–259. http://dx.doi.org/10.1016/j.semnephrol.2007.02.001.

Gagliardini, E., & Benigni, A. (2006). Role of anti-TGF-beta antibodies in the treatment of renal injury. *Cytokine & Growth Factor Reviews, 17*(1–2), 89–96. http://dx.doi.org/10.1016/j.cytogfr.2005.09.005.

Gerhart, J. (1999). 1998 Warkany lecture: Signaling pathways in development. *Teratology, 60*(4), 226–239. http://dx.doi.org/10.1002/(SICI)1096-9926(199910)60:4<226::AID-TERA7>3.0.CO;2-W.

Godin, R. E., Robertson, E. J., & Dudley, A. T. (1999). Role of BMP family members during kidney development. *The International Journal of Developmental Biology*, *43*(5), 405–411.
Godin, R. E., Takaesu, N. T., Robertson, E. J., & Dudley, A. T. (1998). Regulation of BMP7 expression during kidney development. *Development*, *125*(17), 3473–3482.
Goncalves, A., & Zeller, R. (2011). Genetic analysis reveals an unexpected role of BMP7 in initiation of ureteric bud outgrowth in mouse embryos. *PLoS One*, *6*(4), e19370. http://dx.doi.org/10.1371/journal.pone.0019370.
Gould, S. E., Day, M., Jones, S. S., & Dorai, H. (2002). BMP-7 regulates chemokine, cytokine, and hemodynamic gene expression in proximal tubule cells. *Kidney International*, *61*(1), 51–60. http://dx.doi.org/10.1046/j.1523-1755.2002.00103.x.
Green, M. C. (1968). Mechanism of the pleiotropic effects of the short-ear mutant gene in the mouse. *The Journal of Experimental Zoology*, *167*(2), 129–150. http://dx.doi.org/10.1002/jez.1401670202.
Greenwald, J., Groppe, J., Gray, P., Wiater, E., Kwiatkowski, W., Vale, W., et al. (2003). The BMP7/ActRII extracellular domain complex provides new insights into the cooperative nature of receptor assembly. *Molecular Cell*, *11*(3), 605–617.
Gregory, K. E., Ono, R. N., Charbonneau, N. L., Kuo, C. L., Keene, D. R., Bachinger, H. P., et al. (2005). The prodomain of BMP-7 targets the BMP-7 complex to the extracellular matrix. *The Journal of Biological Chemistry*, *280*(30), 27970–27980. http://dx.doi.org/10.1074/jbc.M504270200.
Griffith, D. L., Keck, P. C., Sampath, T. K., Rueger, D. C., & Carlson, W. D. (1996). Three-dimensional structure of recombinant human osteogenic protein 1: Structural paradigm for the transforming growth factor beta superfamily. *Proceedings of the National Academy of Sciences of the United States of America*, *93*(2), 878–883.
Groppe, J., Greenwald, J., Wiater, E., Rodriguez-Leon, J., Economides, A. N., Kwiatkowski, W., et al. (2002). Structural basis of BMP signalling inhibition by the cystine knot protein Noggin. *Nature*, *420*(6916), 636–642. http://dx.doi.org/10.1038/nature01245.
Guo, J. K., & Cantley, L. G. (2010). Cellular maintenance and repair of the kidney. *Annual Review of Physiology*, *72*, 357–376. http://dx.doi.org/10.1146/annurev.physiol.010908.163245.
Hamer, R. A., & El Nahas, A. M. (2006). The burden of chronic kidney disease. *BMJ*, *332*(7541), 563–564. http://dx.doi.org/10.1136/bmj.332.7541.563.
Hartwig, S., Bridgewater, D., Di Giovanni, V., Cain, J., Mishina, Y., & Rosenblum, N. D. (2008). BMP receptor ALK3 controls collecting system development. *Journal of the American Society of Nephrology*, *19*(1), 117–124. http://dx.doi.org/10.1681/ASN.2007010080.
Havasi, A., & Borkan, S. C. (2011). Apoptosis and acute kidney injury. *Kidney International*, *80*(1), 29–40. http://dx.doi.org/10.1038/ki.2011.120.
Hayashi, H., Abdollah, S., Qiu, Y., Cai, J., Xu, Y. Y., Grinnell, B. W., et al. (1997). The MAD-related protein Smad7 associates with the TGFbeta receptor and functions as an antagonist of TGFbeta signaling. *Cell*, *89*(7), 1165–1173.
Hershkovitz, D., Burbea, Z., Skorecki, K., & Brenner, B. M. (2007). Fetal programming of adult kidney disease: Cellular and molecular mechanisms. *Clinical Journal of the American Society of Nephrology*, *2*(2), 334–342. http://dx.doi.org/10.2215/CJN.03291006.
Hinck, A. P. (2012). Structural studies of the TGF-betas and their receptors—Insights into evolution of the TGF-beta superfamily. *FEBS Letters*, *586*(14), 1860–1870. http://dx.doi.org/10.1016/j.febslet.2012.05.028.
Hirschberg, R. (2005). Wound healing in the kidney: Complex interactions in renal interstitial fibrogenesis. *Journal of the American Society of Nephrology*, *16*(1), 9–11. http://dx.doi.org/10.1681/ASN.2004110901.

Hocevar, B. A., Brown, T. L., & Howe, P. H. (1999). TGF-beta induces fibronectin synthesis through a c-Jun N-terminal kinase-dependent, Smad4-independent pathway. *The EMBO Journal, 18*(5), 1345–1356. http://dx.doi.org/10.1093/emboj/18.5.1345.

Hruska, K. A., Guo, G., Wozniak, M., Martin, D., Miller, S., Liapis, H., et al. (2000). Osteogenic protein-1 prevents renal fibrogenesis associated with ureteral obstruction. *American Journal of Physiology. Renal Physiology, 279*(1), F130–F143.

Hu, M. C., Piscione, T. D., & Rosenblum, N. D. (2003). Elevated SMAD1/beta-catenin molecular complexes and renal medullary cystic dysplasia in ALK3 transgenic mice. *Development, 130*(12), 2753–2766.

Hu, M. C., Wasserman, D., Hartwig, S., & Rosenblum, N. D. (2004). p38MAPK acts in the BMP7-dependent stimulatory pathway during epithelial cell morphogenesis and is regulated by Smad1. *The Journal of Biological Chemistry, 279*(13), 12051–12059. http://dx.doi.org/10.1074/jbc.M310526200.

Humphreys, B. D., & Bonventre, J. V. (2007). The contribution of adult stem cells to renal repair. *Néphrologie & Thérapeutique, 3*(1), 3–10. http://dx.doi.org/10.1016/j.nephro.2006.12.002.

Humphreys, B. D., & Bonventre, J. V. (2008). Mesenchymal stem cells in acute kidney injury. *Annual Review of Medicine, 59*, 311–325. http://dx.doi.org/10.1146/annurev.med.59.061506.154239.

Humphreys, B. D., Lin, S. L., Kobayashi, A., Hudson, T. E., Nowlin, B. T., Bonventre, J. V., et al. (2010). Fate tracing reveals the pericyte and not epithelial origin of myofibroblasts in kidney fibrosis. *The American Journal of Pathology, 176*(1), 85–97. http://dx.doi.org/10.2353/ajpath.2010.090517.

Humphreys, B. D., Valerius, M. T., Kobayashi, A., Mugford, J. W., Soeung, S., Duffield, J. S., et al. (2008). Intrinsic epithelial cells repair the kidney after injury. *Cell Stem Cell, 2*(3), 284–291. http://dx.doi.org/10.1016/j.stem.2008.01.014.

Hwang, D. Y., Dworschak, G. C., Kohl, S., Saisawat, P., Vivante, A., Hilger, A. C., et al. (2014). Mutations in 12 known dominant disease-causing genes clarify many congenital anomalies of the kidney and urinary tract. *Kidney International, 85*(6), 1429–1433. http://dx.doi.org/10.1038/ki.2013.508.

Ikeda, Y., Jung, Y. O., Kim, H., Oda, T., Lopez-Guisa, J., Maruvada, R., et al. (2004). Exogenous bone morphogenetic protein-7 fails to attenuate renal fibrosis in rats with overload proteinuria. *Nephron. Experimental Nephrology, 97*(4), e123–e135. http://dx.doi.org/10.1159/000079177.

Imamura, T., Takase, M., Nishihara, A., Oeda, E., Hanai, J., Kawabata, M., et al. (1997). Smad6 inhibits signalling by the TGF-beta superfamily. *Nature, 389*(6651), 622–626. http://dx.doi.org/10.1038/39355.

Inazaki, K., Kanamaru, Y., Kojima, Y., Sueyoshi, N., Okumura, K., Kaneko, K., et al. (2004). Smad3 deficiency attenuates renal fibrosis, inflammation, and apoptosis after unilateral ureteral obstruction. *Kidney International, 66*(2), 597–604. http://dx.doi.org/10.1111/j.1523-1755.2004.00779.x.

Irish, V. F., & Gelbart, W. M. (1987). The decapentaplegic gene is required for dorsal-ventral patterning of the Drosophila embryo. *Genes & Development, 1*(8), 868–879.

Isaka, Y., Tsujie, M., Ando, Y., Nakamura, H., Kaneda, Y., Imai, E., et al. (2000). Transforming growth factor-beta 1 antisense oligodeoxynucleotides block interstitial fibrosis in unilateral ureteral obstruction. *Kidney International, 58*(5), 1885–1892. http://dx.doi.org/10.1111/j.1523-1755.2000.00360.x.

Ishibe, S., & Cantley, L. G. (2008). Epithelial-mesenchymal-epithelial cycling in kidney repair. *Current Opinion in Nephrology and Hypertension, 17*(4), 379–385. http://dx.doi.org/10.1097/MNH.0b013e3283046507.

Israel, D. I., Nove, J., Kerns, K. M., Kaufman, R. J., Rosen, V., Cox, K. A., et al. (1996). Heterodimeric bone morphogenetic proteins show enhanced activity in vitro and in vivo. *Growth Factors, 13*(3–4), 291–300.

Iwano, M., Plieth, D., Danoff, T. M., Xue, C., Okada, H., & Neilson, E. G. (2002). Evidence that fibroblasts derive from epithelium during tissue fibrosis. *The Journal of Clinical Investigation*, *110*(3), 341–350. http://dx.doi.org/10.1172/JCI15518.

Jain, S., Suarez, A. A., McGuire, J., & Liapis, H. (2007). Expression profiles of congenital renal dysplasia reveal new insights into renal development and disease. *Pediatric Nephrology*, *22*(7), 962–974. http://dx.doi.org/10.1007/s00467-007-0466-6.

Kavsak, P., Rasmussen, R. K., Causing, C. G., Bonni, S., Zhu, H., Thomsen, G. H., et al. (2000). Smad7 binds to Smurf2 to form an E3 ubiquitin ligase that targets the TGF beta receptor for degradation. *Molecular Cell*, *6*(6), 1365–1375.

Kim, J., Johnson, K., Chen, H. J., Carroll, S., & Laughon, A. (1997). Drosophila Mad binds to DNA and directly mediates activation of vestigial by Decapentaplegic. *Nature*, *388*(6639), 304–308. http://dx.doi.org/10.1038/40906.

Kirsch, T., Sebald, W., & Dreyer, M. K. (2000). Crystal structure of the BMP-2-BRIA ectodomain complex. *Nature Structural Biology*, 7(6), 492–496. http://dx.doi.org/10.1038/75903.

Knaus, P., & Sebald, W. (2001). Cooperativity of binding epitopes and receptor chains in the BMP/TGFbeta superfamily. *Biological Chemistry*, *382*(8), 1189–1195. http://dx.doi.org/10.1515/BC.2001.149.

Koenig, B. B., Cook, J. S., Wolsing, D. H., Ting, J., Tiesman, J. P., Correa, P. E., et al. (1994). Characterization and cloning of a receptor for BMP-2 and BMP-4 from NIH 3 T3 cells. *Molecular and Cellular Biology*, *14*(9), 5961–5974.

Kohlhase, J., Wischermann, A., Reichenbach, H., Froster, U., & Engel, W. (1998). Mutations in the SALL1 putative transcription factor gene cause Townes-Brocks syndrome. *Nature Genetics*, *18*(1), 81–83. http://dx.doi.org/10.1038/ng0198-81.

Kopp, J. B., Factor, V. M., Mozes, M., Nagy, P., Sanderson, N., Bottinger, E. P., et al. (1996). Transgenic mice with increased plasma levels of TGF-beta 1 develop progressive renal disease. *Laboratory Investigation*, *74*(6), 991–1003.

Kusanagi, K., Inoue, H., Ishidou, Y., Mishima, H. K., Kawabata, M., & Miyazono, K. (2000). Characterization of a bone morphogenetic protein-responsive Smad-binding element. *Molecular Biology of the Cell*, *11*(2), 555–565.

Larrain, J., Oelgeschlager, M., Ketpura, N. I., Reversade, B., Zakin, L., & De Robertis, E. M. (2001). Proteolytic cleavage of Chordin as a switch for the dual activities of Twisted gastrulation in BMP signaling. *Development*, *128*(22), 4439–4447.

Lee, M. K., Pardoux, C., Hall, M. C., Lee, P. S., Warburton, D., Qing, J., et al. (2007). TGF-beta activates Erk MAP kinase signalling through direct phosphorylation of ShcA. *The EMBO Journal*, *26*(17), 3957–3967. http://dx.doi.org/10.1038/sj.emboj.7601818.

Lelongt, B., & Ronco, P. (2003). Role of extracellular matrix in kidney development and repair. *Pediatric Nephrology*, *18*(8), 731–742. http://dx.doi.org/10.1007/s00467-003-1153-x.

Leung-Hagesteijn, C., Hu, M. C., Mahendra, A. S., Hartwig, S., Klamut, H. J., Rosenblum, N. D., et al. (2005). Integrin-linked kinase mediates bone morphogenetic protein 7-dependent renal epithelial cell morphogenesis. *Molecular and Cellular Biology*, *25*(9), 3648–3657. http://dx.doi.org/10.1128/MCB.25.9.3648-3657.2005.

Li, R. X., Yiu, W. H., Wu, H. J., Wong, D. W., Chan, L. Y., Lin, M., et al. (2015). BMP7 reduces ****inflammation and oxidative stress in diabetic tubulopathy. *Clinical Science (London)*, *128*(4), 269–280. http://dx.doi.org/10.1042/CS20140401.

Lin, S. L., Kisseleva, T., Brenner, D. A., & Duffield, J. S. (2008). Pericytes and perivascular fibroblasts are the primary source of collagen-producing cells in obstructive fibrosis of the kidney. *The American Journal of Pathology*, *173*(6), 1617–1627. http://dx.doi.org/10.2353/ajpath.2008.080433.

Lin, X., Liang, M., & Feng, X. H. (2000). Smurf2 is a ubiquitin E3 ligase mediating proteasome-dependent degradation of Smad2 in transforming growth factor-beta signaling. *The Journal of Biological Chemistry*, *275*(47), 36818–36822. http://dx.doi.org/10.1074/jbc.C000580200.

Lin, J., Patel, S. R., Cheng, X., Cho, E. A., Levitan, I., Ullenbruch, M., et al. (2005). Kielin/chordin-like protein, a novel enhancer of BMP signaling, attenuates renal fibrotic disease. *Nature Medicine*, *11*(4), 387–393. http://dx.doi.org/10.1038/nm1217.

Little, M. H. (2006). Regrow or repair: Potential regenerative therapies for the kidney. *Journal of the American Society of Nephrology*, *17*(9), 2390–2401. http://dx.doi.org/10.1681/ASN.2006030218.

Little, S. C., & Mullins, M. C. (2009). Bone morphogenetic protein heterodimers assemble heteromeric type I receptor complexes to pattern the dorsoventral axis. *Nature Cell Biology*, *11*(5), 637–643. http://dx.doi.org/10.1038/ncb1870.

Liu, F., Ventura, F., Doody, J., & Massague, J. (1995). Human type II receptor for bone morphogenic proteins (BMPs): Extension of the two-kinase receptor model to the BMPs. *Molecular and Cellular Biology*, *15*(7), 3479–3486.

Lo, R. S., Chen, Y. G., Shi, Y., Pavletich, N. P., & Massague, J. (1998). The L3 loop: A structural motif determining specific interactions between SMAD proteins and TGF-beta receptors. *The EMBO Journal*, *17*(4), 996–1005. http://dx.doi.org/10.1093/emboj/17.4.996.

Long, J., Badal, S. S., Wang, Y., Chang, B. H., Rodriguez, A., & Danesh, F. R. (2013). MicroRNA-22 is a master regulator of bone morphogenetic protein-7/6 homeostasis in the kidney. *The Journal of Biological Chemistry*, *288*(51), 36202–36214. http://dx.doi.org/10.1074/jbc.M113.498634.

Luo, G., Hofmann, C., Bronckers, A. L., Sohocki, M., Bradley, A., & Karsenty, G. (1995). BMP-7 is an inducer of nephrogenesis, and is also required for eye development and skeletal patterning. *Genes & Development*, *9*(22), 2808–2820.

Lv, S., Liu, G., Sun, A., Wang, J., Cheng, J., Wang, W., et al. (2014). Mesenchymal stem cells ameliorate diabetic glomerular fibrosis in vivo and in vitro by inhibiting TGF-beta signalling via secretion of bone morphogenetic protein 7. *Diabetes & Vascular Disease Research*, *11*(4), 251–261. http://dx.doi.org/10.1177/1479164114531300.

Manson, S. R., Niederhoff, R. A., Hruska, K. A., & Austin, P. F. (2011a). The BMP-7-Smad1/5/8 pathway promotes kidney repair after obstruction induced renal injury. *The Journal of Urology*, *185*(6 Suppl.), 2523–2530. http://dx.doi.org/10.1016/j.juro.2011.01.034.

Manson, S. R., Niederhoff, R. A., Hruska, K. A., & Austin, P. F. (2011b). Endogenous BMP-7 is a critical molecular determinant of the reversibility of obstruction-induced renal injuries. *American Journal of Physiology. Renal Physiology*, *301*(6), F1293–F1302. http://dx.doi.org/10.1152/ajprenal.00071.2011.

Manson, S. R., Song, J. B., Guo, Q., Liapis, H., & Austin, P. F. (2015). Cell type-specific changes in BMP-7 expression contribute to the pathogenesis of kidney disease in patients with obstructive uropathy. *The Journal of Urology*, *193*(5 Suppl.), 1860–1869.

Manson, S. R., Song, J. B., Hruska, K. A., & Austin, P. F. (2014). HDAC dependent transcriptional repression of Bmp-7 potentiates TGF-beta mediated renal fibrosis in obstructive uropathy. *The Journal of Urology*, *191*(1), 242–252. http://dx.doi.org/10.1016/j.juro.2013.06.110.

Massague, J., Seoane, J., & Wotton, D. (2005). Smad transcription factors. *Genes & Development*, *19*(23), 2783–2810. http://dx.doi.org/10.1101/gad.1350705.

Meng, X. M., Nikolic-Paterson, D. J., & Lan, H. Y. (2014). Inflammatory processes in renal fibrosis. *Nature Reviews. Nephrology*, *10*(9), 493–503. http://dx.doi.org/10.1038/nrneph.2014.114.

Michos, O. (2009). Kidney development: From ureteric bud formation to branching morphogenesis. *Current Opinion in Genetics & Development*, *19*(5), 484–490. http://dx.doi.org/10.1016/j.gde.2009.09.003.

Michos, O., Goncalves, A., Lopez-Rios, J., Tiecke, E., Naillat, F., Beier, K., et al. (2007). Reduction of BMP4 activity by gremlin 1 enables ureteric bud outgrowth and GDNF/WNT11 feedback signalling during kidney branching morphogenesis. *Development*, *134*(13), 2397–2405. http://dx.doi.org/10.1242/dev.02861.

Michos, O., Panman, L., Vintersten, K., Beier, K., Zeller, R., & Zuniga, A. (2004). Gremlin-mediated BMP antagonism induces the epithelial-mesenchymal feedback signaling controlling metanephric kidney and limb organogenesis. *Development*, *131*(14), 3401–3410. http://dx.doi.org/10.1242/dev.01251.

Mitu, G. M., Wang, S., & Hirschberg, R. (2007). BMP7 is a podocyte survival factor and rescues podocytes from diabetic injury. *American Journal of Physiology. Renal Physiology*, *293*(5), F1641–F1648. http://dx.doi.org/10.1152/ajprenal.00179.2007.

Miyajima, A., Chen, J., Lawrence, C., Ledbetter, S., Soslow, R. A., Stern, J., et al. (2000). Antibody to transforming growth factor-beta ameliorates tubular apoptosis in unilateral ureteral obstruction. *Kidney International*, *58*(6), 2301–2313. http://dx.doi.org/10.1046/j.1523-1755.2000.00414.x.

Miyazaki, Y., Oshima, K., Fogo, A., Hogan, B. L., & Ichikawa, I. (2000). Bone morphogenetic protein 4 regulates the budding site and elongation of the mouse ureter. *The Journal of Clinical Investigation*, *105*(7), 863–873. http://dx.doi.org/10.1172/JCI8256.

Miyazono, K., Maeda, S., & Imamura, T. (2005). BMP receptor signaling: Transcriptional targets, regulation of signals, and signaling cross-talk. *Cytokine & Growth Factor Reviews*, *16*(3), 251–263. http://dx.doi.org/10.1016/j.cytogfr.2005.01.009.

Miyazono, K., & Miyazawa, K. (2002). Id: A target of BMP signaling. *Science's STKE*, *2002*(151), pe40. http://dx.doi.org/10.1126/stke.2002.151.pe40.

Morrissey, J., Hruska, K., Guo, G., Wang, S., Chen, Q., & Klahr, S. (2002). Bone morphogenetic protein-7 improves renal fibrosis and accelerates the return of renal function. *Journal of the American Society of Nephrology*, *13*(Suppl. 1), S14–S21.

Moustakas, A., & Heldin, C. H. (2005). Non-Smad TGF-beta signals. *Journal of Cell Science*, *118*(Pt. 16), 3573–3584. http://dx.doi.org/10.1242/jcs.02554.

Mueller, T. D., & Nickel, J. (2012). Promiscuity and specificity in BMP receptor activation. *FEBS Letters*, *586*(14), 1846–1859. http://dx.doi.org/10.1016/j.febslet.2012.02.043.

Mulder, K. M., & Morris, S. L. (1992). Activation of p21ras by transforming growth factor beta in epithelial cells. *The Journal of Biological Chemistry*, *267*(8), 5029–5031.

Murakami, G., Watabe, T., Takaoka, K., Miyazono, K., & Imamura, T. (2003). Cooperative inhibition of bone morphogenetic protein signaling by Smurf1 and inhibitory Smads. *Molecular Biology of the Cell*, *14*(7), 2809–2817. http://dx.doi.org/10.1091/mbc.E02-07-0441.

Nangaku, M. (2004). Mechanisms of tubulointerstitial injury in the kidney: Final common pathways to end-stage renal failure. *Internal Medicine*, *43*(1), 9–17.

National Kidney Foundation. (2002). K/DOQI clinical practice guidelines for chronic kidney disease: Evaluation, classification, and stratification. *American Journal of Kidney Diseases*, *39*(2 Suppl. 1), S1–S266.

Neuwirt, H., Perco, P., Kainz, A., Muhlberger, I., Leierer, J., Braniff, S. J., et al. (2014). A 3-biomarker-panel predicts renal outcome in patients with proteinuric renal diseases. *BMC Medical Genomics*, 7(1), 75. http://dx.doi.org/10.1186/s12920-014-0075-8.

Nishinakamura, R., Matsumoto, Y., Nakao, K., Nakamura, K., Sato, A., Copeland, N. G., et al. (2001). Murine homolog of SALL1 is essential for ureteric bud invasion in kidney development. *Development*, *128*(16), 3105–3115.

Nohe, A., Hassel, S., Ehrlich, M., Neubauer, F., Sebald, W., Henis, Y. I., et al. (2002). The mode of bone morphogenetic protein (BMP) receptor oligomerization determines different BMP-2 signaling pathways. *The Journal of Biological Chemistry*, *277*(7), 5330–5338. http://dx.doi.org/10.1074/jbc.M102750200.

Onichtchouk, D., Chen, Y. G., Dosch, R., Gawantka, V., Delius, H., Massague, J., et al. (1999). Silencing of TGF-beta signalling by the pseudoreceptor BAMBI. *Nature, 401*(6752), 480–485. http://dx.doi.org/10.1038/46794.

Otani, H., Otsuka, F., Inagaki, K., Takeda, M., Miyoshi, T., Suzuki, J., et al. (2007). Antagonistic effects of bone morphogenetic protein-4 and -7 on renal mesangial cell proliferation induced by aldosterone through MAPK activation. *American Journal of Physiology. Renal Physiology, 292*(5), F1513–F1525. http://dx.doi.org/10.1152/ajprenal.00402.2006.

Oxburgh, L., Brown, A. C., Fetting, J., & Hill, B. (2011). BMP signaling in the nephron progenitor niche. *Pediatric Nephrology, 26*(9), 1491–1497. http://dx.doi.org/10.1007/s00467-011-1819-8.

Oxburgh, L., Dudley, A. T., Godin, R. E., Koonce, C. H., Islam, A., Anderson, D. C., et al. (2005). BMP4 substitutes for loss of BMP7 during kidney development. *Developmental Biology, 286*(2), 637–646. http://dx.doi.org/10.1016/j.ydbio.2005.08.024.

Ozkaynak, E., Rueger, D. C., Drier, E. A., Corbett, C., Ridge, R. J., Sampath, T. K., et al. (1990). OP-1 cDNA encodes an osteogenic protein in the TGF-beta family. *The EMBO Journal, 9*(7), 2085–2093.

Ozkaynak, E., Schnegelsberg, P. N., & Oppermann, H. (1991). Murine osteogenic protein (OP-1): High levels of mRNA in kidney. *Biochemical and Biophysical Research Communications, 179*(1), 116–123.

Pires-daSilva, A., & Sommer, R. J. (2003). The evolution of signalling pathways in animal development. *Nature Reviews. Genetics, 4*(1), 39–49. http://dx.doi.org/10.1038/nrg977.

Piscione, T. D., Phan, T., & Rosenblum, N. D. (2001). BMP7 controls collecting tubule cell proliferation and apoptosis via Smad1-dependent and -independent pathways. *American Journal of Physiology. Renal Physiology, 280*(1), F19–F33.

Piscione, T. D., Yager, T. D., Gupta, I. R., Grinfeld, B., Pei, Y., Attisano, L., et al. (1997). BMP-2 and OP-1 exert direct and opposite effects on renal branching morphogenesis. *The American Journal of Physiology, 273*(6 Pt. 2), F961–F975.

Pouponnot, C., Jayaraman, L., & Massague, J. (1998). Physical and functional interaction of SMADs and p300/CBP. *The Journal of Biological Chemistry, 273*(36), 22865–22868.

Qu, Y., Du, E., Zhang, Y., Li, S., Han, R., & Qiu, M. (2012). Changes in the expression of bone morphogenetic protein 7 and tamm- horsfall protein in the early stages of diabetic nephropathy. *Nephro-Urology Monthly, 4*(2), 466–469. http://dx.doi.org/10.5812/numonthly.2124.

Ramel, M. C., & Hill, C. S. (2012). Spatial regulation of BMP activity. *FEBS Letters, 586*(14), 1929–1941. http://dx.doi.org/10.1016/j.febslet.2012.02.035.

Ramoshebi, L. N., & Ripamonti, U. (2000). Osteogenic protein-1, a bone morphogenetic protein, induces angiogenesis in the chick chorioallantoic membrane and synergizes with basic fibroblast growth factor and transforming growth factor-beta1. *The Anatomical Record, 259*(1), 97–107.

Rentzsch, F., Zhang, J., Kramer, C., Sebald, W., & Hammerschmidt, M. (2006). Crossveinless 2 is an essential positive feedback regulator of Bmp signaling during zebrafish gastrulation. *Development, 133*(5), 801–811. http://dx.doi.org/10.1242/dev.02250.

Ricardo, S. D., van Goor, H., & Eddy, A. A. (2008). Macrophage diversity in renal injury and repair. *The Journal of Clinical Investigation, 118*(11), 3522–3530. http://dx.doi.org/10.1172/JCI36150.

Ronco, P., & Chatziantoniou, C. (2008). Matrix metalloproteinases and matrix receptors in progression and reversal of kidney disease: Therapeutic perspectives. *Kidney International, 74*(7), 873–878. http://dx.doi.org/10.1038/ki.2008.349.

Ruberte, E., Marty, T., Nellen, D., Affolter, M., & Basler, K. (1995). An absolute requirement for both the type II and type I receptors, punt and thick veins, for dpp signaling in vivo. *Cell, 80*(6), 889–897.

Ruppert, R., Hoffmann, E., & Sebald, W. (1996). Human bone morphogenetic protein 2 contains a heparin-binding site which modifies its biological activity. *European Journal of Biochemistry, 237*(1), 295–302.

Sampath, T. K., Coughlin, J. E., Whetstone, R. M., Banach, D., Corbett, C., Ridge, R. J., et al. (1990). Bovine osteogenic protein is composed of dimers of OP-1 and BMP-2A, two members of the transforming growth factor-beta superfamily. *The Journal of Biological Chemistry, 265*(22), 13198–13205.

Sanderson, N., Factor, V., Nagy, P., Kopp, J., Kondaiah, P., Wakefield, L., et al. (1995). Hepatic expression of mature transforming growth factor beta 1 in transgenic mice results in multiple tissue lesions. *Proceedings of the National Academy of Sciences of the United States of America, 92*(7), 2572–2576.

Satow, R., Kurisaki, A., Chan, T. C., Hamazaki, T. S., & Asashima, M. (2006). Dullard promotes degradation and dephosphorylation of BMP receptors and is required for neural induction. *Developmental Cell, 11*(6), 763–774. http://dx.doi.org/10.1016/j.devcel.2006.10.001.

Schedl, A. (2007). Renal abnormalities and their developmental origin. *Nature Reviews. Genetics, 8*(10), 791–802. http://dx.doi.org/10.1038/nrg2205.

Schrimpf, C., Teebken, O. E., Wilhelmi, M., & Duffield, J. S. (2014). The role of pericyte detachment in vascular rarefaction. *Journal of Vascular Research, 51*(4), 247–258. http://dx.doi.org/10.1159/000365149.

Self, M., Lagutin, O. V., Bowling, B., Hendrix, J., Cai, Y., Dressler, G. R., et al. (2006). Six2 is required for suppression of nephrogenesis and progenitor renewal in the developing kidney. *The EMBO Journal, 25*(21), 5214–5228. http://dx.doi.org/10.1038/sj.emboj.7601381.

Shi, Y., Wang, Y. F., Jayaraman, L., Yang, H., Massague, J., & Pavletich, N. P. (1998). Crystal structure of a Smad MH1 domain bound to DNA: Insights on DNA binding in TGF-beta signaling. *Cell, 94*(5), 585–594.

Shibuya, H., Iwata, H., Masuyama, N., Gotoh, Y., Yamaguchi, K., Irie, K., et al. (1998). Role of TAK1 and TAB1 in BMP signaling in early Xenopus development. *The EMBO Journal, 17*(4), 1019–1028. http://dx.doi.org/10.1093/emboj/17.4.1019.

Shim, J. H., Greenblatt, M. B., Xie, M., Schneider, M. D., Zou, W., Zhai, B., et al. (2009). TAK1 is an essential regulator of BMP signalling in cartilage. *The EMBO Journal, 28*(14), 2028–2041. http://dx.doi.org/10.1038/emboj.2009.162.

Shim, J. H., Xiao, C., Paschal, A. E., Bailey, S. T., Rao, P., Hayden, M. S., et al. (2005). TAK1, but not TAB1 or TAB2, plays an essential role in multiple signaling pathways in vivo. *Genes & Development, 19*(22), 2668–2681. http://dx.doi.org/10.1101/gad.1360605.

Sieber, C., Kopf, J., Hiepen, C., & Knaus, P. (2009). Recent advances in BMP receptor signaling. *Cytokine & Growth Factor Reviews, 20*(5-6), 343–355. http://dx.doi.org/10.1016/j.cytogfr.2009.10.007.

Simon, M., Feliers, D., Arar, M., Bhandari, B., & Abboud, H. E. (2002). Cloning of the 5′-flanking region of the murine bone morphogenetic protein-7 gene. *Molecular and Cellular Biochemistry, 233*(1-2), 31–37.

Simon, M., Maresh, J. G., Harris, S. E., Hernandez, J. D., Arar, M., Olson, M. S., et al. (1999). Expression of bone morphogenetic protein-7 mRNA in normal and ischemic adult rat kidney. *The American Journal of Physiology, 276*(3 Pt. 2), F382–F389.

Sirard, C., Kim, S., Mirtsos, C., Tadich, P., Hoodless, P. A., Itie, A., et al. (2000). Targeted disruption in murine cells reveals variable requirement for Smad4 in transforming growth factor beta-related signaling. *The Journal of Biological Chemistry, 275*(3), 2063–2070.

Sorrentino, A., Thakur, N., Grimsby, S., Marcusson, A., von Bulow, V., Schuster, N., et al. (2008). The type I TGF-beta receptor engages TRAF6 to activate TAK1 in a receptor kinase-independent manner. *Nature Cell Biology, 10*(10), 1199–1207. http://dx.doi.org/10.1038/ncb1780.

Spanjol, J., Djordjevic, G., Markic, D., Fuckar, D., Krpina, K., & Bobinac, D. (2010). Bone morphogenetic protein-7 expression in human pyelonephritis. *Collegium Antropologicum, 34*(Suppl. 2), 61–64.

Sugimoto, H., Grahovac, G., Zeisberg, M., & Kalluri, R. (2007). Renal fibrosis and glomerulosclerosis in a new mouse model of diabetic nephropathy and its regression by bone morphogenic protein-7 and advanced glycation end product inhibitors. *Diabetes, 56*(7), 1825–1833. http://dx.doi.org/10.2337/db06-1226.

Sugimoto, H., LeBleu, V. S., Bosukonda, D., Keck, P., Taduri, G., Bechtel, W., et al. (2012). Activin-like kinase 3 is important for kidney regeneration and reversal of fibrosis. *Nature Medicine, 18*(3), 396–404. http://dx.doi.org/10.1038/nm.2629.

Suzuki, Y., Yandell, M. D., Roy, P. J., Krishna, S., Savage-Dunn, C., Ross, R. M., et al. (1999). A BMP homolog acts as a dose-dependent regulator of body size and male tail patterning in *Caenorhabditis elegans*. *Development, 126*(2), 241–250.

Takeda, M., Mizuide, M., Oka, M., Watabe, T., Inoue, H., Suzuki, H., et al. (2004). Interaction with Smad4 is indispensable for suppression of BMP signaling by c-Ski. *Molecular Biology of the Cell, 15*(3), 963–972. http://dx.doi.org/10.1091/mbc.E03-07-0478.

Teichert-Kuliszewska, K., Kutryk, M. J., Kuliszewski, M. A., Karoubi, G., Courtman, D. W., Zucco, L., et al. (2006). Bone morphogenetic protein receptor-2 signaling promotes pulmonary arterial endothelial cell survival: Implications for loss-of-function mutations in the pathogenesis of pulmonary hypertension. *Circulation Research, 98*(2), 209–217. http://dx.doi.org/10.1161/01.RES.0000200180.01710.e6.

ten Dijke, P., Yamashita, H., Sampath, T. K., Reddi, A. H., Estevez, M., Riddle, D. L., et al. (1994). Identification of type I receptors for osteogenic protein-1 and bone morphogenetic protein-4. *The Journal of Biological Chemistry, 269*(25), 16985–16988.

Tomita, M., Asada, M., Asada, N., Nakamura, J., Oguchi, A., Higashi, A. Y., et al. (2013). Bmp7 maintains undifferentiated kidney progenitor population and determines nephron numbers at birth. *PLoS One, 8*(8), e73554. http://dx.doi.org/10.1371/journal.pone.0073554.

Tuglular, S., Gogas Yavuz, D., Cakalagaoglu, F., Citak, L., Arikan, H., Kocak, H., et al. (2004). Cyclosporine-A induced nephrotoxicity is associated with decreased renal bone morphogenetic protein-7 expression in rats. *Transplantation Proceedings, 36*(1), 131–133. http://dx.doi.org/10.1016/j.transproceed.2003.11.018.

Turk, T., Leeuwis, J. W., Gray, J., Torti, S. V., Lyons, K. M., Nguyen, T. Q., et al. (2009). BMP signaling and podocyte markers are decreased in human diabetic nephropathy in association with CTGF overexpression. *The Journal of Histochemistry and Cytochemistry, 57*(7), 623–631. http://dx.doi.org/10.1369/jhc.2009.953224.

Ueda, H., Miyazaki, Y., Matsusaka, T., Utsunomiya, Y., Kawamura, T., Hosoya, T., et al. (2008). Bmp in podocytes is essential for normal glomerular capillary formation. *Journal of the American Society of Nephrology, 19*(4), 685–694. http://dx.doi.org/10.1681/ASN.2006090983.

Varley, J. E., McPherson, C. E., Zou, H., Niswander, L., & Maxwell, G. D. (1998). Expression of a constitutively active type I BMP receptor using a retroviral vector promotes the development of adrenergic cells in neural crest cultures. *Developmental Biology, 196*(1), 107–118. http://dx.doi.org/10.1006/dbio.1998.8853.

Varley, J. E., Wehby, R. G., Rueger, D. C., & Maxwell, G. D. (1995). Number of adrenergic and islet-1 immunoreactive cells is increased in avian trunk neural crest cultures in the presence of human recombinant osteogenic protein-1. *Developmental Dynamics, 203*(4), 434–447. http://dx.doi.org/10.1002/aja.1002030406.

Vukicevic, S., Basic, V., Rogic, D., Basic, N., Shih, M. S., Shepard, A., et al. (1998). Osteogenic protein-1 (bone morphogenetic protein-7) reduces severity of injury after ischemic acute renal failure in rat. *The Journal of Clinical Investigation, 102*(1), 202–214. http://dx.doi.org/10.1172/JCI2237.

Vukicevic, S., Kopp, J. B., Luyten, F. P., & Sampath, T. K. (1996). Induction of nephrogenic mesenchyme by osteogenic protein 1 (bone morphogenetic protein 7). *Proceedings of the National Academy of Sciences of the United States of America, 93*(17), 9021–9026.

Vukicevic, S., Latin, V., Chen, P., Batorsky, R., Reddi, A. H., & Sampath, T. K. (1994). Localization of osteogenic protein-1 (bone morphogenetic protein-7) during human embryonic development: High affinity binding to basement membranes. *Biochemical and Biophysical Research Communications, 198*(2), 693–700. http://dx.doi.org/10.1006/bbrc.1994.1100.

Walsh, D. W., Godson, C., Brazil, D. P., & Martin, F. (2010). Extracellular BMP-antagonist regulation in development and disease: Tied up in knots. *Trends in Cell Biology, 20*(5), 244–256. http://dx.doi.org/10.1016/j.tcb.2010.01.008.

Wang, S., Chen, Q., Simon, T. C., Strebeck, F., Chaudhary, L., Morrissey, J., et al. (2003). Bone morphogenic protein-7 (BMP-7), a novel therapy for diabetic nephropathy. *Kidney International, 63*(6), 2037–2049. http://dx.doi.org/10.1046/j.1523-1755.2003.00035.x.

Wang, S., de Caestecker, M., Kopp, J., Mitu, G., Lapage, J., & Hirschberg, R. (2006). Renal bone morphogenetic protein-7 protects against diabetic nephropathy. *Journal of the American Society of Nephrology, 17*(9), 2504–2512. http://dx.doi.org/10.1681/ASN.2006030278.

Wang, S., & Hirschberg, R. (2003). BMP7 antagonizes TGF-beta -dependent fibrogenesis in mesangial cells. *American Journal of Physiology. Renal Physiology, 284*(5), F1006–F1013. http://dx.doi.org/10.1152/ajprenal.00382.2002.

Wang, S., & Hirschberg, R. (2004). Bone morphogenetic protein-7 signals opposing transforming growth factor beta in mesangial cells. *The Journal of Biological Chemistry, 279*(22), 23200–23206. http://dx.doi.org/10.1074/jbc.M311998200.

Wang, S. N., Lapage, J., & Hirschberg, R. (2001). Loss of tubular bone morphogenetic protein-7 in diabetic nephropathy. *Journal of the American Society of Nephrology, 12*(11), 2392–2399.

Wang, W., Mariani, F. V., Harland, R. M., & Luo, K. (2000). Ski represses bone morphogenic protein signaling in Xenopus and mammalian cells. *Proceedings of the National Academy of Sciences of the United States of America, 97*(26), 14394–14399. http://dx.doi.org/10.1073/pnas.97.26.14394.

Weber, S., Taylor, J. C., Winyard, P., Baker, K. F., Sullivan-Brown, J., Schild, R., et al. (2008). SIX2 and BMP4 mutations associate with anomalous kidney development. *Journal of the American Society of Nephrology, 19*(5), 891–903. http://dx.doi.org/10.1681/ASN.2006111282.

Weiskirchen, R., & Meurer, S. K. (2013). BMP-7 counteracting TGF-beta1 activities in organ fibrosis. *Frontiers in Bioscience (Landmark Edition), 18*, 1407–1434.

Wharton, K. A., Thomsen, G. H., & Gelbart, W. M. (1991). Drosophila 60A gene, another transforming growth factor beta family member, is closely related to human bone morphogenetic proteins. *Proceedings of the National Academy of Sciences of the United States of America, 88*(20), 9214–9218.

White, A. P., Vaccaro, A. R., Hall, J. A., Whang, P. G., Friel, B. C., & McKee, M. D. (2007). Clinical applications of BMP-7/OP-1 in fractures, nonunions and spinal fusion. *International Orthopaedics, 31*(6), 735–741. http://dx.doi.org/10.1007/s00264-007-0422-x.

Witzgall, R., Brown, D., Schwarz, C., & Bonventre, J. V. (1994). Localization of proliferating cell nuclear antigen, vimentin, c-Fos, and clusterin in the postischemic kidney. Evidence for a heterogenous genetic response among nephron segments, and a large pool

of mitotically active and dedifferentiated cells. *The Journal of Clinical Investigation*, *93*(5), 2175–2188. http://dx.doi.org/10.1172/JCI117214.
Wong, M. G., Perkovic, V., Woodward, M., Chalmers, J., Li, Q., Hillis, G. S., et al. (2013). Circulating bone morphogenetic protein-7 and transforming growth factor-beta1 are better predictors of renal end points in patients with type 2 diabetes mellitus. *Kidney International*, *83*(2), 278–284. http://dx.doi.org/10.1038/ki.2012.383.
Wozney, J. M., Rosen, V., Celeste, A. J., Mitsock, L. M., Whitters, M. J., Kriz, R. W., et al. (1988). Novel regulators of bone formation: Molecular clones and activities. *Science*, *242*(4885), 1528–1534.
Wu, C. F., Chiang, W. C., Lai, C. F., Chang, F. C., Chen, Y. T., Chou, Y. H., et al. (2013). Transforming growth factor beta-1 stimulates profibrotic epithelial signaling to activate pericyte-myofibroblast transition in obstructive kidney fibrosis. *The American Journal of Pathology*, *182*(1), 118–131. http://dx.doi.org/10.1016/j.ajpath.2012.09.009.
Xu, L., Alarcon, C., Col, S., & Massague, J. (2003). Distinct domain utilization by Smad3 and Smad4 for nucleoporin interaction and nuclear import. *The Journal of Biological Chemistry*, *278*(43), 42569–42577. http://dx.doi.org/10.1074/jbc.M307601200.
Xu, L., Chen, Y. G., & Massague, J. (2000). The nuclear import function of Smad2 is masked by SARA and unmasked by TGFbeta-dependent phosphorylation. *Nature Cell Biology*, *2*(8), 559–562. http://dx.doi.org/10.1038/35019649.
Xu, X., Han, J., Ito, Y., Bringas, P., Jr., Deng, C., & Chai, Y. (2008). Ectodermal Smad4 and p38 MAPK are functionally redundant in mediating TGF-beta/BMP signaling during tooth and palate development. *Developmental Cell*, *15*(2), 322–329. http://dx.doi.org/10.1016/j.devcel.2008.06.004.
Xu, Y., Wan, J., Jiang, D., & Wu, X. (2009). BMP-7 counteracts TGF-beta1-induced epithelial-to-mesenchymal transition in human renal proximal tubular epithelial cells. *Journal of Nephrology*, *22*(3), 403–410.
Yamada, S., Nakamura, J., Asada, M., Takase, M., Matsusaka, T., Iguchi, T., et al. (2014). Twisted gastrulation, a BMP antagonist, exacerbates podocyte injury. *PLoS One*, *9*(2), e89135. http://dx.doi.org/10.1371/journal.pone.0089135.
Yamaguchi, K., Nagai, S., Ninomiya-Tsuji, J., Nishita, M., Tamai, K., Irie, K., et al. (1999). XIAP, a cellular member of the inhibitor of apoptosis protein family, links the receptors to TAB1-TAK1 in the BMP signaling pathway. *The EMBO Journal*, *18*(1), 179–187. http://dx.doi.org/10.1093/emboj/18.1.179.
Yamashita, M., Fatyol, K., Jin, C., Wang, X., Liu, Z., & Zhang, Y. E. (2008). TRAF6 mediates Smad-independent activation of JNK and p38 by TGF-beta. *Molecular Cell*, *31*(6), 918–924. http://dx.doi.org/10.1016/j.molcel.2008.09.002.
Yanagita, M. (2005). BMP antagonists: Their roles in development and involvement in pathophysiology. *Cytokine & Growth Factor Reviews*, *16*(3), 309–317. http://dx.doi.org/10.1016/j.cytogfr.2005.02.007.
Yanagita, M., Okuda, T., Endo, S., Tanaka, M., Takahashi, K., Sugiyama, F., et al. (2006). Uterine sensitization-associated gene-1 (USAG-1), a novel BMP antagonist expressed in the kidney, accelerates tubular injury. *The Journal of Clinical Investigation*, *116*(1), 70–79. http://dx.doi.org/10.1172/JCI25445.
Yang, L., Humphreys, B. D., & Bonventre, J. V. (2011). Pathophysiology of acute kidney injury to chronic kidney disease: Maladaptive repair. *Contributions to Nephrology*, *174*, 149–155. http://dx.doi.org/10.1159/000329385.
Yeh, C. H., Chang, C. K., Cheng, M. F., Lin, H. J., & Cheng, J. T. (2009). Decrease of bone morphogenetic protein-7 (BMP-7) and its type II receptor (BMP-RII) in kidney of type 1-like diabetic rats. *Hormone and Metabolic Research*, *41*(8), 605–611. http://dx.doi.org/10.1055/s-0029-1220736.
Ying, Q. L., Nichols, J., Chambers, I., & Smith, A. (2003). BMP induction of Id proteins suppresses differentiation and sustains embryonic stem cell self-renewal in collaboration with STAT3. *Cell*, *115*(3), 281–292.

Yosypiv, I. V. (2012). Congenital anomalies of the kidney and urinary tract: A genetic disorder? *International Journal of Nephrology*, *2012*, 909083. http://dx.doi.org/10.1155/2012/909083.

Zeisberg, M., Bottiglio, C., Kumar, N., Maeshima, Y., Strutz, F., Muller, G. A., et al. (2003). Bone morphogenic protein-7 inhibits progression of chronic renal fibrosis associated with two genetic mouse models. *American Journal of Physiology. Renal Physiology*, *285*(6), F1060–F1067. http://dx.doi.org/10.1152/ajprenal.00191.2002.

Zeisberg, M., Hanai, J., Sugimoto, H., Mammoto, T., Charytan, D., Strutz, F., et al. (2003). BMP-7 counteracts TGF-beta1-induced epithelial-to-mesenchymal transition and reverses chronic renal injury. *Nature Medicine*, *9*(7), 964–968. http://dx.doi.org/10.1038/nm888.

Zeisberg, M., Shah, A. A., & Kalluri, R. (2005). Bone morphogenic protein-7 induces mesenchymal to epithelial transition in adult renal fibroblasts and facilitates regeneration of injured kidney. *The Journal of Biological Chemistry*, *280*(9), 8094–8100. http://dx.doi.org/10.1074/jbc.M413102200.

Zhang, Y. E. (2009). Non-Smad pathways in TGF-beta signaling. *Cell Research*, *19*(1), 128–139. http://dx.doi.org/10.1038/cr.2008.328.

Zhen-Qiang, F., Bing-Wei, Y., Yong-Liang, L., Xiang-Wei, W., Shan-Hong, Y., Yuan-Ning, Z., et al. (2012). Localized expression of human BMP-7 by BM-MSCs enhances renal repair in an in vivo model of ischemia-reperfusion injury. *Genes to Cells*, *17*(1), 53–64. http://dx.doi.org/10.1111/j.1365-2443.2011.01572.x.

CHAPTER FOUR

The Role of BMP Signaling and NF-κB Signaling on Osteoblastic Differentiation, Cancer Development, and Vascular Diseases—Is the Activation of NF-κB a Friend or Foe of BMP Function?

Eijiro Jimi[1]
Division of Molecular Signaling and Biochemistry, Department of Health Promotion, Center for Oral Biological Research, Kyushu Dental University, Kitakyushu, Fukuoka, Japan
[1]Corresponding author: e-mail address: ejimi@kyu-dent.ac.jp

Contents

Vitamins and Hormones, Volume 99
ISSN 0083-6729
http://dx.doi.org/10.1016/bs.vh.2015.05.002

Abstract

Bone morphogenetic proteins (BMPs), members of the transforming growth factor-β family, were first identified as potent inducers of ectopic bone formation when implanted into muscle tissue. Subsequent studies have demonstrated that BMPs play important roles during developmental processes, including cell proliferation, differentiation, and apoptosis. Furthermore, recent studies have shown that BMPs are also involved in the initiation and/or progression of various diseases, such as skeletal diseases, cancer, and vascular diseases. Nuclear factor κ light-chain enhancer of activated B cells (NF-κB) was originally identified as a transcription factor that bound to the enhancer region of the immunoglobulin κ light-chain promoter in B cells. A wide range of stimuli, including inflammatory cytokines and bacterial and viral products, activate the NF-κB pathway, leading to the expression of NF-κB target genes. NF-κB also has functions in multiple biological processes, such as immune and inflammatory responses, cell differentiation, cellular stress responses, and cancer development. Recent findings have demonstrated that BMP and NF-κB signaling agonistically or antagonistically regulate bone development, cancer development, and vascular diseases. This review describes the role of BMPs and NF-κB in bone development, cancer development, and vascular diseases with special attention given to concepts that have emerged in recent years.

1. INTRODUCTION

Bone morphogenetic proteins (BMPs) are multifunctional growth factors that belong to the transforming growth factor-β (TGF-β) superfamily (Katagiri, Suda, & Miyazono, 2008; Urist, 1965; Wozney et al., 1988; Katagiri & Tsukamoto, 2013). BMPs were first identified in bone matrix extracts that could induce ectopic bone formation when implanted subcutaneously in rats (Urist, 1965). BMPs have been widely recognized for roles in bone formation during mammalian development, and they exhibit versatile functions in the body (Katagiri et al., 2008). The unique and specific ability of BMPs to induce bone formation was expected to be useful for the development of bone regeneration treatments; however, BMPs cannot generate enough of a clinical response to be used in this context (Buma, Arts, Gardeniers, Verdonschot, & Schreurs, 2008; Kusumoto et al., 2002; Vaidya et al., 2008). A possible reason might be that inflammatory cytokines, such as tumor necrosis factor α (TNFα) or interleukin 1β (IL-1β), inhibit the bone formation and osteoblast differentiation induced by BMPs (Gilbert et al., 2000; Mukai et al., 2007; Nakase et al., 1997).

The stimulation of TNFα or IL-1β induces various immune or inflammatory responses via several intracellular signaling pathways. Among these intracellular molecules, the transcription factor NF-κB is activated by inflammatory cytokines and bacterial and viral products (DiDonato, Mercurio, & Karin, 2012; Hayden & Ghosh, 2012, 2014). In this pathway, activated NF-κB, mainly p50 and p65 heterodimers, enters the nucleus and binds to NF-κB-responsive elements to regulate the expression of genes that are involved in the regulation of immune and inflammatory responses, cell proliferation, tumorigenesis, and antiapoptosis (DiDonato et al., 2012; Hayden & Ghosh, 2012, 2014).

Inflammatory cytokines contribute to local and systemic bone loss in inflammatory bone diseases, such as rheumatoid arthritis and periodontitis, as well as in estrogen deficiency (Ammann et al., 1997; Choy & Panayi, 2001; Pacifici et al., 1991). Anti-TNF drugs, such as infliximab, etanercept, and adalimumab, have been shown to not only diminish the signs and symptoms of disease but also prevent joint damage (Ranganathan, 2008). Further support for this inhibitory effect was provided by TNFα transgenic mice, which exhibit the well-characterized features of rheumatoid arthritis and develop the symptoms of general osteoporosis (Kaneki et al., 2006). In addition, osteoblastic cells derived from TNFα transgenic mice form significantly fewer and smaller nodules under basal conditions and in the presence of BMP2. Subsequent studies have demonstrated that TNFα inhibits BMP-induced osteoblast differentiation via Smad-dependent or -independent mechanisms (Kaneki et al., 2006).

Recent genetic studies in humans and mice have shown that perturbations in BMP signaling lead not only to skeletal diseases but also to cancer and vascular diseases (Cai, Pardali, Sánchez-Duffhues, & ten Dijke, 2012; Goumans & Mummery, 2000; Lowery & de Caestecker, 2010). In clinical investigations, the expression levels of BMPs or BMP receptors were correlated with tumor grade, malignant characteristics, or prognosis (Ye, Mason, & Jiang, 2011). The NF-κB signaling pathway is also activated in many cancers (Ben-Neriah & Karin, 2011; DiDonato et al., 2012). Mutations in BMPRII have been linked to pulmonary arterial hypertension (PAH), and BMPs are involved in the promotion of vascular calcification and tumor angiogenesis (Cai et al., 2012; Lowery & de Caestecker, 2010). The activation of NF-κB by proinflammatory cytokines or high pressure plays a part in the promotion or worsening of these BMP-related diseases. Although different BMPs can exert distinct effects depending on the cell type, the stage of cell differentiation and the presence of other cytokines (Anderson et al., 2010), BMP, and NF-κB may regulate cancer progression and vascular diseases as well as bone metabolism (Csiszar, Labinskyy, Jo,

Ballabh, & Ungvari, 2008; Csiszar et al., 2005; Graham et al., 2009; Lai, Fong, Fu, Yang, & Tang, 2008). This review will focus on the cross talk between BMP and NF-κB signaling in osteoblast differentiation and in the pathology of cancer and vascular diseases.

2. ACTIVATION OF NF-κB SIGNALING

2.1 Structures of NF-κB/Rel Family, IκB Family, and IKK Complex

NF-κB was originally identified as a transcription factor that bound to the enhancer region of the immunoglobulin κ light-chain promoter in B cells (Sen & Baltimore, 1986). The NF-κB family of ubiquitously expressed transcription factors consists of five members in mammals: p65 (RelA), c-Rel, RelB, NF-κB1 (p105/p50), and NF-κB2 (p100/p52), which form various homo- and heterodimers (Fig. 1) (DiDonato et al., 2012; Hayden & Ghosh, 2012, 2014). All five members share an N-terminal domain of 300 amino acids, designated the Rel homology domain (RHD), with the retroviral oncoprotein v-Rel; the RHD is responsible for DNA binding, dimerization, and the interaction with IκB (inhibitor of κB) proteins. Three members, p65, c-Ral, and RelB, contain C-terminal transcriptional activation domains (TADs) that are crucial for their ability to induce targeted gene expression. p65, c-Rel, and RelB are synthesized as mature proteins, whereas NF-κB1 and NF-κB2 are synthesized as the large precursors p105 and p100, which undergo processing to generate the mature NF-κB subunits p50 and p52, respectively. The C-terminal regions of p105 and p100 containing ankyrin repeats are selectively degraded by the ubiquitin/proteasome pathway. The processing of p100 to p52 is tightly regulated by this process, while p50 is generated from the constitutive processing of p105. Thus, the homodimers of p50 and p52 lack TADs and therefore have no intrinsic ability to drive transcription as transcriptional repressors. However, the p65:p50, c-Rel:p50, and RelB:p52 heterodimers work as transcriptional activators (DiDonato et al., 2012; Hayden & Ghosh, 2012, 2014).

The activation of NF-κB signaling is regulated by IκB proteins, such as IκBα, IκBβ, IκBε, IκBζ, BCL-3, and IκBNS (IκBδ), and the precursors p105 (NF-κB1) and p100 (NF-κB2), which are characterized by the presence of multiple ankyrin repeat domains and the ability to bind NF-κB dimers (Ghosh & Hayden, 2012; Hinz, Arslan, & Scheidereit, 2012). In unstimulated cells, NF-κB dimers are maintained in the cytosol in complex

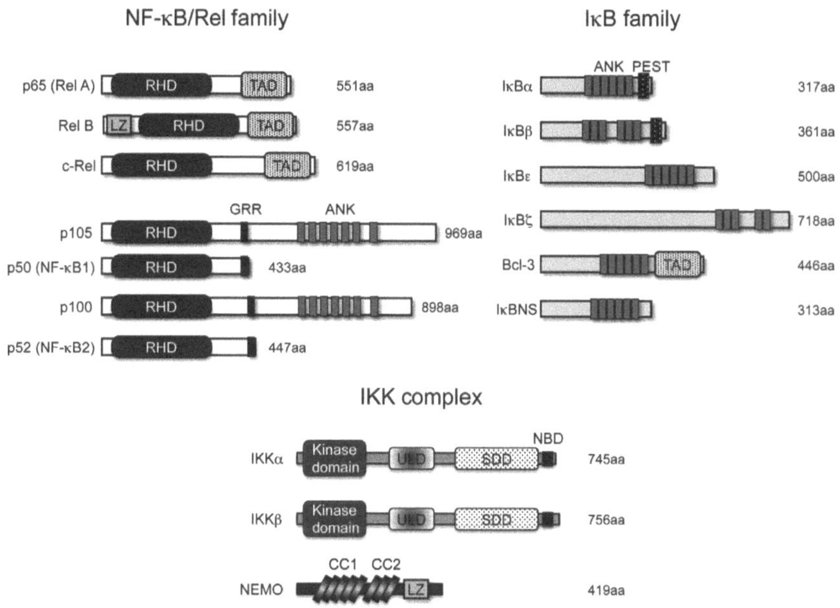

Figure 1 Schematic representation of the NF-κB/IκB protein family and the IKK family. Members of the NF-κB/IκB protein family and the IKK family are shown. The number of amino acids in each protein is indicated on the right. Presumed sites of cleavage for p105 (amino acid 433) and p100 (amino acid 447) are shown. RHD, Rel homology domain; TAD, transcriptional activation domain; LZ, leucine zipper; GRR, glycine-rich repeat; ANK, ankyrin repeat; PEST, PEST domain; CC, coiled-coil domain; ULD, ubiquitin-like domain; SDD, scaffolding and dimerization domain; NBD, NEMO-binding domain.

with IκB proteins (Fig. 1). The activation of NF-κB is achieved through the phosphorylation of IκBs on conserved serine residues, so-called destruction box serine residues (DSGXXS), leading to recognition by βTrCP proteins. The phosphorylation induces K48-linked polyubiquitination by the Skp1-Culin-Roc1/Rbx1/Hrt-1-F-box (SCF or SCRF) family of E3 ubiquitin ligases acting coordinately with the E2 enzyme UbcH5. The typical and most extensively studied member of the family is IκBα. IκBα associates with NF-κB dimers, mainly p65:p50, and inhibits their nuclear export by masking the nuclear localization sequence of p65. The phosphorylation of IκB is a critical step in the NF-κB signaling pathway.

The IκB kinase (IKK) is an enzyme complex that is involved in modulating the cellular response (DiDonato et al., 2012; Hayden & Ghosh, 2012, 2014). The IKK complex consists of three subunits that are each encoded by a separate gene, IKKα (also known as IKK1), IKKβ (also known as IKK2), and NF-κB essential modulator (NEMO) (also known as IKKγ) (Fig. 1). The α- and β-subunits together are catalytically active, whereas NEMO serves a regulatory function. IKKα and IKKβ, 85 and 87 kDa, respectively, have high sequence homology and contain an N-terminal kinase domain, a dimerization domain, and a C-terminal NEMO-binding domain (May et al., 2000). The kinase activity critically depends on lysine 44 and phosphorylated T-loop serines S176 and 180 for IKKα and S177 and 181 for IKKβ. IKKβ is required mainly for the canonical NF-κB pathway that depends on IκB degradation. Despite the structural similarity of IKKα and IKKβ, biological and genetic studies have indicated that IKKβ is the dominant kinase involved in IκB phosphorylation. IKKβ-deficient mice present a phenotype similar to that of p65-deficient mice, which die at E13.5 from severe liver damage due to massive apoptosis, suggesting the importance of IKKβ in IκB phosphorylation (Beg, Sha, Bronson, Ghosh, & Baltimore, 1995; Doi, Takahashi, Taguchi, Azuma, & Obata, 1997; Li, Chu, et al., 1999; Li, Van Antwerp, Mercurio, Lee, & Verma, 1999; Tanaka et al., 1999). However, IKKα-deficient mice die perinatally with multiple morphological defects (Hu et al., 1999; Li, Lu, et al., 1999). Although initial studies have demonstrated few roles for IKKα in NF-κB activation, recent studies have shown that IKKα is involved in an alternative NF-κB pathway that regulates the RelB/p52 dimer (Sun, 2011). Mice, deficient in the third component, NEMO, died at E12.5–E13.0 from severe liver damage due to massive apoptosis, suggesting that NEMO is indispensable for the activation of NF-κB signaling (Schmidt-Supprian et al., 2000).

2.2 Regulatory Mechanisms of NF-κB Signaling

NF-κB is activated by two distinct pathways, termed "classical or canonical" and "alternative or noncanonical" NF-κB activation pathways (Fig. 2) (DiDonato et al., 2012; Hayden & Ghosh, 2012, 2014). These pathways are differentially regulated by IKKα and IKKβ. The classical NF-κB signaling pathway is activated by a broad range of receptors via the activation of the members of the mitogen-activated protein kinase kinase kinase (MAPKKK) superfamily, such as MAPK/extracellular signal-related kinase kinase (MEKK) and TGF-β-activated kinase-1. In response to a variety of stimuli, such as TNFα, IL-1β, or LPS, IκBs, mainly IκBα and IκBβ, are phosphorylated (Ser 32 and Ser 36 for IκBα and Ser 19 and Ser 21 for IκBβ) by the activated IKK complex, mainly IKKβ, leading to their ubiquitination

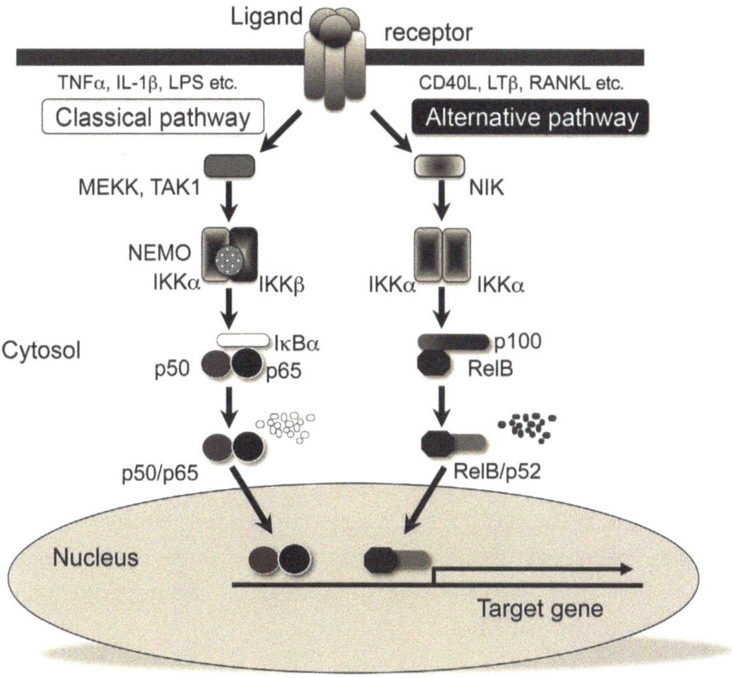

Figure 2 The classical and alternative NF-κB signaling pathways. The classical (canonical) pathway is activated by a large number of agonists, such as TNFα, IL-1β, lipopolysaccharide, and T-cell receptors. The activation of this pathway depends on the inhibitor of κB (IκB) kinase (IKK) complex, which phosphorylates IκBα to induce rapid degradation. This pathway is essential for immune responses, inflammation, tumorigenesis, and cell survival. The alternative (noncanonical) pathway is activated by a limited number of agonists, which are involved in secondary lymphoid organogenesis, mature B-cell function, and adaptive immunity. This pathway requires NF-κB inducing kinase (NIK) and IKKα to induce the slow processing of p52 from p100, resulting in the dimerization and activation of the p52/RelB heterodimer.

and degradation by the 26S proteasome. This degradation allows NF-κB dimers, mainly p65:p50, to translocate to the nucleus and regulate gene expression.

By contrast, the alternative NF-κB pathway depends on IKKα and is independent of NEMO (Sun, 2011). The alternative NF-κB pathway is activated by a select group of TNFR superfamily members, including CD40, lymphotoxin-β receptor (LTβR), and receptor activator of NF-κB (RANK) (Heusch, Lin, Geleziunas, & Greene, 1999; Sun, 2011). The NF-κB-inducing kinase (NIK) phosphorylation of IKKα leads to the phosphorylation of p100 at serines 866 and 870 in p100: RelB heterodimers and the processing to p52. NIK also phosphorylates p100 through a kinase-independent mechanism. Unlike canonical signaling to IκBα degradation, it was shown that the NIK- and IKKα-dependent pathway leading to p100 processing to p52 does not require NEMO or IKKβ. Similar to the process with IκBα, the phosphorylation of p100 leads to recognition by βTrCP. Subsequently, the polyubiquitination of Lys 855 in a region with sequence homology to Lys22 of IκBα is mediated by the SCF complex, leading to subsequent proteasomal processing of p100 to p52 by an as-yet undefined mechanism (Sun, 2011). Another structural element of p100 that is also involved in p105 processing to p50 and that may contribute to selective processing is a glycine-rich repeat (GRR) that is thought to prevent complete degradation and promote the release of active RelB: p52 dimers. Indeed, p100 lacking GRR (p100ΔGRR) undergoes NIK-induced degradation rather than processing to p52 (Heusch et al., 1999). Given that IKKα can contribute to some classical pathways and augment alternative signaling through the induction of p100 expression, the most useful distinction between classical and alternative signaling remains the dependence of the signaling pathway on NEMO.

3. BMP REGULATES OSTEOBLAST DIFFERENTIATION AND BONE FORMATION

Bone consists of hydroxyapatite crystals and various extracellular matrix proteins, including type I collagen, osteocalcin, osteopontin, bone sialoprotein, and proteoglycans. Most of these bone matrix proteins are secreted and deposited by mature osteoblasts, which are aligned on the bone surface. The formation of hydroxyapatite crystals in osteoid is also regulated by osteoblasts. The expression of a number of bone-related extracellular matrix proteins, high enzymatic activity of alkaline phosphatase (ALP), and the responsiveness to osteotropic hormones and cytokines are believed to be major characteristics of osteoblasts (Mundlos & Olsen, 1997; Robey et al., 1993; Young, Kerr, Ibaraki, Heegaard, & Robey, 1992).

During embryogenesis, bone tissues are formed through two distinct pathways: intramembranous ossification and endochondral ossification (Karsenty, Kronenberg, & Settembre, 2009; Komori, 2010; Kronenberg, 2003). In the case of intramembranous ossification, osteoblasts differentiate directly from mesenchymal cell condensations, whereas in the case of endochondral ossification, the condensed mesenchymal cells differentiate into chondrocytes and form a cartilaginous template. Then, the surrounding mesenchymal cells immediately differentiate into osteoblasts after the maturation of hypertrophic chondrocytes in the template, suggesting that osteoblasts and chondrocytes are derived from a common progenitor cell.

Several cytokines and hormones, such as BMP, TGF-β, Wnt, hedgehog, fibroblast growth factors, and estrogen, modulate the regulation of mesenchymal cell differentiation by stimulating intracellular signaling pathways. Among them, BMP is one of the most powerful inducers of ectopic bone formation and strongly promotes the differentiation of mesenchymal cells into osteoblasts (Katagiri et al., 2008; Komori, 2010; Lee et al., 2000). The biological effects of recombinant BMP proteins on osteoblastic differentiation have been studied *in vitro* using cell lines and primary cells. In cultures of osteoblast lineage cells, various BMPs induced the expression of ALP, parathyroid hormone (PTH)/PTH-related protein receptor, type I collagen, and osteocalcin (Aoki et al., 2001; Ebisawa et al., 1999). BMPs also stimulated the formation of mineralized bone-like nodules in these cells. BMPs can also induce osteoblastic differentiation in several other types of cells in culture and some pluripotent cell lines, such as C3H10T1/2 cells (Ahrens et al., 1993; Katagiri et al., 1990; Wang, Israel, Kelly, & Luxenberg, 1993). BMP2 inhibited the myogenic differentiation of a mouse myoblast cell line, C2C12, and instead diverted their differentiation pathway into osteoblasts (Katagiri et al., 1994; Yamaguchi et al., 1991).

3.1 Activation of NF-κB Signaling in an Inflammatory Environment Suppresses BMP-Induced Osteoblast Differentiation

The involvement of NF-κB in osteoclastogenesis has been well investigated (Abu-Amer, 2013; Novack, 2011). For example, the inhibition of NF-κB pathways using transgenic mice expressing the active form of IKKβ in myeloid cells induced severe bone loss (Otero, Chen, Zhang, & Abu-Amer, 2012), whereas IKKβ-specific peptides suppresses inflammatory bone loss by inhibiting osteoclastogenesis in an arthritis model, suggesting that NF-κB is a major target of inflammatory bone diseases (Dai, Hirayama, Abbas, & Abu-Amer, 2004; Jimi et al., 2004). However, the importance of NF-κB in BMP-induced osteoblast differentiation or bone formation in an inflammatory environment has been revealed in recent studies (Krum, Chang, Miranda-Carboni, & Wang, 2010).

Blocking the activation of NF-κB by TNFα using a cell-permeable NF-κB activation antagonist, but not the inactive mutant, prevented TNFα from suppressing BMP-2-induced Runx2 mRNA expression and osteoblast differentiation in MC3T3-E1 cells (Li et al., 2010). Zinc also antagonized NF-κB activation driven by TNFα and stimulated BMP2-induced osteoblastogenesis (Yamaguchi & Weitzmann, 2011). UBS109, a novel curcumin analogue, potently stimulated BMP2-induced Smad-luciferase activity and osteoblastogenesis by inhibiting TNFα-induced NF-κB activity (Yamaguchi, Moore, Sun, Snyder, & Shoji, 2012). Treatment with pyrrolidine dithiocarbamate (PDTC), an NF-κB inhibitor, partially rescued BMP2-induced Smad phosphorylation and osteoblastogenesis of bone marrow mesenchymal stem cells from systemic *lupus erythematosus* patients with an increased risk for osteoporosis by suppressing TNFα-induced NF-κB activation in these cells (Tang et al., 2013). These results indicate that the inhibition of NF-κB could suppress the antiosteoblastogenic activity of TNFα. In addition to TNFα, lipopolysaccharide (LPS) activates NF-κB via a Toll-like receptor 4/Myd88-dependent manner and inhibited BMP2-induced Smad1/5 phosphorylation and osteoblastogenesis (Huang et al., 2014). The blocking of NF-κB signaling by pretreatment with specific inhibitors, such as BAY11-7082, tosyl phenylalanyl chloromethyl ketone, and PDTC, or by transfection of p65 siRNA could significantly reverse the inhibitory effects of BMP2-induced BMP/Smad signaling and osteogenic differentiation.

We have also shown that TNFα inhibited ALP activity and markedly reduced BMP2- and Smad-induced reporter activity without affecting Smad1/5 phosphorylation and the nuclear translocation of the Smad1–Smad4 complex in MC3T3-E1 cells (Yamazaki et al., 2009). In mouse embryonic fibroblasts (MEFs) prepared from mice deficient in p65, the main subunit of NF-κB, the overexpression of p65 inhibited BMP2- and Smad-induced reporter activity in a dose-dependent manner. Chromatin immunoprecipitation and an electrophoresis mobility shift assay showed that TNFα suppressed the binding of Smad proteins to the target gene DNA. The pharmacological NF-κB inhibitor BAY11-7082 restored the inhibitory effects of TNFα on both BMP-induced DNA binding of Smad proteins and ALP activity.

3.2 Classical NF-κB Signaling Modulates BMP-Induced Osteoblastogenesis and Bone Formation

Recently, the inhibition of NF-κB in mature mouse osteoblasts expressing a dominant-negative form of IKKβ (OB-IKKβDN) showed increased bone mineral density and bone volume due to the increased activity of osteoblasts with no effects on osteoclast activity (Chang et al., 2009). Osteoblasts derived from OB-IKKβDN mice were more responsive to BMP2. These results suggest that the inhibition of NF-κB also enhances BMP-induced osteoblastogenesis and bone formation in a physiological environment. The inhibition of NF-κB by the overexpression of a dominant-negative form of IκBα (IκBαDN) leads to the induction of osteoblast differentiation (Eliseev et al., 2006). The overexpression of IκBαDN in osteosarcoma Saos2 cells resulted in the higher expression of osteoblastic phenotypic genes, such as ALP, Runx2, and osteocalcin, and in more responsiveness to BMP2 compared with wild-type Saos2 cells or empty vector-transfected Saos2 cells. We have also shown that a selective inhibitor of the classic NF-κB pathway BAY11-7082 enhanced BMP2-induced ectopic bone formation *in vivo* (Hirata-Tsuchiya et al., 2014). In p65 −/− MEFs, BMP2 induced osteoblastic differentiation via the Smad complex to a greater extent than that in wild-type MEFs because the BMP2-activated Smad complex bound much more stably to the target element than it did in wild-type MEFs. The C-terminal TA2 domain of p65 is associated with the MH1 domain of Smad4 but not with Smad1. Our results indicate that p65 inhibits BMP signaling by blocking the DNA binding of the Smad complex via an interaction with Smad4.

However, contrary to these results, a few studies have demonstrated that NF-κB enhances BMP-induced osteoblastic differentiation from human mesenchymal stem cells or MC3T3 subclone 4 (Hess, Ushmorov, Fiedler, Brenner, & Wirth, 2009; Kim, Lee, Kim, Min, & Kim, 2008). These seemingly contradictory findings may be due to cell type-specific or differentiation stage-specific effects.

3.3 Alternative NF-κB Signaling Negatively Regulates BMP-Induced Osteoblastogenesis and Bone Formation

Alymphoplasia (*aly/aly*) mice have a natural loss-of-function mutation in the gene encoding the NIK, a kinase essential for the processing of p100 to p52 in the alternative NF-κB pathway (Shinkura et al., 1999). The alternative NF-κB signaling pathway is inhibited downstream of NIK in these mice. We have recently reported that *aly/aly* mice exhibited mild osteopetrosis with significantly reduced osteoclast number and increased bone formation (Maruyama et al., 2010; Soysa et al., 2010). By contrast, p100-deficient (p100 −/−) mice, which carry a homozygous deletion of the COOH-terminal ankyrin repeats of p100 but still express functional p52 protein (Ishikawa, Carrasco, Claudio, Ryseck, & Bravo, 1997), have precisely the opposite bone phenotype of *aly/aly* mice: osteopenia with increased numbers of osteoclasts and decreased numbers of osteoblasts (Soysa et al., 2010).

The ectopic bone formation induced by BMP2 was larger in *aly/aly* mice than in wild-type (WT) mice *in vivo*. BMP2-induced osteoblast differentiation, as indicated by ALP activity, osteocalcin mRNA expression, and phosphorylation of Smad1/5 were enhanced in primary osteoblasts (POBs) from *aly/aly* mice. By contrast, BMP2-induced ALP activity and Smad1/5 phosphorylation were decreased in POBs from p100 −/− mice compared with WT mice. These results suggest that the inhibition of the alternative NF-κB signaling pathway by the prevention of p100 and p52 processing positively regulates osteoblastogenesis. Furthermore, p100ΔGRR, which is not processed into p52 due to substitutions at the critical processing site, and p52 interacted with the BMP receptor ALK2 and changed the ALK2 protein levels in opposite directions: p100ΔGRR increased ALK2 protein levels and p52 decreased them (Seo et al., 2012).

Take together, inhibition of the classical NF-κB signaling and the alternative NF-κB signaling pathway by selective inhibitor for IKKβ and *aly* mutant, respectively, might be useful for BMP-induced bone formation and regeneration (Fig. 3).

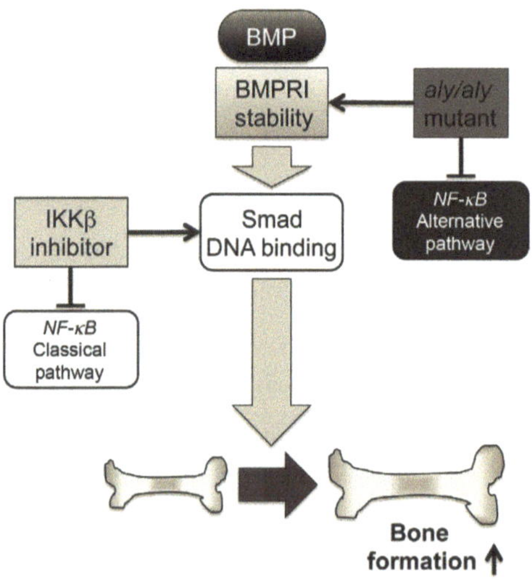

Figure 3 Inhibition of NF-κB is useful for bone formation and regeneration. Inhibition of the classical and alternative NF-κB pathways enhances osteoblastic bone formation by modulating BMP/Smad signaling through two different mechanisms: extension of BMP-induced Smad DNA binding and stabilization of BMP receptors.

4. THE ROLE OF NF-κB IN BMP-INDUCED CHONDROGENIC DIFFERENTIATION

The cartilaginous structure of a growth plate consists of four zones, the resting zone, the proliferative zone, the prehypertrophic zone, and the hypertrophic zone, that contain chondrocytes at different stages of differentiation (Kronenberg, 2003; Nishimura, Hata, Matsubara, Wakabayashi, & Yoneda, 2012). BMPs are involved in all phases of chondrogenesis and are expressed in a gradient in the growth plate, with the greatest BMP signaling occurring in the hypertrophic zone and the least in the resting zone. Consistent with this finding, phosphorylated Smad1/5 localizes with increasing distance from the epiphysis in the growth plate (Yoon et al., 2006). These expression patterns might contribute to the progressive differentiation of resting to proliferative to hypertrophic chondrocytes. BMP2 promotes the hypertrophy of resting zone chondrocytes in an organ culture model. Mice expressing the constitutively active form of BMPr1a showed accelerating hypertrophic differentiation with no effects on proliferation (Kobayashi, Lyons, McMahon, & Kronenberg, 2005). The requirement of BMPs for chondrocyte differentiation is supported by evidence that the loss of Smad1 and 5 or the inhibition of the BMP/Smad1/5 axis blocks the differentiation of chondrocytes and leads to severe cartilage defects (Retting, Song, Yoon, & Lyons, 2009). BMPs also play an important role in the regeneration of cartilage, in processes such as chondrocyte hypertrophy and matrix degradation (Mariani, Pulsatelli, & Facchini, 2014). Furthermore, BMP2 promotes matrix synthesis and the expression of SRY-related high-mobility-group box transcription factor (SOX) 9, which is required for chondrocyte differentiation (Healy, Uwanogho, & Sharpe, 1999; Lui, Nilsson, & Baron, 2014).

NF-κB has been recognized as a key developmental signaling mediator that regulates endochondral ossification. Early bone fracture healing by endochondral ossification depends on a hematoma-induced inflammatory environment, and several NF-κB target genes are involved in bone fracture repair (Mariani et al., 2014). Furthermore, in chick embryos, the overexpression of the dominant-negative form of IκBα, which blocks NF-κB activation, leads to abnormalities in limb bud development and the downregulation of BMP signaling (Kanegae, Tavares, Izpisúa Belmonte, & Verma, 1997).

Several groups have clarified the functional interaction between NF-κB and BMP2 in chondrocyte differentiation (Feng et al., 2003; Wu, Flint,

Rezvani, & De Luca, 2007). The expression of p65 was observed throughout the growth plate, predominantly in the resting and hypertrophic zones. The inhibition of NF-κB signaling using PDTC, BAY11-7082, or p65 siRNA suppressed metatarsal liner growth using cultured rat metatarsal bones and growth plate chondrocytes by decreasing chondrocyte proliferation and differentiation and increasing apoptosis due to suppressed BMP2 expression (Wu et al., 2007). By contrast, the overexpression of p65 in cultured chondrocytes induced chondrocyte proliferation and differentiation and decreased apoptosis by increasing BMP2 expression (Feng et al., 2003). Two putative NF-κB-responsive elements were found in the promoter region of the *BMP2* gene. These elements were functional in chondrocytes, and NF-κB induced BMP2 expression through these elements. Consistent with these results, the expression of BMP2 in growth plates chondrocytes is dramatically reduced, which leads to a significant decrease in chondrocyte proliferation in NF-κB1 and NF-κB2 double knock-out mice, suggesting that in postnatal growth plate chondrocytes, the expression of BMP2 is regulated by NF-κB, which plays an important role in chondrogenesis (Feng et al., 2003).

In addition, recent findings have shown the importance of transient NF-κB activation on the initiation of chondrogenic differentiation using ATDC5, a chondrogenic cell line (Caron et al., 2012). BMP2 induced transient activation of NF-κB during the very first hours of chondroprogenic differentiation, and p65 siRNA suppressed BMP2-induced Sox9 expression in ATDC cells, suggesting that BMP2 actions in the early chondrogenic phase of endochondral ossification explain, in part, early transient NF-κB activation and Sox9 expression.

5. THE ROLE OF BMP AND NF-κB SIGNALING ON CANCER PROGRESSION

Bone is a common site of certain solid tumors, such as breast cancer, prostate cancer, and lung cancer (Sterling, Edwards, Martin, & Mundy, 2011). Although metastasis is the major cause of mortality for cancer patients, it is still unclear how the cancer cells acquire capacities and characteristics, leading to the spread and development of bone metastasis due to a complex and multistage process in which secondary tumors are formed in distant sites.

BMPs and the signaling of their receptor have been involved in the development and progression of a variety of solid tumors, including prostate cancer, breast cancer, and lung cancer (Ye, Bokobza, & Jiang, 2009; Ye, Lewis-Russell, Kyanaston, & Jiang, 2007). Prostate cancer and breast cancer are the most common malignancies with the highest incidence of bone metastasis and are also representative of two distinct types of metastatic bone lesions: osteoblastic and osteolytic lesions. The NF-κB signaling pathway is also activated in many cancers, including breast cancer and liver cancer, contributing to the acquisition of malignant characteristics, such as the increased invasion, survival, chemoresistance, and angiogenesis of cancers (Ben-Neriah & Karin, 2011; DiDonato et al., 2012).

The expression of BMPRIA, BMPRIB, and BMPRII in human prostate cancer tissues has been investigated and was found to correlate with tumor grade (Kim et al., 2000, 2004). Integrins are major adhesive molecules in mammalian cells that have been associated with prostate cancer metastasis (Jacquemet, Humphries, & Caswell, 2013). The conditioned media from osteoblast cultures (OBCM) increased the migration of human prostate cancer cells by inducing β1 and β3 integrin expression via NF-κB activation (Lai et al., 2008). BMP2 siRNA specifically reduced OBCM-induced migration and integrin upregulation by inhibiting Akt/ERK and NF-κB activation. These results suggest that osteoblast-derived BMP2 acts through Akt and ERK, which in turn activate NF-κB, resulting in the activation of β1 and β3 integrin and contributing to the migration of prostate cancer cells, whereas it was reported that PI3K/AKT–NF-κB axis promotes bone metastasis by prostate cancer cells in part by activating BMP2/Smad signaling (Graham et al., 2009).

Gastric cancer is a malignancy derived from gastric mucosal epithelial cells and accounts for 95% of gastric malignancies. Gastric cancer has become a common malignancy threatening human health (Shim et al., 2013). BMP2 induced the phosphorylation of ERK and Akt and the activation of NF-κB in gastric cancer cells. BMP2 also induced motility and invasiveness in association with the induction of matrix metalloprotease (MMP) 9 expression and its activity (Kang et al., 2011). The blockade of PI3K/Akt and ERK signaling using LY294002 and PD98059, respectively, significantly suppressed BMP2-induced MMP9 expression and its activity by inhibiting NF-κB activation, suggesting that BMP2 signaling enhances tumor metastasis in gastric cancer by the sequential activation of PI3K/Akt and ERK followed by NF-κB activation and MMP9 activity.

Bone tumors are usually associated with disturbed bone metabolic activities. Among different types of bone tumors, chondrosarcoma is the second most common type of bone tumor and a malignant primary bone tumor with a poor response to currently used chemotherapy or radiation treatments (Chow, 2007). BMP2 promotes cell migration and induces the expression of β1 integrin on the cell surface of a chondrosarcoma cell line, JJ012. Pretreatment with either PI3K (LY294002) or PDTC inhibited BMP2-mediated cell migration and integrin upregulation (Fong et al., 2008). Similar results were also reported for BMP7, which is expressed at significantly higher levels in chondrosarcoma, but it is not detected or is found at very low expression levels in normal cartilage; BMP7 also enhances cell migration and αvβ3 integrin via the c-Src–NF-κB-dependent pathway in human chondrosarcoma cells (Chen et al., 2014). These results suggest that the BMPs activate NF-κB, resulting in the upregulation of integrins and contributing to the migration of chondrosarcoma cells.

6. BMP SIGNALING AND NF-κB SIGNALING IN VASCULAR DISEASES

6.1 Pulmonary Arterial Hypertension

Genetic studies in humans and mice have shown that the inhibition of BMP signaling leads to vascular diseases as well as skeletal diseases and cancer (Cai et al., 2012; Goumans & Mummery, 2000; Lowery & de Caestecker, 2010). The genetic deletion or misexpression of different components of BMP signaling leads to embryonic lethality due to cardiovascular malformations and defects in vascular remodeling. Furthermore, adequate BMP signaling in both epithelial cells and smooth muscle cells has been shown to be required for appropriate vasculogenesis and angiogenesis (Cai et al., 2012).

PAH is a disease characterized by progressively increasing pulmonary vascular resistance, leading to right heart failure and death (Machado & Farber, 2013). Two types of PAH have been reported: sporadic or idiopathic PAH (IPAH) and familial or hereditary PAH (HPAH) (Cogan et al., 2005; Lane et al., 2000). Mutations in BMPRII are responsible for the majority of HPAH cases and loss of Smad8 in mice led to increased vascular inflammation (Huang, Wang, Ihida-Stansbury, Jones, & Martin, 2008; Shintani, Yagi, Nakayama, Saji, & Matsuoka, 2009). Furthermore, NF-κB is activated in the pulmonary vessels of patients with end-stage IPAH (Davies et al., 2012; Eickelberg et al., 1999). These reports indicate that NF-κB signaling may be involved in BMPRII-related PAH. Mice expressing BMPR2^{R899X}, a Bmpr2 mutation that is truncated in the cytoplasmic tail domain, or BMPR2^{delx4}, a Bmpr2 mutant with an inserted "T" at base 504 that leads to a premature stop 18 amino acids into the kinase domain, develop

PAH. Both mutations blocked Smad downstream signaling (Machado et al., 2001; Tada et al., 2007). Although NF-κB is activated in the lungs in end-stage disease in mice, the BMPR2 mutation did not activate NF-κB in early disease development (Talati et al., 2014). Thus, BMPR2 indirectly activates NF-κB after pulmonary vascular pressers have already been increased. NF-κB activation leads to the further inhibition of BMP signaling followed by worsening muscularization and pulmonary pressures.

6.2 Vascular Calcification

Vascular calcification commonly occurs with advancing age, chronic kidney disease, diabetes mellitus, and atherosclerosis and is closely associated with cardiovascular morbidity and mortality (Abedin, Tintut, & Demer, 2004; Mazzini & Schulze, 2006). BMPs play important roles in vascular calcification and atherosclerosis (Cai et al., 2012) NF-κB activation was observed in the endothelial cells of atherosclerotic plaques (Brand et al., 1996).

Oxidized (ox)-low-density lipoprotein (LDL) participates in the initiation and development of atherosclerotic plaques and has been associated with subclinical atherosclerosis. Ox-LDL induced both the expression of BMP2 mRNA and the production of BMP2 in human umbilical vein endothelial cells (HUVECs) (Zhang et al., 2008). Ox-LDL also modulates the oxidative status in HUVECs by suppressing the activity of an antioxidative enzyme superoxide dismutase and increasing malondialdehyde, a marker of oxidative stress. Pretreatment with either PDTC or atorvastatin inhibited ox-LDL-induced BMP2 expression by inhibiting NF-κB activation in HUVECs. Similar events were also observed in angiotensin II (AngII)-treated HUVECs. AngII is essential for the regulation of blood pressure and is also involved in arterial wall remodeling. AngII also induces BMP2 expression in HUVECs, and pretreatment with either PDTC or probucol, an antihyperlipidemic drug, inhibited AngII-induced BMP2 expression by inhibiting NF-κB activation in HUVECs (Zhang et al., 2013). Free fatty acids play a crucial role in the development of atherosclerotic cardiovascular disease. Palmitic acid induced BMP2 expression by long-chain acyl-CoA synthetase 3-induced NF-κB activation in human aortic smooth muscle cells. Eicosapentaenoic acid inhibited these effects (Kageyama et al., 2013).

Proinflammatory cytokines and high intraluminal pressure elicit vascular oxidative stress (Ungvari et al., 2003). In particular, an increased level of reactive oxygen species, including increased H_2O_2 production, was mediated by NF-κB activation. TNFα induced NF-κB activation and elicited a significant increase in BMP2 mRNA and protein in primary coronary arterial endothelial cells and HUVECs that was suppressed by NF-κB inhibitors (PDTC, SN-50), p65 siRNA, or catalase. The addition of H_2O_2 also induced both NF-κB activation and BMP2 production (Csiszar et al., 2005). The exposure of rat arteries to high pressure induced H_2O_2 production, NF-κB activation, and BMP2 production *ex vivo*. These results suggest that vascular BMP2 expression is regulated by H_2O_2-mediated NF-κB both from inflammatory stimuli and from high intravascular pressure.

By contrast, a study showed that certain reduced BMP signaling causes the progression and worsening of vascular calcification and atherosclerosis (Kim et al., 2013). The knockdown of BMPRII but not other BMP receptors in HUVECs induced endothelial inflammation in a ligand-independent manner by a mechanism mediated by reactive oxygen species-induced NF-κB activation. BMPRII+/−ApoE+/− mice developed accelerated atherosclerosis compared with BMPRII+/+ApoE+/− mice. The loss of BMPRII in mice or human coronary arteries in clinical samples advanced atherosclerotic plaque formation caused by NF-κB-induced expressions of adhesion molecules followed by monocyte infiltration, which become foam cells.

7. CONCLUSIONS AND FUTURE DIRECTIONS

Although BMPs were originally identified as potent inducers of ectopic bone formation *in vivo* and of osteoblastogenesis in several types of cells *in vitro* (Katagiri et al., 2008; Urist, 1965; Wozney et al., 1988; Katagiri & Tsukamoto, 2013), genetic studies in humans and mice have shown that the inhibition of BMP signaling revealed not only skeletal diseases but also vascular diseases and cancer (Cai et al., 2012; Goumans & Mummery, 2000; Lowery & de Caestecker, 2010). However, NF-κB plays important roles in immune and inflammatory responses and tumorigenesis (Ben-Neriah & Karin, 2011; DiDonato et al., 2012; Hayden & Ghosh, 2012, 2014). Several lines of evidence have shown that BMP/Smad and NF-κB are antagonistic during osteoblast differentiation and bone formation. However, BMP and NF-κB cooperatively regulate chondrocyte differentiation. Furthermore, the expression of BMP is in part regulated by NF-κB in chondrocytes. In clinical investigations and mouse models, reduced BMP signals promote vascular diseases and cancer by activating the NF-κB signaling induced by several stimuli.

Thousands of experiments using distinct and/or different approaches have provided a variety of answers to how the BMP/Smad pathway interacts with the NF-κB pathway. These studies often show contradictory results that are not easy explainable. Although these discrepancies could have arisen from the variations in experimental conditions, they may also reflect the true adaptability that an organism must possess to survive in a constantly changing environment. It is very difficult to simplify the role of the cross talk between BMP and NF-κB signaling using cultured cells *in vitro* because various types of cells at different developmental stages exist and are in changing physiological or pathological condition *in vivo*. In conclusion, because BMP and NF-κB pathways regulate bone formation, vascular diseases, and tumorigenesis either as agonists or as antagonists under pathological conditions, the fine-tuning or balance between BMP and NF-κB pathways might be useful for therapeutic means in certain diseases.

ACKNOWLEDGMENT

This work was supported by the Ministry of Education, Culture, Sports, Science and Technology of Japan (to E.J.: 26293396).

REFERENCES

Abedin, M., Tintut, Y., & Demer, L. L. (2004). Vascular calcification: Mechanisms and clinical ramifications. *Arteriosclerosis, Thrombosis, and Vascular Biology*, *24*, 1161–1170.

Abu-Amer, Y. (2013). NF-κB signaling and bone resorption. *Osteoporosis International*, *24*, 2377–2386.

Ahrens, M., Ankenbauer, T., Schröder, D., Hollnagel, A., Mayer, H., & Gross, G. (1993). Expression of human bone morphogenetic proteins-2 or -4 in murine mesenchymal progenitor C3H10T1/2 cells induces differentiation into distinct mesenchymal cell lineages. *DNA and Cell Biology*, *12*, 871–880.

Ammann, P., Rizzoli, R., Bonjour, J. P., Bourrin, S., Meyer, J. M., Vassalli, P., et al. (1997). Transgenic mice expressing soluble tumor necrosis factor-receptor are protected against bone loss caused by estrogen deficiency. *Journal of Clinical Investigation*, *99*, 1699–1703.

Anderson, L., Lowery, J. W., Frank, D. B., Novitskaya, T., Jones, M., Mortlock, D. P., et al. (2010). Bmp2 and Bmp4 exert opposing effects in hypoxic pulmonary hypertension. *The American Journal of Physiology—Regulatory, Integrative and Comparative Physiology*, *298*, R833–R842.

Aoki, H., Fujii, M., Imamura, T., Yagi, K., Takehara, K., Kato, M., et al. (2001). Synergistic effects of different bone morphogenetic protein type I receptors on alkaline phosphatase induction. *Journal of Cell Science*, *114*, 1483–1489.

Beg, A. A., Sha, W. C., Bronson, R. T., Ghosh, S., & Baltimore, D. (1995). Embryonic lethality and liver degeneration in mice lacking the RelA component of NF-κB. *Nature*, *376*, 167–170.

Ben-Neriah, Y., & Karin, M. (2011). Inflammation meets cancer, with NF-κB as the matchmaker. *Nature Immunology*, *12*, 715–723.

Brand, K., Page, S., Rogler, G., Bartsch, A., Brandl, R., Knuechel, R., et al. (1996). Activated transcription factor nuclear factor-κB is present in the atherosclerotic lesion. *Journal of Clinical Investigation*, *97*, 1715–1722.

Buma, P., Arts, J. J., Gardeniers, J. W., Verdonschot, N., & Schreurs, B. W. (2008). No effect of bone morphogenetic protein-7 (OP-1) on the incorporation of impacted bone grafts in a realistic acetabular model. *Journal of Biomedical Materials Research. Part B, Applied Biomaterials*, *84*, 231–239.

Cai, J., Pardali, E., Sánchez-Duffhues, G., & ten Dijke, P. (2012). BMP signaling in vascular diseases. *FEBS Letters*, *586*, 1993–2002.

Caron, M. M., Emans, P. J., Surtel, D. A., Cremers, A., Voncken, J. W., Welting, T. J., et al. (2012). Activation of NF-κB/p65 facilitates early chondrogenic differentiation during endochondral ossification. *PLoS One*, 7, e33467.

Chang, J., Wang, Z., Tang, E., Fan, Z., McCauley, L., Franceschi, R., et al. (2009). Novel functions for NF-κB: Inhibition of bone formation. *Nature Medicine*, *15*, 682–689.

Chen, J. C., Yang, S. T., Lin, C. Y., Hsu, C. J., Tsai, C. H., Su, J. L., et al. (2014). BMP-7 enhances cell migration and αvβ3 integrin expression via a c-Src-dependent pathway in human chondrosarcoma cells. *PLoS One*, *9*, e112636.

Chow, W. A. (2007). Update on chondrosarcomas. *Current Opinion in Oncology*, *19*, 371–376.

Choy, E. H., & Panayi, G. S. (2001). Cytokine pathways and joint inflammation in rheumatoid arthritis. *The New England Journal of Medicine*, *344*, 907–916.

Cogan, J. D., Vnencak-Jones, C. L., Phillips, J. A., 3rd., Lane, K. B., Wheeler, L. A., Robbins, I. M., et al. (2005). BMPR2 gene rearrangements constitute a new cause for primary pulmonary hypertension. *Genetics in Medicine*, 7, 169–174.

Csiszar, A., Labinskyy, N., Jo, H., Ballabh, P., & Ungvari, Z. (2008). Differential proinflammatory and prooxidant effects of bone morphogenetic protein-4 in coronary and pulmonary arterial endothelial cells. *The American Journal of Physiology—Heart and Circulatory Physiology*, *295*, H569–H577.

Csiszar, A., Smith, K. E., Koller, A., Kaley, G., Edwards, J. G., & Ungvari, Z. (2005). Regulation of bone morphogenetic protein-2 expression in endothelial cells: Role of nuclear factor-κB activation by tumor necrosis factor-α, H_2O_2, and high intravascular pressure. *Circulation*, *111*, 2364–2372.

Dai, S., Hirayama, T., Abbas, S., & Abu-Amer, Y. (2004). The IκB kinase (IKK) inhibitor, NEMO-binding domain peptide, blocks osteoclastogenesis and bone erosion in inflammatory arthritis. *The Journal of Biological Chemistry*, *279*, 37219–37222.

Davies, R. J., Holmes, A. M., Deighton, J., Long, L., Yang, X., Barker, L., et al. (2012). BMP type II receptor deficiency confers resistance to growth inhibition by TGF-β in pulmonary artery smooth muscle cells: Role of proinflammatory cytokines. *The American Journal of Physiology—Lung Cellular and Molecular Physiology*, *302*, L604–L615.

DiDonato, J. A., Mercurio, F., & Karin, M. (2012). NF-κB and the link between inflammation and cancer. *Immunological Reviews*, *246*, 379–400.

Doi, T. S., Takahashi, T., Taguchi, O., Azuma, T., & Obata, Y. (1997). NF-κB RelA-deficient lymphocytes: Normal development of T cells and B cells, impaired production of IgA and IgG1 and reduced proliferative responses. *Journal of Experimental Medicine*, *185*, 953–961.

Ebisawa, T., Tada, K., Kitajima, I., Tojo, K., Sampath, T. K., Kawabata, M., et al. (1999). Characterization of bone morphogenetic protein-6 signaling pathways in osteoblast differentiation. *Journal of Cell Science*, *112*, 3519–3527.

Eickelberg, O., Roth, M., Mussmann, R., Rüdiger, J. J., Tamm, M., Perruchoud, A. P., et al. (1999). Calcium channel blockers activate the interleukin-6 gene via the transcription factors NF-IL6 and NF-κB in primary human vascular smooth muscle cells. *Circulation*, *99*, 2276–2282.

Eliseev, R. A., Schwarz, E. M., Zuscik, M. J., O'Keefe, R. J., Drissi, H., & Rosier, R. N. (2006). Smad7 mediates inhibition of Saos2 osteosarcoma cell differentiation by NF-κB. *Experimental Cell Research*, *312*, 40–50.

Feng, J. Q., Xing, L., Zhang, J. H., Zhao, M., Horn, D., Chan, J., et al. (2003). NF-κB specifically activates BMP-2 gene expression in growth plate chondrocytes *in vivo* and in a chondrocyte cell line *in vitro*. *The Journal of Biological Chemistry*, *278*, 29130–29135.

Fong, Y. C., Li, T. M., Wu, C. M., Hsu, S. F., Kao, S. T., Chen, R. J., et al. (2008). BMP-2 increases migration of human chondrosarcoma cells via PI3K/Akt pathway. *The Journal of Cellular Physiology*, *217*, 846–855.

Ghosh, S., & Hayden, M. S. (2012). Celebrating 25 years of NF-κB research. *Immunological Reviews*, *246*, 5–13.

Gilbert, L., He, X., Farmer, P., Boden, S., Kozlowski, M., Rubin, J., et al. (2000). Inhibition of osteoblast differentiation by tumor necrosis factor-α. *Endocrinology*, *141*, 3956–3964.

Goumans, M. J., & Mummery, C. (2000). Functional analysis of the TGFβ receptor/Smad pathway through gene ablation in mice. *The International Journal of Developmental Biology*, *44*, 253–265.

Graham, T. R., Odero-Marah, V. A., Chung, L. W., Agrawal, K. C., Davis, R., & Abdel-Mageed, A. B. (2009). PI3K/Akt-dependent transcriptional regulation and activation of BMP-2-Smad signaling by NF-κB in metastatic prostate cancer cells. *Prostate*, *69*, 168–180.

Hayden, M. S., & Ghosh, S. (2012). NF-κB, the first quarter-century: Remarkable progress and outstanding questions. *Genes & Development*, *26*, 203–234.

Hayden, M. S., & Ghosh, S. (2014). Regulation of NF-κB by TNF family cytokines. *Seminars in Immunology*, *26*, 253–266.

Healy, C., Uwanogho, D., & Sharpe, P. T. (1999). Regulation and role of Sox9 in cartilage formation. *Developmental Dynamics*, *215*, 69–78.

Hess, K., Ushmorov, A., Fiedler, J., Brenner, R. E., & Wirth, T. (2009). TNFα promotes osteogenic differentiation of human mesenchymal stem cells by triggering the NF-κB signaling pathway. *Bone*, *45*, 367–376.

Heusch, M., Lin, L., Geleziunas, R., & Greene, W. C. (1999). The generation of nfkb2 p52: Mechanism and efficiency. *Oncogene*, *18*, 6201–6208.

Hinz, M., Arslan, S. Ç., & Scheidereit, C. (2012). It takes two to tango: IκBs, the multifunctional partners of NF-κB. *Immunological Reviews*, *246*, 59–76.

Hirata-Tsuchiya, S., Fukushima, H., Katagiri, T., Ohte, S., Shin, M., Nagano, K., et al. (2014). Inhibition of BMP2-induced bone formation by the p65 subunit of NF-κB via an interaction with Smad4. *Molecular Endocrinology*, *28*, 1460–1470.

Hu, Y., Baud, V., Delhase, M., Zhang, P., Deerinck, T., Ellisman, M., et al. (1999). Abnormal morphogenesis but intact IKK activation in mice lacking the IKKα subunit of IκB kinase. *Science*, *284*, 316–320.

Huang, Z., Wang, D., Ihida-Stansbury, K., Jones, P. L., & Martin, J. F. (2008). Defective pulmonary vascular remodeling in Smad8 mutant mice. *Human Molecular Genetics*, *18*, 2791–2801.

Huang, R. L., Yuan, Y., Zou, G. M., Liu, G., Tu, J., & Li, Q. (2014). LPS-stimulated inflammatory environment inhibits BMP-2-induced osteoblastic differentiation through crosstalk between TLR4/MyD88/NF-κB and BMP/Smad signaling. *Stem Cells and Development*, *23*, 277–289.

Ishikawa, H., Carrasco, D., Claudio, E., Ryseck, R. P., & Bravo, R. (1997). Gastric hyperplasia and increased proliferative responses of lymphocytes in mice lacking the COOH-terminal ankyrin domain of NF-κB2. *The Journal of Experimental Medicine*, *186*, 999–1014.

Jacquemet, G., Humphries, M. J., & Caswell, P. T. (2013). Role of adhesion receptor trafficking in 3D cell migration. *Current Opinion in Cell Biology*, *25*, 627–632.

Jimi, E., Aoki, K., Saito, H., D'Acquisto, F., May, M. J., Nakamura, I., et al. (2004). Selective inhibition of NF-κB blocks osteoclastogenesis and prevents inflammatory bone destruction *in vivo*. *Nature Medicine*, *10*, 617–624.

Kageyama, A., Matsui, H., Ohta, M., Sambuichi, K., Kawano, H., & Notsu, T. (2013). Palmitic acid induces osteoblastic differentiation in vascular smooth muscle cells through ACSL3 and NF-κB, novel targets of eicosapentaenoic acid. *PLoS One*, *8*, e68197.

Kanegae, Y., Tavares, A. T., Izpisúa Belmonte, J. C., & Verma, I. M. (1997). Role of Rel/NF-κB transcription factors during the outgrowth of the vertebrate limb. *Nature*, *392*, 611–614.

Kaneki, H., Guo, R., Chen, D., Yao, Z., Schwarz, E. M., Zhang, Y. E., et al. (2006). Tumor necrosis factor promotes Runx2 degradation through up-regulation of Smurf1 and Smurf2 in osteoblasts. *The Journal of Biological Chemistry*, *281*, 4326–4333.

Kang, M. H., Oh, S. C., Lee, H. J., Kang, H. N., Kim, J. L., Kim, J. S., et al. (2011). Metastatic function of BMP-2 in gastric cancer cells: The role of PI3K/AKT, MAPK, the NF-κB pathway, and MMP-9 expression. *Experimental Cell Research*, *317*, 1746–1762.

Karsenty, G., Kronenberg, H. M., & Settembre, C. (2009). Genetic control of bone formation. *The Annual Review of Cell and Developmental Biology*, *25*, 629–648.

Katagiri, T., Suda, T., & Miyazono, K. (2008). The bone morphogenetic proteins. In R. Derynck, & K. Miyazono (Eds.), *The TGF-β family* (pp. 121–149). New York: Cold Spring Harbor Monograph.

Katagiri, T., & Tsukamoto, S. (2013). The unique activity of bone morphogenetic proteins in bone: A critical role of the Smad signaling pathway. *Biological Chemistry*, *394*, 614–619.

Katagiri, T., Yamaguchi, A., Ikeda, T., Yoshiki, S., Wozney, J. M., Rosen, V., et al. (1990). The non-osteogenic mouse pluripotent cell line, C3H10T1/2, is induced to differentiate into osteoblastic cells by recombinant human bone morphogenetic protein-2. *Biochemical and Biophysical Research Communications*, *172*, 295–299.

Katagiri, T., Yamaguchi, A., Komaki, M., Abe, E., Takahashi, N., Ikeda, T., et al. (1994). Bone morphogenetic protein-2 converts the differentiation pathway of C2C12 myoblasts into the osteoblast lineage. *The Journal of Cell Biology*, *127*, 1755–1766.

Kim, I. Y., Lee, D. H., Ahn, H. J., Tokunaga, H., Song, W., Devereaux, L. M., et al. (2000). Expression of bone morphogenetic protein receptors type-IA, -IB and -II correlates with tumor grade in human prostate cancer tissues. *Cancer Research*, *60*, 2840–2844.

Kim, J. M., Lee, S. U., Kim, Y. S., Min, Y. K., & Kim, S. H. (2008). Baicalein stimulates osteoblast differentiation via coordinating activation of MAP kinases and transcription factors. *Journal of Cellular Biochemistry*, *104*, 1906–1917.

Kim, I. Y., Lee, D. H., Lee, D. K., Ahn, H. J., Kim, M. M., Kim, S. J., et al. (2004). Loss of expression of bone morphogenetic protein receptor type II in human prostate cancer cells. *Oncogene*, *23*, 7651–7659.

Kim, C. W., Song, H., Kumar, S., Nam, D., Kwon, H. S., Chang, K. H., et al. (2013). Anti-inflammatory and antiatherogenic role of BMP receptor II in endothelial cells. *Arteriosclerosis, Thrombosis, and Vascular Biology*, *33*, 1350–1359.

Kobayashi, T., Lyons, K. M., McMahon, A. P., & Kronenberg, H. M. (2005). BMP signaling stimulates cellular differentiation at multiple steps during cartilage development. *Proceedings of the National Academy of Sciences of the United States of America*, *102*, 18023–18027.

Komori, T. (2010). Regulation of bone development and extracellular matrix protein genes by RUNX2. *Cell and Tissue Research*, *339*, 189–195.

Kronenberg, H. M. (2003). Developmental regulation of the growth plate. *Nature*, *423*, 332–336.

Krum, S. A., Chang, J., Miranda-Carboni, G., & Wang, C. Y. (2010). Novel functions for NF-κB: Inhibition of bone formation. *Nature Reviews. Rheumatology*, *6*, 607–611.

Kusumoto, K., Bessho, K., Fujimura, K., Akioka, J., Okubo, Y., Wang, Y., et al. (2002). Osteoinduction by recombinant human bone morphogenetic protein-2 in muscles of non-human primates. *The Journal of International Medical Research, 30*, 251–259.

Lai, T. H., Fong, Y. C., Fu, W. M., Yang, R. S., & Tang, C. H. (2008). Osteoblasts-derived BMP-2 enhances the motility of prostate cancer cells via activation of integrins. *Prostate, 68*, 1341–1353.

Lane, K. B., Machado, R. D., Pauciulo, M. W., Thomson, J. R., Phillips, J. A., 3rd., Loyd, J. E., et al. (2000). Heterozygous germline mutations in BMPR2, encoding a TGF-β receptor, cause familial primary pulmonary hypertension. *Nature Genetics, 26*, 81–84.

Lee, K. S., Kim, H. J., Li, Q. L., Chi, X. Z., Ueta, C., Komori, T., et al. (2000). Runx2 is a common target of transforming growth factor β1 and bone morphogenetic protein 2, and cooperation between Runx2 and Smad5 induces osteoblast-specific gene expression in the pluripotent mesenchymal precursor cell line C2C12. *Molecular and Cellular Biology, 20*(8783–87), 92.

Li, Z. W., Chu, W., Hu, Y., Delhase, M., Deerinck, T., Ellisman, M., et al. (1999). The IKKβ subunit of IκB kinase (IKK) is essential for nuclear factor κB activation and prevention of apoptosis. *The Journal of Experimental Medicine, 189*, 1839–1845.

Li, Ql, Lu, Q., Hwang, J. Y., Büscher, D., Lee, K. F., Izpisua-Belmonte, J. C., et al. (1999). IKK1-deficient mice exhibit abnormal development of skin and skeleton. *Genes & Development, 13*, 322–328.

Li, Q., Van Antwerp, D., Mercurio, F., Lee, K. F., & Verma, I. M. (1999). Severe liver degeneration in mice lacking the IκB kinase 2 gene. *Science, 284*, 321–325.

Li, W., Yu, B., Li, M., Sun, D., Hu, Y., Zhao, M., et al. (2010). NEMO-binding domain peptide promotes osteoblast differentiation impaired by tumor necrosis factor α. *Biochemical and Biophysical Research Communications, 391*, 1228–1233.

Lowery, J. W., & de Caestecker, M. P. (2010). BMP signaling in vascular development and disease. *Cytokine & Growth Factor Reviews, 21*, 287–298.

Lui, J. C., Nilsson, O., & Baron, J. (2014). Recent research on the growth plate: Recent insights into the regulation of the growth plate. *Journal of Molecular Endocrinology, 53*, T1–T9.

Machado, R. F., & Farber, H. W. (2013). Pulmonary hypertension associated with chronic hemolytic anemia and other blood disorders. *Clinics in Chest Medicine, 34*, 739–752.

Machado, R. D., Pauciulo, M. W., Thomson, J. R., Lane, K. B., Morgan, N. V., Wheeler, L., et al. (2001). BMPR2 haploinsufficiency as the inherited molecular mechanism for primary pulmonary hypertension. *The American Journal of Human Genetics, 68*, 92–102.

Mariani, E., Pulsatelli, L., & Facchini, A. (2014). Signaling pathways in cartilage repair. *International Journal of Molecular Sciences, 15*, 8667–8698.

Maruyama, T., Fukushima, H., Nakao, K., Shin, M., Yasuda, H., Weih, F., et al. (2010). Processing of the NF-κB2 precursor p100 to p52 is critical for RANKL-induced osteoclast differentiation. *Journal of Bone and Mineral Research, 25*, 1058–1067.

May, M. J., D'Acquisto, F., Madge, L. A., Glöckner, J., Pober, J. S., & Ghosh, S. (2000). Selective inhibition of NF-κB activation by a peptide that blocks the interaction of NEMO with the IκB kinase complex. *Science, 289*, 1550–1554.

Mazzini, M. J., & Schulze, P. C. (2006). Proatherogenic pathways leading to vascular calcification. *European Journal of Radiology, 57*, 384–389.

Mukai, T., Otsuka, F., Otani, H., Yamashita, M., Takasugi, K., Inagaki, K., et al. (2007). TNF-α inhibits BMP-induced osteoblast differentiation through activating SAPK/JNK signaling. *Biochemical and Biophysical Research Communications, 356*, 1004–1010.

Mundlos, S., & Olsen, B. R. (1997). Heritable diseases of the skeleton. Part II: Molecular insights into skeletal development-matrix components and their homeostasis. *FASEB Journal*, *11*, 227–233.
Nakase, T., Takaoka, K., Masuhara, K., Shimizu, K., Yoshikawa, H., & Ochi, T. (1997). Interleukin-1β enhances and tumor necrosis factor-α inhibits bone morphogenetic protein-2-induced alkaline phosphatase activity in MC3T3-E1 osteoblastic cells. *Bone*, *21*, 17–21.
Nishimura, R., Hata, K., Matsubara, T., Wakabayashi, M., & Yoneda, T. (2012). Regulation of bone and cartilage development by network between BMP signalling and transcription factors. *The Journal of Biochemistry*, *151*, 247–254.
Novack, D. V. (2011). Role of NF-κB in the skeleton. *Cell Research*, *21*, 169–182.
Otero, J. E., Chen, T., Zhang, K., & Abu-Amer, Y. (2012). Constitutively active canonical NF-κB pathway induces severe bone loss in mice. *PLoS One*, *7*, e38694.
Pacifici, R., Brown, C., Puscheck, E., Friedrich, E., Slatopolsky, E., Maggio, D., et al. (1991). Effect of surgical menopause and estrogen replacement on cytokine release from human blood mononuclear cells. *Proceedings of the National Academy of Sciences of the United States of America*, *88*, 5134–5138.
Ranganathan, P. (2008). An update on pharmacogenomics in rheumatoid arthritis with a focus on TNF-blocking agents. *Current Opinion in Molecular Therapeutics*, *10*, 562–567.
Retting, K. N., Song, B., Yoon, B. S., & Lyons, K. M. (2009). BMP canonical Smad signaling through Smad1 and Smad5 is required for endochondral bone formation. *Development*, *136*, 1093–1104.
Robey, P. G., Fedarko, N. S., Hefferan, T. E., Bianco, P., Vetter, U. K., Grzesik, W., et al. (1993). Structure and molecular regulation of bone matrix proteins. *Journal of Bone and Mineral Research*, *2*, S483–S487.
Schmidt-Supprian, M., Bloch, W., Courtois, G., Addicks, K., Israël, A., Rajewsky, K., et al. (2000). NEMO/IKKγ-deficient mice model incontinentia pigmenti. *Molecular Cell*, *5*, 981–992.
Sen, R., & Baltimore, D. (1986). Inducibility of κ immunoglobulin enhancer-binding protein NF-κB by a posttranslational mechanism. *Cell*, *47*, 921–928.
Seo, Y., Fukushima, H., Maruyama, T., Kuroishi, K. N., Osawa, K., Nagano, K., et al. (2012). Accumulation of p100, a precursor of NF-κB2, enhances osteoblastic differentiation *in vitro* and bone formation *in vivo* in *aly/aly* mice. *Molecular Endocrinology*, *26*, 414–422.
Shim, J. H., Yoo, H. M., Oh, S. I., Nam, M. J., Jeon, H. M., Park, C. H., et al. (2013). Various types of intracorporeal esophagojejunostomy after laparoscopic total gastrectomy for gastric cancer. *Gastric Cancer*, *16*, 420–427.
Shinkura, R., Kitada, K., Matsuda, F., Tashiro, K., Ikuta, K., Suzuki, M., et al. (1999). Alymphoplasia is caused by a point mutation in the mouse gene encoding Nf-κb-inducing kinase. *Nature Genetics*, *22*, 74–77.
Shintani, M., Yagi, H., Nakayama, T., Saji, T., & Matsuoka, R. (2009). A new nonsense mutation of SMAD8 associated with pulmonary arterial hypertension. *Journal of Medical Genetics*, *46*, 331–337.
Soysa, N. S., Alles, N., Weih, D., Lovas, A., Mian, A. H., Shimokawa, H., et al. (2010). The pivotal role of the alternative NF-κB pathway in maintenance of basal bone homeostasis and osteoclastogenesis. *Journal of Bone and Mineral Research*, *25*, 809–818.
Sterling, J. A., Edwards, J. R., Martin, T. J., & Mundy, G. R. (2011). Advances in the biology of bone metastasis: How the skeleton affects tumor behavior. *Bone*, *48*, 6–15.
Sun, S. C. (2011). Non-canonical NF-κB signaling pathway. *Cell Research*, *21*, 71–85.
Tada, Y., Majka, S., Carr, M., Harral, J., Crona, D., Kuriyama, T., et al. (2007). Molecular effects of loss of BMPR2 signaling in smooth muscle in a transgenic mouse model of

PAH. *The American Journal of Physiology—Lung Cellular and Molecular Physiology, 292*, L1556–L1563.

Talati, M., Mutlak, H., Lane, K. B., Han, W., Hemnes, A., Mutlak, O., et al. (2014). NF-κB activation exacerbates, but is not required for murine Bmpr2-related pulmonary hypertension. *Diseases, 2*, 148–167.

Tanaka, M., Fuentes, M. E., Yamaguchi, K., Durnin, M. H., Dalrymple, S. A., Hardy, K. L., et al. (1999). Embryonic lethality, liver degeneration, and impaired NF-κB activation in IKK-β-deficient mice. *Immunity, 10*, 421–429.

Tang, Y., Xie, H., Chen, J., Geng, L., Chen, H., Li, X., et al. (2013). Activated NF-κB in bone marrow mesenchymal stem cells from systemic lupus erythematosus patients inhibits osteogenic differentiation through downregulating Smad signaling. *Stem Cells and Development, 22*, 668–678.

Ungvari, Z., Csiszar, A., Edwards, J. G., Kaminski, P. M., Wolin, M. S., Kaley, G., et al. (2003). Increased superoxide production in coronary arteries in hyperhomocysteinemia: Role of tumor necrosis factor-α, NAD(P)H oxidase, and inducible nitric oxide synthase. *Arteriosclerosis, Thrombosis, and Vascular Biology, 23*, 418–424.

Urist, M. R. (1965). Bone: Formation by autoinduction. *Science, 150*, 893–899.

Vaidya, R., Sethi, A., Bartol, S., Jacobson, M., Coe, C., & Craig, J. G. (2008). Complications in the use of rhBMP-2 in PEEK cages for interbody spinal fusions. *Journal of Spinal Disorders & Techniques, 21*, 557–562.

Wang, E. A., Israel, D. I., Kelly, S., & Luxenberg, D. P. (1993). Bone morphogenetic protein-2 causes commitment and differentiation in C3H10T1/2 and 3T3 cells. *Growth Factors, 9*, 57–71.

Wozney, J. M., Rosen, V., Celeste, A. J., Mitsock, L. M., Whitters, M. J., Kriz, R. W., et al. (1988). Novel regulators of bone formation: Molecular clones and activities. *Science, 242*, 1528–1534.

Wu, S., Flint, J. K., Rezvani, G., & De Luca, F. (2007). Nuclear factor-κB p65 facilitates longitudinal bone growth by inducing growth plate chondrocyte proliferation and differentiation and by preventing apoptosis. *The Journal of Biological Chemistry, 282*, 33698–33706.

Yamaguchi, A., Katagiri, T., Ikeda, T., Wozney, J. M., Rosen, V., Wang, E. A., et al. (1991). Recombinant human bone morphogenetic protein-2 stimulates osteoblastic maturation and inhibits myogenic differentiation *in vitro*. *The Journal of Cell Biology, 113*, 681–687.

Yamaguchi, M., Moore, T. W., Sun, A., Snyder, J. P., & Shoji, M. (2012). Novel curcumin analogue UBS109 potently stimulates osteoblastogenesis and suppresses osteoclastogenesis: Involvement in Smad activation and NF-κB inhibition. *Integrative Biology: Quantitative Biosciences from Nano to Macro, 4*, 905–913.

Yamaguchi, M., & Weitzmann, M. N. (2011). Zinc stimulates osteoblastogenesis and suppresses osteoclastogenesis by antagonizing NF-κB activation. *Molecular and Cellular Biochemistry, 355*, 179–186.

Yamazaki, M., Fukushima, H., Shin, M., Katagiri, T., Doi, T., Takahashi, T., et al. (2009). Tumor necrosis factor α represses bone morphogenetic protein (BMP) signaling by interfering with the DNA binding of Smads through the activation of NF-κB. *The Journal of Biological Chemistry, 284*, 35987–35995.

Ye, L., Bokobza, S. M., & Jiang, W. G. (2009). Bone morphogenetic proteins in development and progression of breast cancer and therapeutic potential. *International Journal of Molecular Medicine, 24*, 591–597.

Ye, L., Lewis-Russell, J. M., Kyanaston, H. G., & Jiang, W. G. (2007). Bone morphogenetic proteins and their receptor signaling in prostate cancer. *Histology and Histopathology, 22*, 1129–1147.

Ye, L., Mason, M. D., & Jiang, W. G. (2011). Bone morphogenetic protein and bone metastasis, implication and therapeutic potential. *Frontiers in Bioscience (Landmark Edition)*, *16*, 865–897.

Yoon, B. S., Pogue, R., Ovchinnikov, D. A., Yoshii, I., Mishina, Y., Behringer, R. R., et al. (2006). BMPs regulate multiple aspects of growth-plate chondrogenesis through opposing actions on FGF pathways. *Development*, *133*, 4667–4678.

Young, M. F., Kerr, J. M., Ibaraki, K., Heegaard, A. M., & Robey, P. G. (1992). Structure, expression, and regulation of the major noncollagenous matrix proteins of bone. *Clinical Orthopedics Related Research*, *281*, 275–294.

Zhang, M., Wang, J., Liu, J. H., Chen, S. J., Zhen, B., Wang, C. H., et al. (2013). Effects of probucol on angiotensin II-induced BMP-2 expression in human umbilical vein endothelial cells. *Molecular Medicine Reports*, *7*, 177–182.

Zhang, M., Zhou, S. H., Li, X. P., Shen, X. Q., Fang, Z. F., Liu, Q. M., et al. (2008). Atorvastatin downregulates BMP-2 expression induced by oxidized low-density lipoprotein in human umbilical vein endothelial cells. *Circulation Journal*, *72*, 807–812.

CHAPTER FIVE

Bone Morphogenetic Protein Signaling Regulates Development and Activation of $CD4^+$ T Cells

Michal Kuczma*, Piotr Kraj[†,1]
*Cancer Center, Georgia Regents University, Augusta, Georgia, USA
[†]Department of Biological Sciences, Old Dominion University, Norfolk, Virginia, USA
[1]Corresponding author: e-mail address: pkraj@odu.edu

Contents

Abstract

Bone morphogenetic proteins (BMPs) are growth factors belonging to the TGF-β (transforming growth factor β) superfamily. BMPs were found to regulate multiple cell processes such as proliferation, survival, differentiation, and apoptosis. They were originally described to play a pivotal role in inducing bone, cartilage, ligament, and tendon formation at both heterotopic and orthotopic sites but were found to play a significant role in embryogenesis and development of multiple tissues and organs. Activities of BMPs are regulated by a number of secreted proteins, which modulate their availability to bind cellular receptors. The functions of individual BMPs are highly redundant due to binding the same receptors and inducing overlapping signal transduction pathways. Recently, BMPs were found to regulate cells of the innate and adaptive immune system. BMPs are involved in thymic development of T cells at the early, double negative, as well as later, double positive, stages of thymopoesis. They specifically modulate thymic development of regulatory T cells (T_{reg}). In the periphery, BMPs affect T cell activation, promoting generation of T_{reg} cells. We found that mice deficient for one of the receptors activated by BMPs demonstrated slower growth of transplantable melanoma tumors.

Vitamins and Hormones, Volume 99
ISSN 0083-6729
http://dx.doi.org/10.1016/bs.vh.2015.05.001

1. INTRODUCTION

Bone Morphogenetic Proteins (BMPs) are a protein family acting during vertebrate embryogenesis and organogenesis by regulating tissue patterning. They determine cell fate during lineage determination and in adults regulate self-renewal of tissues of multiple organs including hematopoietic system. The term "BMP" was proposed in late 1960s when it was observed that proteins contained in the demineralized bone segments implanted into muscles were able to induce ectopic bone formation (Urist, 1965). It is now apparent that besides inducing differentiation of bone-forming cells—osteoblasts, BMPs regulate stem cell renewal and epithelial–mesenchymal transition. BMPs are involved in differentiation and maintenance of multiple tissues and organs including bones, teeth, cartilage, and nonosseous tissues of lungs, heart, kidneys, and vessels (Ferguson & Anderson, 1992; Zou & Niswander, 1996). BMPs' role in organogenesis is dependent on their potential to control multiple cell functions: proliferation, adhesion, migration, apoptosis, and differentiation.

Despite ubiquitous expression and general significance, it took the next 20 years for BMPs to be cloned and characterized (Wozney et al., 1988). The role of BMPs in individual organs is subject of active research and their involvement in regulating the development, homeostasis, and functions of the immune system started to be appreciated only recently. Since BMPs and their receptors are expressed by cells of innate and adaptive immune system as well as nonimmune cells, they are involved in communication between organ stromal cells and infiltrating hematopoietic cells. Defects in BMP production/signaling were found to lead to autoimmunity, inflammation, and even cancer. Examining the role of BMPs in the immune system may explain the basis of many pathological conditions and lead to the design of novel therapeutic strategies.

2. BONE MORPHOGENETIC PROTEINS AND THEIR RECEPTORS

BMPs represent the largest subgroup of the transforming growth factor β (TGF-β) superfamily which also includes activins, nodal proteins, Mullerian-inhibiting substance, growth and differentiation factors, and TGF-β itself (Carreira, Alves, Zambuzzi, Sogayar, & Granjeiro, 2014). BMPs are further subdivided into four subfamilies based on amino acid sequence homology. Despite structural similarities, individual BMPs may exert opposing influences. For example, BMP2, 4, and 7 are known for their osteogenic function, while BMP3 decreases bone density and BMP13 strongly inhibits bone formation (Mueller & Nickel, 2012). To date, 20 BMPs were discovered and characterized in humans. BMPs are formed as large (400–500 amino acids) precursor proteins (Xiao, Xiang, & Shao, 2007). Their N-terminal fragment is a signal peptide important for the folding of mature protein; this fragment is cleaved during BMP activation (i.e., BMP4 is activated by a protease furin) (Nelsen & Christian, 2009). BMPs are synthesized as large, inactive precursors, which undergo posttranslational processing and act as disulfide-linked homo- and heterodimer glycoproteins. Once secreted, BMPs may associate with molecules present in extracellular space which include soluble proteins, noggin, chordin, gremlin, or follistatin; membrane/matrix-associated proteins like chondroitin sulfate small leucine-rich proteins; coreceptors and pseudoreceptors like BAMBI (BMP and activin membrane-bound inhibitor); and matrix protein fibrin (Umulis, O'Connor, & Blair, 2009). These interactions regulate BMP availability to bind membrane receptors and control timing of BMP signals. Matrix proteins may also stimulate BMP signaling by increasing activation and release of BMP, increasing movement of BMPs, protecting them from endocytosis and degradation, acting as shuttles to redistribute BMP ligands and increase their concentration far from sites where these proteins are secreted. This complex system of BMP-binding proteins regulating their maturation by proteolytic cleavage, degradation, availability, diffusion ensures that appropriate BMP gradient is established in tissues.

BMPs and other members of TGF-β family bind two types of membrane receptors, which have cytoplasmic domains with activity of serine/threonine kinases. Type I receptors are 50 kDa proteins, while type II receptors are larger (70–80 kDa). There are seven type I receptors of TGF-β family: ALK1 (ACVRL1), ALK2 (ACVR1), ALK3 (BMPR1A), ALK4 (ACVR1B), ALK5 (TGFβR1), ALK6 (BMPR1B), and ALK7 (ACVR1C) and four type II receptors: BMPR2, ACTR2A, ACTR2B, and TGFβR2. Receptor signaling is initiated by binding of a ligand with one of type I receptors (except BMP6 which first binds type II receptor). Ligand binding induces association with type II receptor which leads to a conformational change of a heteromeric receptor—ligand complex and transphosphorylation of type I receptor which induces kinase activity of type I receptors. Activated type I receptors in turn phosphorylate receptor-regulated transcription factors, R-Smads. Binding of TGF-β or activins results in phosphorylation of Smad2 and 3, while binding of a BMP ligand leads to phosphorylation of Smad1, 5, and 8. SARA (Smad anchor for receptor activation) binds Smad2/3 and facilitates their interaction with receptors. Phosphorylated Smads associate with a co-Smad, Smad4, and the whole complex translocates to the nucleus where it acts as a transcription factor by further interaction with multiple coactivators or corepressors to regulate gene expression (Akhurst & Hata, 2012).

BMPs and other members of TGF-β family, e.g., activins, were found to associate with multiple receptors, and individual receptors may bind more than one ligand. Three of type I receptors, BMPR1A (ALK3), BMPR1B (ALK6), and ACVR2A (activin receptor, ALK2), bind BMPs. Of four TGF-β type II receptors, three, BMPR2, ACVR2A, and ACVR2B, bind BMP ligands. Type I receptors BMPR1A, BMPR1B and type II receptor BMPR2 are specific for BMPs, and ACVR1A, ACVR2A, and ACVR2B can bind both BMPs and activins (Kawabata, Imamura, & Miyazono, 1998). This promiscuity of ligand–receptor interactions contributes to redundant functions of BMPs but also makes it difficult to define roles of individual ligand–receptor complexes.

3. SIGNALING PATHWAYS

BMP signaling is highly dependent on cellular and environmental context and interactions with multiple proteins modulate availability and functional concentrations of BMPs and ensure that gradients of BMPs that regulate tissue patterning in embryogenesis and organogenesis are established and maintained. Once BMPs bind their receptors, they initiate signaling through canonical pathway, involving phosphorylation of Smads, and through multiple noncanonical pathways. In addition to canonical and noncanonical pathways, BMP signaling is influenced by cross talk with different signaling pathways, including the PI3K–AKT, WNT, Hedgehog, Notch, interferon (IFN), TNF, and RAS pathways. Often signaling modules involving TGFβ/BMPs/activins are studied only in certain cell types or activation states making it difficult to put together a comprehensive picture of signaling circuits and assessing their significance.

Binding of BMPs and activins to BMP receptors induces phosphorylation of Smad1/5/8 or Smad2/3, respectively. Upon phosphorylation, both subsets of R-Smads associate with Smad4 and translocate into nucleus to regulate gene expression. DNA motif GTCT/AGAC bound by Smad2/3 (Smad-binding element) and GGCGCC/GGAGCC motif bound by

Smad1/5/8 (BMP-binding element) are present in promoters or enhancers of many genes, including genes regulating lineage commitment of T cell subsets and T cell functions like Runx, Schnurri, C/EBPa, Foxo, and Foxp3 (Itoh et al., 2004; Korchynskyi & ten, 2002; Kusanagi et al., 2000; Lin, Lerch, Cook, Jardetzky, & Woodruff, 2006; Morikawa et al., 2011). Id gene (inhibitors of DNA binding) family is one of the most studied genes downstream of BMP receptors (Miyazono, Kamiya, & Morikawa, 2010). The Id proteins inhibit lineage commitment by binding and sequestering basic helix–loop–helix transcription factors and have been implicated in regulating invasion, proliferation, and survival of various cell types. Id3-deficient T cells were found to poorly upregulate Foxp3 that led to defective generation of regulatory $CD4^{+}Foxp3^{+}$ cells (T_{reg}) and accumulation of Th17 cells and autoimmune disease (Maruyama et al., 2011). Another target of BMP-induced Smads is Runx family of transcription factors important for regulating T cell functions (Javed et al., 2008). Runx1–3 are Runt-related heteromeric transcription factors consisting of a DNA-binding α-subunits specific for individual proteins and a common non-DNA-binding β subunit. In T cells, Runx1 and 3 were found to attenuate Th2 cell differentiation and mice with $CD4^{+}$ T cells deficient in these proteins are prone to Th2-mediated diseases like asthma (Naoe et al., 2007). Transgenic expression of Runx3 augments Th1 cytokine production (Kohu et al., 2009). In regulatory T cells (T_{reg}) cells, Runx1 associates with Foxp3 controlling their suppressor function (Ono et al., 2007; Rudra et al., 2012). It is not been extensively studied to what extent Smad proteins responding to BMP signaling bind to Runx family proteins in T cells but such interactions were found in other cell types (Javed et al., 2008). BMP-specific R-Smads1/5/8 are involved in feedback regulation of transcription of inhibitory Smad6, Smad7, and a decoy receptor BAMBI. Signal transduction mediated by activated R-Smads may be modulated by interactions with inhibitory Smad6 or Smad7 (I-Smads), which block their association with common Smad4 and nuclear translocation. Expression of Smad7 is increased by IFNγ (acting through Jak1/STAT4) and proinflammatory cytokines TNFα and IL-1β (acting through NFκB/RelA).

Another mechanism which antagonizes TGF-β/BMP signaling involves interactions with other signaling pathways, like phosphorylation of linker region of R-Smads by CDK8/9, activation of mitogen-activated protein kinases ERK1/2 or GSK-3β (Guo et al., 2008; Kretzschmar, Doody, & Massague, 1997). ERK1/2 may act downstream of Ras, which is activated by T-cell antigen receptor (TCR) signaling (Kretzschmar, Doody,

Timokhina, & Massague, 1999). Reduction in the expression of Axin/GSK-3β leads to increased Smad3 stability and transcriptional activity. Wnt signaling, which is known to inactivate GSK-3β, reduces Smad ubiquitination and stabilizes the protein. Linker region phosphorylation (different from phosphorylation by BMP receptors) marks R-Smads for recognition by HECT type E3 ubiquitin ligases SMURF1, SMURF2/NEDD4L, NEDD4-2 which mediate their polyubiquitination and degradation (Sapkota, Alarcon, Spagnoli, Brivanlou, & Massague, 2007). We have found that NEDD4 is expressed in activated Foxp3$^+$ T_{reg} cells. OTUB1 (ubiquitin thioesterase) protects pSmad2/3 from ubiquitination and proteolytic degradation (Herhaus, Al-Salihi, Macartney, Weidlich, & Sapkota, 2013). In summary, phosphorylation linker region is an important determinant of steady-state stability of Smads which determines cellular sensitivity to TGF-β/BMP and integrates multiple signaling pathways. AKT (protein kinase B) may restrict Smad3 activity through a kinase-independent mechanism involving the direct binding of AKT to Smad3, which blocks activation of Smad3 by sequestering Smad3 from TGFβRI (Conery et al., 2004). PI3K (phosphatidylinositide 3-kinase)/AKT may also inhibit TGF-β/BMP signaling through a kinase-dependent mechanism involving mTOR (Song, Wang, Krebs, & Danielpour, 2006; Wahdan-Alaswad et al., 2012). AKT is a strong repressor of induction of the T_{reg} phenotype *in vitro* and *in vivo* so and inhibiting TGF-β/BMP signaling could complement reported mTOR-dependent inhibition of Foxp3 expression (Haxhinasto, Mathis, & Benoist, 2008; Sauer et al., 2008). Recently reported association of Smad and Foxo family transcription factors opens one more avenue how AKT could impact BMP signaling (Neurath et al., 2002; Seoane, Le, Shen, Anderson, & Massague, 2004). Foxo transcription factors control survival and homing of naive T cells by inducing expression of IL-7Rα, adhesion molecule CD62L, and cytokine receptor CCR7 (Kerdiles et al., 2009). They also control differentiation and function of T_{reg} cells (Kerdiles et al., 2010; Ouyang, Beckett, Ma, Paik, et al., 2010; Ouyang et al., 2012). Activity of Foxo proteins is regulated by their subcellular localization, they complex with 14-3-3 proteins, and phosphorylation diminishes their translocation into nucleus and binding to DNA. PI3K through PDK1 (phosphoinositide-dependent kinase-1) activates AKT and this phosphorylates Foxo proteins barring them, and associated Smads, from entering nucleus and regulating target genes.

In addition to Smad-mediated signaling, BMP receptor binding triggers activation of alternative, nonconventional pathways. BMPR signaling was found to activate MAP-kinase-mediated signaling cascades. BMP2/4 binding to receptors led to phosphorylation of MKK4–JNK (and AP-1) and MKK3–p38. TAK1 (TGF-β-activated kinase 1) was identified to directly phosphorylate MAP kinases (Mao et al., 2011; Shim et al., 2009). TAK1 was proposed to act downstream of TRAF6 (TNF receptor-associated factor 6) which moderates signaling of TGF-β/BMP receptors and receptors for TNF, IL-1, and Toll-like receptors (Ohkura et al., 2012). TRAF6 is an E3 ubiquitin ligase, which synthesizes polyubiquitin chains, by self-polyubiquitination of Lys63, which do not target protein for degradation but serve as docking sites to assemble signaling complexes (Yamashita et al., 2008). The significance of p38 signaling in peripheral T cells was demonstrated by showing that it is necessary for conversion to induced T_{reg} (iT_{reg}) cells (Huber et al., 2008). This conversion requires TGF-β signaling but is enhanced by BMPs and activin A which is particularly important in inflammatory responses when TGF-β concentration could be low. Interestingly, activin A has a minimal effect on Foxp3 induction in thymic-derived T_{reg} cells, highlighting a possibility of different regulation of Foxp3 gene expression in the thymus and peripheral T_{reg} cells (Huber et al., 2009). The synergistic effect of TGFβ and BMP/activins was dependent on activation of ERK1/2 and JNK and could be blocked by specific small molecule inhibitors (Lu, Ma, et al., 2010). Analysis of Smad2/3-deficient mice only partially abrogated generation of iT_{reg} and Th17 cells. The importance of noncanonical TGF-β/BMP signaling was further revealed by demonstrating that a similar proportion of Smad2/3-deficient and -sufficient mice succumbed to experimentally induced EAE, suggesting that generation of Th17 cells *in vivo* is TGF-β dependent but Smad independent (Lu, Wang, et al., 2010). While the molecular mechanisms of BMP signaling were not studied extensively in T cells, they are known to utilize signaling pathways acted on by BMPs in other cell types what suggests that they may play more important role than currently appreciated. For example, in colon cancer cells, BMPs were found to suppress PTEN (phosphatase and tensin homolog) expression via RAS/ERK pathway (Beck & Carethers, 2007). In T cells, PTEN is regulated in response to TCR signaling and affects AKT/mTOR axis, which has a major effect on effector and T_{reg} cell differentiation in response to antigen (Powell & Delgoffe, 2010). In another example, BMP-induced Smad1 interacted with β-catenin in bone marrow stromal cells impeding its transfer into nucleus (Liu, Tang, Qiu, Cao, & Clemens, 2006). A cross talk between BMP and Wnt signaling pathways in T cells is known to regulate Th17, T_{reg}, and CD8 cells (Ding, Shen,

Lino, Curotto de Lafaille, & Lafaille, 2008; Muranski et al., 2011; Zhao et al., 2010). Finally, TGFβ/BMP pathways communicate with Notch signaling pathway and synergistically regulate target genes in many cell types (Guo & Wang, 2009). Activation of Notch is critical at many stages of T cell ontogeny, including original T cell lineage specification (Radtke et al., 1999). While molecular mechanisms of downstream mediators of BMP signaling with Notch are not yet known in T cells, interaction of Notch1 ICD and Smad3 was found to regulate Foxp3 transcription factor in T_{reg} cells (Asano, Watanabe, Kitani, Fuss, & Strober, 2008; Samon et al., 2008). In summary, binding of BMPs to their receptors initiates complex signaling events, and biological outcome is highly context dependent and relies on extensive communication with other signaling pathways.

4. PATHOLOGY ASSOCIATED WITH IMPAIRED OR ABNORMAL BMP SIGNALING

BMPs are important in virtually all aspects of the organism's development and function. Therefore, tight regulation of BMP signaling is crucial to maintain integrity and homeostasis of tissues and organs. This is originated in extracellular space by regulation of BMP availability for binding to receptors. Altered concentration of natural BMP antagonists, like noggin, gremlin, or chordin during embryogenesis, was shown to have detrimental developmental outcomes, including lethality (Bachiller et al., 2000). The most apparent pathology affected skeletal and nervous systems. Deletion of noggin gene caused early mortality as a result of multiple defects (failure to close neural tube, loss of caudal vertebrae) attributed to uncontrolled signaling of BMP2 and BMP4 (McMahon et al., 1998). BMPs were found to sustain stem cell renewal and regulate their differentiation, including tissue specific and cancer stem cells (Chen et al., 2008; He et al., 2004; Reya, Morrison, Clarke, & Weissman, 2001; Ying, Nichols, Chambers, & Smith, 2003). Cancer stem cells were defined as self-renewing cells that can "phenocopy" the tumor they originate from and are considered a cellular component essential for tumor persistence. Recent reports showed that tumors, e.g., gliomas, colorectal cancers, developed mechanisms to tune BMP signaling by secreting gremlin. This regulated tumor invasiveness and supported persistence of nondifferentiating and hence self-renewing cells (Karagiannis, Berk, Dimitromanolakis, & Diamandis, 2013).

Attenuated BMP signaling is also associated with severe developmental and functional abnormalities. Deletion of BMP2 or BMP4 genes in experimental mice leads to embryonic lethality because of defects in mesoderm formation and BMP7-knockout mice die shortly after birth (Dudley, Lyons, & Robertson, 1995; Zhang & Evans, 1996). BMP2, 4, and 7 are ligands of BMPR1A and constitutive knockout of this receptor was found lethal due to defect of mesoderm formation in early gestation (Mishina, Suzuki, Ueno, & Behringer, 1995). Using conditional knockout and mutant mice, BMPs and their receptors were found essential in postnatal ontogeny. For example, mutations in Smad4 and BMPR1A have been linked to the development of juvenile polyposis leading to formation of polyps in duodenum and stomach (Takaku et al., 1999). Unlike the BMPR1A-knockout mice, animals deficient in BMPR1B or BMPR2 gene functions survive past neonatal period but are prone to develop pulmonary hypertension (Chida et al., 2012; Soubrier et al., 2013).

5. BMPs REGULATE DEVELOPMENT AND FUNCTION OF IMMUNE SYSTEM CELLS

While BMP signaling was thoroughly investigated in multiple systems, their role in immune system, particularly in regulating T cell differentiation and function of mature T cells, is only emerging. The main features of BMP signaling, acting in concert and in context of local mediators, including other members of TGF-β family, make it difficult to dissect the functions of individual BMPs. Accumulating data (including our own study) suggest that BMPs and their receptors have a significant role in regulating immune responses of $CD4^+$ T cells, particularly in tumor environment. Here, we will summarize the current knowledge on the role of BMPs in T cell ontogeny.

6. BMPs ARE INVOLVED IN THYMIC DEVELOPMENT OF T CELLS

T cells mature and differentiate in the thymus from bone marrow-derived common lymphoid progenitors, a precursor cells capable of differentiating into various subsets of lymphoid cells including T and B lymphocytes and NK cells (Kondo, Scherer, King, Manz, & Weissman, 2001; Kuo & Leiden, 1999). Earliest T cell precursors do not express markers of mature helper and cytotoxic lymphocytes, CD4 and CD8, respectively, and are called double negative (DN) cells (Fig. 1). DN cells are further subdivided into four developmental stages based on the surface expression of CD117 (c-kit), CD44, and CD25 (IL-2 receptor α-chain). Development

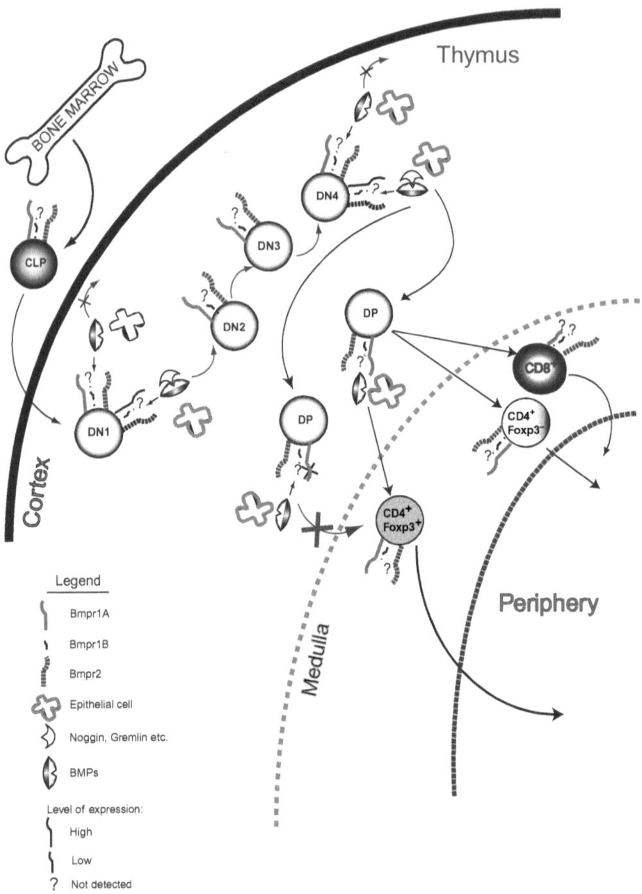

Figure 1 Bone morphogenetic proteins (BMPs) affect thymopoiesis at multiple stages. Bone marrow-derived common lymphoid progenitor (CLP) cell colonizes thymus and undergoes series of changes to become mature T cell. Upon binding of BMPs (mainly BMP2 and 4) produced by multiple cells in the thymus (i.e., epithelial cells) at double negative (DN) stages 1 and 4, thymocyte development in the cortex is arrested. Natural inhibitors of BMP signaling (i.e., Noggin, Gremlin) bind to BMPs sequestering them from their type I (mainly BMPR1A, expression of BMPR1B was not reported in the thymus) and type II (BMPR2) receptors. Regulatory T cells ($CD4^{+}Foxp3^{+}$) develop only upon signal mediated by BMPs in the context of BMPR1A, and their generation is largely affected when DP (double positive) thymocytes lack its expression. $CD8^{+}$ T cells express very low level of BMPR1A, and their development and function is minimally affected by its deletion. (See the color plate.)

within this population progresses from the CD117$^+$CD44$^+$CD25$^-$ cells (DN1) to the CD117$^-$CD44$^+$CD25$^+$ pro-T cells (DN2), and to CD117$^-$CD44$^-$CD25$^+$ pre-T cells (DN3), and finally to the CD117$^-$44$^-$25$^-$ DN4 stage. DN2 pro-T cells initiate rearrangement of the β-chain of TCR, which is completed at the DN3 stage. Productive rearrangement of the β-chain allows for expression of pre-TCR complexes, proliferation and expansion of DN3 thymocytes which progress through the DN4 stage, upregulate expression of CD4 and CD8, and become double positive (DP) thymocytes (Davis & Littman, 1994; von Boehmer et al., 1999). DP thymocytes rearrange their TCRα locus, express a mature TCRα/β complex, and undergo processes of positive and negative selection that generate CD4$^+$ and CD8$^+$ single-positive (SP) thymocytes. A fraction of SP CD4$^+$ thymocytes upregulates Foxp3 transcription factor and generates a population of thymic T_{reg} cells, called natural T_{reg} cells. This cell population is critical for maintenance of immune system homeostasis, immune tolerance and control of immune responses to antigens (Sakaguchi, Yamaguchi, Nomura, & Ono, 2008). However, dysregulated suppressor function of these cells is attributed to incomplete elimination of pathogens in chronic infectious diseases or compromised immune responses in cancer (Josefowicz, Lu, & Rudensky, 2012; Zou, 2006).

Interactions of the developing thymocytes with thymic stromal cells are critical for their differentiation to become DP cells and later, to undergo TCR-mediated selection. Thymic stromal cells were found to produce BMP2/4 and contribute to the specification of T cell lineage. BMPs promote differentiation of early progenitor cells into thymocyte lineage but arrest DN1 cells at the stage where they are still capable of producing T, B, NK, and DC cells (Hager-Theodorides et al., 2002). BMP2/4 signaling is required for T cell progenitor homeostasis and expansion but negatively regulates DN to DP transition. In contrast, BMP4 signaling promotes generation of NK cells by regulating expression of Id3, Nfil3, and CD122 (Nohe et al., 2002). BMP4 added to fetal thymic explants inhibits thymocyte proliferation, enhances early progenitor survival, and arrests thymocyte differentiation at the DN stage. Neutralization of BMP2/4 by treatment with recombinant noggin promotes and accelerates thymocyte differentiation, increasing the expression of CD2 and the proportion of DN4 cells and DP cells. A similar result was obtained by blocking BMP2/4 signaling in epithelium and mesenchymal cells of noggin transgenic mice. Noggin encoding transgene, controlled by FoxN1 promoter, was expressed in epithelial cells of developing thymus and this resulted in marked reduction of thymus volume and its ectopic location in the neck (Bleul & Boehm, 2005). Despite reduced thymus size, newborn thymocytes developed normally. Although splenic T cells were present in these mice, their numbers were over 25-fold reduced compared to wild-type animals. This inhibitory effect of BMPs is most likely mediated by BMPR1A and BMPR2 (BMPR1B was not detected on any thymocyte population). High levels of both receptors were found in DN1 subset, expression decreased in DN2, and again increased from DN3 to DP cells to reach maximum in SP $CD4^+$ cells. However, thymic stroma was also shown to express BMP receptors as well as BMPs and signaling molecules Smad1/5/8, suggesting that it is able to respond to BMPs. *In vitro* analysis of thymic reaggregation cultures also showed that BMP treatment arrests thymocyte precursors at DN1 which is due to signaling both to stromal and hematopoietic cells while the effect of BMP2/4 on transition from DN to DP is mediated by stroma alone (Tsai, Lee, & Wu, 2003). In contrast to earlier results, the most recent analysis of conditional knockout mice, where BMPR1A gene was deleted in hematopoietic cells (by crossing to vav-cre mice) revealed a distinct outcome. Thymic cellularity; proportions of DN, DP, and SP $CD4^+$ and $CD8^+$ T cell subsets; and expression profiles of CD5, CD3, CD24, CD2, and CD69 were normal in adult mice (Hager-Theodorides et al., 2014). Reduced cellularity of DN3 and DN4 populations was observed in day 14.5 of fetal thymi in BMPR1A/vav-cre mice, but cell numbers recovered by day 16.5 when DP cells are first observed. More pronounced effect of BMPR1A deletion in fetal thymocytes could be due to higher receptor levels in fetal thymi. The lack of phenotype in BMPR1A/vav-cre mice shows that earlier observed effects of BMP treatment on DN–DP transition might be indirect and involve thymic stroma. Altogether, despite controversial results, BMP signaling seems to regulate early thymopoiesis, although interpretation of data is confusing since both BMPs and their receptors are expressed by developing thymocytes and stromal elements. Considering that BMPs act locally and their concentration and gradient were found important for the biological outcome in other experimental systems, it might be difficult to dissect the contribution of hematopoietic and stromal elements to T cell generation using *in vitro* systems.

Analyses of fetal thymi or BMPR1A/vav-cre mice provided evidence on the role of BMPs in early stages of thymopoiesis. We have investigated development of T cells in mice, where BMPR1A signaling is abrogated in late stages of thymocyte differentiation ($BMPR1A^{T-}$ mice) (Kuczma, Kurczewska, & Kraj, 2014). Mice where the second exon of the BMPR1A gene, encoding kinase domain of the receptor, is flanked by loxP sites were crossed to CD4cre mice (Lee et al., 2001; Mishina, Hanks, Miura, Tallquist, & Behringer, 2002). BMPR1A function in BMPR1A^{T-} mice is eliminated at the DP stage, in SP CD4$^+$ and CD8$^+$ thymocytes, and in all T cells in the peripheral organs. Analysis of BMPR1A^{T-} mice showed that thymic cellularity and composition of subsets were normal with the exception of a population of Foxp3$^+$ T_{reg} cells which was severely decreased (Kuczma et al., 2014). This suggests that signaling through the BMPR1A has a unique role in the process of generating T_{reg} cells. The mechanistic function of BMPR1A in this process is not currently known. BMPR1A may promote renewal of T_{reg} progenitors and initiate instructive process of precursor commitment to T_{reg} lineage. Alternatively, BMPR1A signaling may synergize with TGF-β to expand and protect T_{reg} precursors from apoptosis (Ouyang, Beckett, Ma, & Li, 2010). Although the Foxp3$^+$ cell numbers were reduced, the BMPR1A^{T-} mice had normal proportion of CD4$^+$CD25$^+$ cells. This suggests that BMPR1A functions in CD25$^+$ thymocytes to upregulate/sustain Foxp3 expression. Analysis of thymic and peripheral CD4$^+$CD25$^+$Foxp3$^-$ cells showed that they do not have suppressor function.

7. BMP SIGNALING MODULATES ACTIVATION OF PERIPHERAL CD4$^+$ T CELLS

Despite well-known role of TGF-β in maintaining immune tolerance but also regulating immune responses, contributions of other members of TGF-β family to control functions of peripheral T cells have been addressed only recently (Zhou et al., 2008). TGF-β may induce upregulation of Foxp3 and skew CD4$^+$ T cell lineage commitment toward iT_{reg} cells when activation occurs in the presence of IL-2 (Davidson, DiPaolo, Andersson, & Shevach, 2007). Alternatively, it may promote Th17 cell generation in the presence of IL-6 (Bettelli et al., 2006). The same combination of TGF-β and IL-6 results in upregulation of ACVR2A specifically in Th17 cells, and not in other CD4 lineages, suggesting the role of activins in Th17 function. The role of BMP signaling in regulating CD4$^+$ T cell activation was analyzed by comparing T cells stimulated *in vitro* in the absence and presence of dorsomorphin (Yoshioka, Ono, Osaki, Konishi, & Sakaguchi, 2012). Dorsomorphin inhibited Smad1/5/8 phosphorylation in Jurkat and mouse T cells, blocked proliferation by suppressing IL-2 production and expression of CD25 and arresting cells in G0/G1 phase.

Dorsomorphin treatment inhibited differentiation of Th1, Th17, and iT_{reg} cells but not Th2 by suppressing expression of lineage-specific transcription factors T-bet, RORγt, and Foxp3 but not GATA3. BMPs control phosphorylation of Runx1 and promote its association with AML, which is known to regulate IL-2 gene expression in concert with Foxp3 (Ono et al., 2007).

Despite similarities in signaling pathways between TGF-β and BMPs, BMP2/4 acting alone were not able to induce Foxp3 expression but increased the ability of TGF-β to promote iT_{reg} development (Lu, Ma, et al., 2010). This synergistic effect was dependent on activation of ERK1/2 and JNK (effect could be blocked by MAP kinase inhibitors PD98059 and SB203580, respectively). Combining BMP treatment with trichostatin A (HDAC inhibitor which promotes expression of Foxp3 and iT_{reg} induction) resulted in increased proportion of iT_{reg} cells in the gut. Another member of TGF-β family, activin A which may bind BMPR1A (ALK3), was found to synergize with TGF-β to promote iT_{reg} generation through Smad3 phosphorylation (Huber et al., 2009). Since activin A production is induced in macrophages stimulated by LPS, different members of the TGF-β family may differentially function in inflammation and steady state. To examine the impact of BMPR1A signaling in peripheral $CD4^+$ T cells, we looked at the expression pattern of this receptor and its ligands. In healthy mice, low levels of BMPR1A are found on activated $CD44^+CD62L^-$ helper T cells and T_{reg} cells expressing high level of Foxp3 (Kuczma et al., 2014). BMPR1A is upregulated upon *in vitro* activation of conventional $CD4^+$ T cells (but not cytotoxic cells), especially in cells converted to become iT_{reg} cells, and in activated T_{reg} cells. This expression pattern suggested that its main function could be to regulate $CD4^+$ T cell lineage commitment. BMPR1A was also expressed in a population of activated human $CD4^+$ T cells, especially in the presence of TGF-β. To further reveal BMPR1A signaling axis, we have examined production of BMPs by T cells. BMPR1A ligands, i.e., BMP2, 4, and 7, were not expressed in naive T cells. BMP2 were produced by Th1 cells, and BMP2 and BMP7 were produced by Th2 cells. We failed to detect BMP4 in activated $CD4^+$ T cells. Expression of BMPR1A ligands in activated T cells (especially Th2 cells) suggests that they may act as autocrine or paracrine factors modulating the outcome of T cell activation.

Analysis of $BMPR1A^{T-}$ mice revealed that they have moderately increased proportion of activated $CD4^+$ cells and decreased proportion of T_{reg} cells in lymph nodes. When stimulated with an antigen, $CD4^+$

T cells from mutant mice were highly activated cells and had proliferative advantage over cells isolated from wild-type animals. Furthermore, we found that signaling through BMPR1A regulates $CD4^+$ cell lineage commitment. Activation of $CD4^+Foxp3^-$ cells in the presence of TGF-β resulted in several fold smaller number of $Foxp3^+$ cells when compared with BMPR1A-sufficient littermates despite the same level of activation measured by CD25, CD44, and CD69 expression. In conclusion, signaling through BMPR1A is essential in peripheral helper cells, stimulated by antigens, to promote development and to sustain population of T_{reg} cells. This finding is consistent with the role of BMPs and other members of TGF-β family revealed in previous *in vitro* studies (Lu, Ma, et al., 2010).

8. BMP SIGNALING IN T CELLS IN CANCER

Besides impaired generation of T_{reg} cells, absence of BMPR1A signaling shifted the phenotype of CD4 cells toward IFN-γ producing Th1 cells. Taking all the information together—Th1 phenotype of activated $CD4^+$ cells, diminished proportion of thymic-derived T_{reg} cells *in situ* and impaired iT_{reg} formation, we decided to analyze tumor growth in $BMPR1A^{T-}$ mice. We used B16-F1 melanoma and showed that the tumor growth and composition of T cells infiltrating the tumor mass (like TIL) were different in $BMPR1A^{T-}$ and wild-type mice. We consistently observed highly impaired kinetics of tumor growth in mice lacking BMPR1A signaling in CD4 cells. Surprisingly, the tumor mass was devoid of T_{reg} cells that were readily found in tumor-draining lymph nodes in the same mice. The lack of T_{reg} cells in tumor infiltrates is not known, but it might be caused by impaired translocation of T_{reg} cells into the tumor mass as a result of lack of BMPR1AA expression, or by a defect in iT_{reg} generation/renewal in tumors.

In vitro T cell activation data suggest that targeting BMPR1A in T cells should result in generation of tumor-specific effector $CD4^+$ cells. Indeed, helper T cells found in the tumors in $BMPR1A^{T-}$ mice expressed higher amounts of IFN-γ. We also observed significantly increased (2–3×) CD8 cell numbers in tumors growing in $BMPR1A^{T-}$ mice. These CD8 cells expressed much higher level of CD44 activation marker than the same population isolated from BMPR1A-sufficient animals. Expanded population of CD8 cells in tumors may be caused by signaling through other BMP receptors, or lack of T_{reg} cells in the tumor and enhanced expression of IFN-γ by

CD4 cells may promote expansion of CD8 cells and increase their antitumor response as it was reported (Song, Song, Tang, & Croft, 2007).

BMPs were recently implicated in the development and function of myeloid-derived suppressor cells (MDSCs). These important cells, together with Foxp3$^+$ T$_{reg}$, constitute the main suppressor populations found in multiple types of cancer, and their presence in the circulating blood was correlated with poor prognosis for cancer patients. MDSC was characterized as CD11b$^+$Gr1$^+$ or CD11b$^+$CD15$^+$CD33$^+$HLA-DR$^-$ in mouse and human, respectively (Gabrilovich & Nagaraj, 2009). MDSC inhibits response of effector helper and cytotoxic T cells using variety of mechanisms including expression of arginase-1, nitric oxide, or indoleamine 2,3-dioxygenase (IDO), production of suppressor cytokines, and upregulation of surface molecules inhibiting immune responses (Serafini, Borrello, & Bronte, 2006; Fig. 2). They were shown to skew immune responses toward generating suppressor T cells. The effect of BMPs on the tumor population of MDSCs is controversial. Some reports suggest that BMPs decreased MDSC population in breast cancer, but others claimed that this treatment increased metastatic potential of human breast cancer lines (Cao et al., 2014; Owens et al., 2012). Interaction between MDSCs and infiltrating T cells is likely shaping adaptive immune responses in tumors. Our preliminary data suggest that a subset of myeloid cells could be major source of BMP2/7 in prostate TRAMP tumors. Thus, BMPs produced by myeloid cells (and prostate tumor cells) could complement known immunosuppressive activity of TGF-β and constitute one of suppressor signaling circuits compromising immune responses in tumors.

9. CONCLUDING REMARKS

For a long time, the importance of BMPs and their receptors in development and differentiation of multiple organs was well known and appreciated. The role of these proteins in controlling immune system functions has started to be uncovered. Reports from several laboratories revealed that BMPs control early stages of thymic development of T cells and regulate functions of mature T cells. The progress to unravel BMP-controlled molecular pathways is hampered by highly redundant functions of individual BMP ligands and their receptors as well as by a context-dependent biological outcome. This has led us to adopt a strategy of eliminating one of the major BMP receptors in T cells in order to reveal its functions. Our data show that lack of BMPR1A signaling led to impaired generation and function of T$_{reg}$.

Figure 2 Effect of BMP signaling in the tumor microenvironment. BMPs act locally to enhance self-renewal/stemness of cancer stem cells (CSCs) by induction of STAT3 expression. This in turn enhances the metastatic potential of tumor cells. BMPs also induce Foxp3 expression in conventional CD4 cells ($CD4^{+}Foxp3^{-}$) to become regulatory T cells (induced T_{reg}, iT_{reg}). T_{reg} (both iT_{reg} and thymic-derived natural T_{reg} (nT_{reg})) and MDSC suppress antitumor response targeting granzyme/perforin producing $CD8^{+}$ T cells, NK cells, and $CD4^{+}$ cells of Th1 lineage by cell-dependent contact and expression of suppressive cytokines or metabolites (TGF-β, IDO, arginase, etc.). MDSC, myeloid-derived suppressor cell; NK, natural killer; B, B cell; MΦ, macrophage; DC, dendritic cell. (See the color plate.)

In addition, conventional T cells deprived of BMPR1A signaling are prone to generate Th1 effector cells when activated. While it is likely to require a significant effort to reveal molecular mechanisms of these processes, the inhibition of tumor growth in BMPR1A^{T-} mice suggests that signaling controlled by this receptor plays an important role in adaptive immune response in cancer.

ACKNOWLEDGMENT

This work was supported by the NIAID R21 AI097600 Grant to P.K.

REFERENCES

Akhurst, R. J., & Hata, A. (2012). Targeting the TGFbeta signalling pathway in disease. *Nature Reviews Drug Discovery*, *11*, 790–811.

Asano, N., Watanabe, T., Kitani, A., Fuss, I. J., & Strober, W. (2008). Notch1 signaling and regulatory T cell function. *Journal of Immunology*, *180*, 2796–2804.

Bachiller, D., Klingensmith, J., Kemp, C., Belo, J. A., Anderson, R. M., May, S. R., et al. (2000). The organizer factors Chordin and Noggin are required for mouse forebrain development. *Nature*, *403*, 658–661.

Beck, S. E., & Carethers, J. M. (2007). BMP suppresses PTEN expression via RAS/ERK signaling. *Cancer Biology and Therapy*, *6*, 1313–1317.

Bettelli, E., Carrier, Y., Gao, W., Korn, T., Strom, T. B., Oukka, M., et al. (2006). Reciprocal developmental pathways for the generation of pathogenic effector TH17 and regulatory T cells. *Nature*, *441*, 235–238.

Bleul, C. C., & Boehm, T. (2005). BMP signaling is required for normal thymus development. *Journal of Immunology*, *175*, 5213–5221.

Cao, Y., Slaney, C. Y., Bidwell, B. N., Parker, B. S., Johnstone, C. N., Rautela, J., et al. (2014). BMP4 inhibits breast cancer metastasis by blocking myeloid-derived suppressor cell activity. *Cancer Research*, *74*, 5091–5102.

Carreira, A. C., Alves, G. G., Zambuzzi, W. F., Sogayar, M. C., & Granjeiro, J. M. (2014). Bone morphogenetic proteins: Structure, biological function and therapeutic applications. *Archives of Biochemistry and Biophysics*, *561*, 64–73.

Chen, X., Xu, H., Yuan, P., Fang, F., Huss, M., Vega, V. B., et al. (2008). Integration of external signaling pathways with the core transcriptional network in embryonic stem cells. *Cell*, *133*, 1106–1117.

Chida, A., Shintani, M., Nakayama, T., Furutani, Y., Hayama, E., Inai, K., et al. (2012). Missense mutations of the BMPR1B (ALK6) gene in childhood idiopathic pulmonary arterial hypertension. *Circulation Journal: Official Journal of the Japanese Circulation Society*, *76*, 1501–1508.

Conery, A. R., Cao, Y., Thompson, E. A., Townsend, C. M., Jr., Ko, T. C., & Luo, K. (2004). Akt interacts directly with Smad3 to regulate the sensitivity to TGF-beta induced apoptosis. *Nature Cell Biology*, *6*, 366–372.

Davidson, T. S., DiPaolo, R. J., Andersson, J., & Shevach, E. M. (2007). Cutting Edge: IL-2 is essential for TGF-beta-mediated induction of Foxp3+ T regulatory cells. *Journal of Immunology*, *178*, 4022–4026.

Davis, C. B., & Littman, D. R. (1994). Thymocyte lineage commitment: Is it instructed or stochastic? *Current Opinion in Immunology*, *6*, 266–272.

Ding, Y., Shen, S., Lino, A. C., Curotto de Lafaille, M. A., & Lafaille, J. J. (2008). Beta-catenin stabilization extends regulatory T cell survival and induces anergy in nonregulatory T cells. *Nature Medicine, 14*, 162–169.

Dudley, A. T., Lyons, K. M., & Robertson, E. J. (1995). A requirement for bone morphogenetic protein-7 during development of the mammalian kidney and eye. *Genes & Development, 9*, 2795–2807.

Ferguson, E. L., & Anderson, K. V. (1992). Localized enhancement and repression of the activity of the TGF-beta family member, decapentaplegic, is necessary for dorsal-ventral pattern formation in the Drosophila embryo. *Development, 114*, 583–597.

Gabrilovich, D. I., & Nagaraj, S. (2009). Myeloid-derived suppressor cells as regulators of the immune system. *Nature Reviews Immunology, 9*, 162–174.

Guo, X., Ramirez, A., Waddell, D. S., Li, Z., Liu, X., & Wang, X. F. (2008). Axin and GSK3- control Smad3 protein stability and modulate TGF- signaling. *Genes and Development, 22*, 106–120.

Guo, X., & Wang, X. F. (2009). Signaling cross-talk between TGF-beta/BMP and other pathways. *Cell Research, 19*, 71–88.

Hager-Theodorides, A. L., Outram, S. V., Shah, D. K., Sacedon, R., Shrimpton, R. E., Vicente, A., et al. (2002). Bone morphogenetic protein 2/4 signaling regulates early thymocyte differentiation. *Journal of Immunology, 169*, 5496–5504.

Hager-Theodorides, A. L., Ross, S. E., Sahni, H., Mishina, Y., Furmanski, A. L., & Crompton, T. (2014). Direct BMP2/4 signaling through BMP receptor IA regulates fetal thymocyte progenitor homeostasis and differentiation to CD4+CD8+ double-positive cell. *Cell Cycle, 13*, 324–333.

Haxhinasto, S., Mathis, D., & Benoist, C. (2008). The AKT-mTOR axis regulates de novo differentiation of CD4+Foxp3+ cells. *Journal of Experimental Medicine, 205*, 565–574.

He, X. C., Zhang, J., Tong, W. G., Tawfik, O., Ross, J., Scoville, D. H., et al. (2004). BMP signaling inhibits intestinal stem cell self-renewal through suppression of Wnt-beta-catenin signaling. *Nature Genetics, 36*, 1117–1121.

Herhaus, L., Al-Salihi, M., Macartney, T., Weidlich, S., & Sapkota, G. P. (2013). OTUB1 enhances TGFbeta signalling by inhibiting the ubiquitylation and degradation of active SMAD2/3. *Nature Communications, 4*, 2519.

Huber, S., Schrader, J., Fritz, G., Presser, K., Schmitt, S., Waisman, A., et al. (2008). P38 MAP kinase signaling is required for the conversion of CD4+CD25- T cells into iTreg. *PLoS One, 3*, e3302.

Huber, S., Stahl, F. R., Schrader, J., Luth, S., Presser, K., Carambia, A., et al. (2009). Activin a promotes the TGF-beta-induced conversion of CD4+CD25- T cells into Foxp3+ induced regulatory T cells. *Journal of Immunology, 182*, 4633–4640.

Itoh, F., Itoh, S., Goumans, M. J., Valdimarsdottir, G., Iso, T., Dotto, G. P., et al. (2004). Synergy and antagonism between Notch and BMP receptor signaling pathways in endothelial cells. *EMBO Journal, 23*, 541–551.

Javed, A., Bae, J. S., Afzal, F., Gutierrez, S., Pratap, J., Zaidi, S. K., et al. (2008). Structural coupling of Smad and Runx2 for execution of the BMP2 osteogenic signal. *The Journal of Biological Chemistry, 283*, 8412–8422.

Josefowicz, S. Z., Lu, L. F., & Rudensky, A. Y. (2012). Regulatory T cells: Mechanisms of differentiation and function. *Annual Review of Immunology, 30*, 531–564.

Karagiannis, G. S., Berk, A., Dimitromanolakis, A., & Diamandis, E. P. (2013). Enrichment map profiling of the cancer invasion front suggests regulation of colorectal cancer progression by the bone morphogenetic protein antagonist, gremlin-1. *Molecular Oncology, 7*, 826–839.

Kawabata, M., Imamura, T., & Miyazono, K. (1998). Signal transduction by bone morphogenetic proteins. *Cytokine & Growth Factor Reviews, 9*, 49–61.

Kerdiles, Y. M., Beisner, D. R., Tinoco, R., Dejean, A. S., Castrillon, D. H., Depinho, R. A., et al. (2009). Foxo1 links homing and survival of naive T cells by regulating L-selectin, CCR7 and interleukin 7 receptor. *Nature Immunology*, *10*, 176–184.

Kerdiles, Y. M., Stone, E. L., Beisner, D. R., McGargill, M. A., Ch'en, I. L., Stockmann, C., et al. (2010). Foxo transcription factors control regulatory T cell development and function. *Immunity*, *33*, 890–904.

Kohu, K., Ohmori, H., Wong, W. F., Onda, D., Wakoh, T., Kon, S., et al. (2009). The Runx3 transcription factor augments Th1 and down-modulates Th2 phenotypes by interacting with and attenuating GATA3. *Journal of Immunology*, *183*, 7817–7824.

Kondo, M., Scherer, D. C., King, A. G., Manz, M. G., & Weissman, I. L. (2001). Lymphocyte development from hematopoietic stem cells. *Current Opinion in Genetics & Development*, *11*, 520–526.

Korchynskyi, O., & ten, D. P. (2002). Identification and functional characterization of distinct critically important bone morphogenetic protein-specific response elements in the Id1 promoter. *The Journal of Biological Chemistry*, *277*, 4883–4891.

Kretzschmar, M., Doody, J., & Massague, J. (1997). Opposing BMP and EGF signalling pathways converge on the TGF-beta family mediator Smad1. *Nature*, *389*, 618–622.

Kretzschmar, M., Doody, J., Timokhina, I., & Massague, J. (1999). A mechanism of repression of TGFbeta/Smad signaling by oncogenic Ras. *Genes and Development*, *13*, 804–816.

Kuczma, M., Kurczewska, A., & Kraj, P. (2014). Modulation of bone morphogenic protein signaling in T-cells for cancer immunotherapy. *Journal of Immunotoxicology*, *11*, 319–327.

Kuo, C. T., & Leiden, J. M. (1999). Transcriptional regulation of T lymphocyte development and function. *Annual Review of Immunology*, *17*, 149–187.

Kusanagi, K., Inoue, H., Ishidou, Y., Mishima, H. K., Kawabata, M., & Miyazono, K. (2000). Characterization of a bone morphogenetic protein-responsive Smad-binding element. *Molecular Biology of the Cell*, *11*, 555–565.

Lee, P. P., Fitzpatrick, D. R., Beard, C., Jessup, H. K., Lehar, S., Makar, K. W., et al. (2001). A critical role for Dnmt1 and DNA methylation in T cell development, function, and survival. *Immunity*, *15*, 763–774.

Lin, S. J., Lerch, T. F., Cook, R. W., Jardetzky, T. S., & Woodruff, T. K. (2006). The structural basis of TGF-beta, bone morphogenetic protein, and activin ligand binding. *Reproduction*, *132*, 179–190.

Liu, Z., Tang, Y., Qiu, T., Cao, X., & Clemens, T. L. (2006). A dishevelled-1/Smad1 interaction couples WNT and bone morphogenetic protein signaling pathways in uncommitted bone marrow stromal cells. *The Journal of Biological Chemistry*, *281*, 17156–17163.

Lu, L., Ma, J., Wang, X., Wang, J., Zhang, F., Yu, J., et al. (2010). Synergistic effect of TGF-beta superfamily members on the induction of Foxp3+ Treg. *European Journal of Immunology*, *40*, 142–152.

Lu, L., Wang, J., Zhang, F., Chai, Y., Brand, D., Wang, X., et al. (2010). Role of SMAD and non-SMAD signals in the development of Th17 and regulatory T cells. *Journal of Immunology*, *184*, 4295–4306.

Mao, R., Fan, Y., Mou, Y., Zhang, H., Fu, S., & Yang, J. (2011). TAK1 lysine 158 is required for TGF-beta-induced TRAF6-mediated Smad-independent IKK/NF-kappaB and JNK/AP-1 activation. *Cellular Signalling*, *23*, 222–227.

Maruyama, T., Li, J., Vaque, J. P., Konkel, J. E., Wang, W., Zhang, B., et al. (2011). Control of the differentiation of regulatory T cells and T(H)17 cells by the DNA-binding inhibitor Id3. *Nature Immunology*, *12*, 86–95.

McMahon, J. A., Takada, S., Zimmerman, L. B., Fan, C. M., Harland, R. M., & McMahon, A. P. (1998). Noggin-mediated antagonism of BMP signaling is required for growth and patterning of the neural tube and somite. *Genes & Development*, *12*, 1438–1452.

Mishina, Y., Hanks, M. C., Miura, S., Tallquist, M. D., & Behringer, R. R. (2002). Generation of Bmpr/Alk3 conditional knockout mice. *Genesis*, *32*, 69–72.

Mishina, Y., Suzuki, A., Ueno, N., & Behringer, R. R. (1995). Bmpr encodes a type I bone morphogenetic protein receptor that is essential for gastrulation during mouse embryogenesis. *Genes & Development*, *9*, 3027–3037.

Miyazono, K., Kamiya, Y., & Morikawa, M. (2010). Bone morphogenetic protein receptors and signal transduction. *Journal of Biochemistry*, *147*, 35–51.

Morikawa, M., Koinuma, D., Tsutsumi, S., Vasilaki, E., Kanki, Y., Heldin, C. H., et al. (2011). ChIP-seq reveals cell type-specific binding patterns of BMP-specific Smads and a novel binding motif. *Nucleic Acids Research*, *39*, 8712–8727.

Mueller, T. D., & Nickel, J. (2012). Promiscuity and specificity in BMP receptor activation. *FEBS Letters*, *586*, 1846–1859.

Muranski, P., Borman, Z. A., Kerkar, S. P., Klebanoff, C. A., Ji, Y., Sanchez-Perez, L., et al. (2011). Th17 cells are long lived and retain a stem cell-like molecular signature. *Immunity*, *35*, 972–985.

Naoe, Y., Setoguchi, R., Akiyama, K., Muroi, S., Kuroda, M., Hatam, F., et al. (2007). Repression of interleukin-4 in T helper type 1 cells by Runx/Cbf beta binding to the Il4 silencer. *Journal of Experimental Medicine*, *204*, 1749–1755.

Nelsen, S. M., & Christian, J. L. (2009). Site-specific cleavage of BMP4 by furin, PC6, and PC7. *The Journal of Biological Chemistry*, *284*, 27157–27166.

Neurath, M. F., Weigmann, B., Finotto, S., Glickman, J., Nieuwenhuis, E., Iijima, H., et al. (2002). The transcription factor T-bet regulates mucosal T cell activation in experimental colitis and Crohn's disease. *Journal of Experimental Medicine*, *195*, 1129–1143.

Nohe, A., Hassel, S., Ehrlich, M., Neubauer, F., Sebald, W., Henis, Y. I., et al. (2002). The mode of bone morphogenetic protein (BMP) receptor oligomerization determines different BMP-2 signaling pathways. *The Journal of Biological Chemistry*, *277*, 5330–5338.

Ohkura, N., Hamaguchi, M., Morikawa, H., Sugimura, K., Tanaka, A., Ito, Y., et al. (2012). T cell receptor stimulation-induced epigenetic changes and Foxp3 expression are independent and complementary events required for Treg cell development. *Immunity*, *37*, 785–799.

Ono, M., Yaguchi, H., Ohkura, N., Kitabayashi, I., Nagamura, Y., Nomura, T., et al. (2007). Foxp3 controls regulatory T-cell function by interacting with AML1/Runx1. *Nature*, *446*, 685–689.

Ouyang, W., Beckett, O., Ma, Q., & Li, M. O. (2010). Transforming growth factor-beta signaling curbs thymic negative selection promoting regulatory T cell development. *Immunity*, *32*, 642–653.

Ouyang, W., Beckett, O., Ma, Q., Paik, J. H., Depinho, R. A., & Li, M. O. (2010). Foxo proteins cooperatively control the differentiation of Foxp3+ regulatory T cells. *Nature Immunology*, *11*, 618–627.

Ouyang, W., Liao, W., Luo, C. T., Yin, N., Huse, M., Kim, M. V., et al. (2012). Novel Foxo1-dependent transcriptional programs control T(reg) cell function. *Nature*, *491*, 554–559.

Owens, P., Pickup, M. W., Novitskiy, S. V., Chytil, A., Gorska, A. E., Aakre, M. E., et al. (2012). Disruption of bone morphogenetic protein receptor 2 (BMPR2) in mammary tumors promotes metastases through cell autonomous and paracrine mediators. *Proceedings of the National Academy of Sciences of the United States of America*, *109*, 2814–2819.

Powell, J. D., & Delgoffe, G. M. (2010). The mammalian target of rapamycin: Linking T cell differentiation, function, and metabolism. *Immunity*, *33*, 301–311.

Radtke, F., Wilson, A., Stark, G., Bauer, M., van Meerwijk, J., MacDonald, H. R., et al. (1999). Deficient T cell fate specification in mice with an induced inactivation of Notch1. *Immunity*, *10*, 547–558.

Reya, T., Morrison, S. J., Clarke, M. F., & Weissman, I. L. (2001). Stem cells, cancer, and cancer stem cells. *Nature*, *414*, 105–111.

Rudra, D., deRoos, P., Chaudhry, A., Niec, R. E., Arvey, A., Samstein, R. M., et al. (2012). Transcription factor Foxp3 and its protein partners form a complex regulatory network. *Nature Immunology*, *13*, 1010–1019.

Sakaguchi, S., Yamaguchi, T., Nomura, T., & Ono, M. (2008). Regulatory T cells and immune tolerance. *Cell*, *133*, 775–787.

Samon, J. B., Champhekar, A., Minter, L. M., Telfer, J. C., Miele, L., Fauq, A., et al. (2008). Notch1 and TGFbeta1 cooperatively regulate Foxp3 expression and the maintenance of peripheral regulatory T cells. *Blood*, *112*, 1813–1821.

Sapkota, G., Alarcon, C., Spagnoli, F. M., Brivanlou, A. H., & Massague, J. (2007). Balancing BMP signaling through integrated inputs into the Smad1 linker. *Molecular Cell*, *25*, 441–454.

Sauer, S., Bruno, L., Hertweck, A., Finlay, D., Leleu, M., Spivakov, M., et al. (2008). T cell receptor signaling controls Foxp3 expression via PI3K, Akt, and mTOR. *Proceedings of the National academy of Sciences of the United States of America*, *105*, 7797–7802.

Seoane, J., Le, H. V., Shen, L., Anderson, S. A., & Massague, J. (2004). Integration of Smad and forkhead pathways in the control of neuroepithelial and glioblastoma cell proliferation. *Cell*, *117*, 211–223.

Serafini, P., Borrello, I., & Bronte, V. (2006). Myeloid suppressor cells in cancer: Recruitment, phenotype, properties, and mechanisms of immune suppression. *Seminars in Cancer Biology*, *16*, 53–65.

Shim, J. H., Greenblatt, M. B., Xie, M., Schneider, M. D., Zou, W., Zhai, B., et al. (2009). TAK1 is an essential regulator of BMP signalling in cartilage. *EMBO Journal*, *28*, 2028–2041.

Song, A., Song, J., Tang, X., & Croft, M. (2007). Cooperation between CD4 and CD8 T cells for anti-tumor activity is enhanced by OX40 signals. *European Journal of Immunology*, *37*, 1224–1232.

Song, K., Wang, H., Krebs, T. L., & Danielpour, D. (2006). Novel roles of Akt and mTOR in suppressing TGF-beta/ALK5-mediated Smad3 activation. *EMBO Journal*, *25*, 58–69.

Soubrier, F., Chung, W. K., Machado, R., Grunig, E., Aldred, M., Geraci, M., et al. (2013). Genetics and genomics of pulmonary arterial hypertension. *Journal of the American College of Cardiology*, *62*, D13–D21.

Takaku, K., Miyoshi, H., Matsunaga, A., Oshima, M., Sasaki, N., & Taketo, M. M. (1999). Gastric and duodenal polyps in Smad4 (Dpc4) knockout mice. *Cancer Research*, *59*, 6113–6117.

Tsai, P. T., Lee, R. A., & Wu, H. (2003). BMP4 acts upstream of FGF in modulating thymic stroma and regulating thymopoiesis. *Blood*, *102*, 3947–3953.

Umulis, D., O'Connor, M. B., & Blair, S. S. (2009). The extracellular regulation of bone morphogenetic protein signaling. *Development*, *136*, 3715–3728.

Urist, M. R. (1965). Bone: Formation by autoinduction. *Science*, *150*, 893–899.

von Boehmer, H., Aifantis, I., Feinberg, J., Lechner, O., Saint-Ruf, C., Walter, U., et al. (1999). Pleiotropic changes controlled by the pre-T-cell receptor. *Current Opinion in Immunology*, *11*, 135–142.

Wahdan-Alaswad, R. S., Bane, K. L., Song, K., Shola, D. T., Garcia, J. A., & Danielpour, D. (2012). Inhibition of mTORC1 kinase activates Smads 1 and 5 but not Smad8 in human prostate cancer cells, mediating cytostatic response to rapamycin. *Molecular Cancer Research*, *10*, 821–833.

Wozney, J. M., Rosen, V., Celeste, A. J., Mitsock, L. M., Whitters, M. J., Kriz, R. W., et al. (1988). Novel regulators of bone formation: Molecular clones and activities. *Science*, *242*, 1528–1534.

Xiao, Y. T., Xiang, L. X., & Shao, J. Z. (2007). Bone morphogenetic protein. *Biochemical and Biophysical Research Communications, 362*, 550–553.

Yamashita, M., Fatyol, K., Jin, C., Wang, X., Liu, Z., & Zhang, Y. E. (2008). TRAF6 mediates Smad-independent activation of JNK and p38 by TGF-beta. *Molecular Cell, 31*, 918–924.

Ying, Q. L., Nichols, J., Chambers, I., & Smith, A. (2003). BMP induction of Id proteins suppresses differentiation and sustains embryonic stem cell self-renewal in collaboration with STAT3. *Cell, 115*, 281–292.

Yoshioka, Y., Ono, M., Osaki, M., Konishi, I., & Sakaguchi, S. (2012). Differential effects of inhibition of bone morphogenic protein (BMP) signalling on T-cell activation and differentiation. *European Journal of Immunology, 42*, 749–759.

Zhang, C., & Evans, T. (1996). BMP-like signals are required after the midblastula transition for blood cell development. *Developmental Genetics, 18*, 267–278.

Zhao, D. M., Yu, S., Zhou, X., Haring, J. S., Held, W., Badovinac, V. P., et al. (2010). Constitutive activation of Wnt signaling favors generation of memory CD8 T cells. *Journal of Immunology, 184*, 1191–1199.

Zhou, L., Lopes, J. E., Chong, M. M., Ivanov, I. I., Min, R., Victora, G. D., et al. (2008). TGF-beta-induced Foxp3 inhibits T(H)17 cell differentiation by antagonizing RORgammat function. *Nature, 453*, 236–240.

Zou, W. (2006). Regulatory T cells, tumour immunity and immunotherapy. *Nature Reviews Immunology, 6*, 295–307.

Zou, H., & Niswander, L. (1996). Requirement for BMP signaling in interdigital apoptosis and scale formation. *Science, 272*, 738–741.

CHAPTER SIX

Bone Morphogenetic Proteins: Inhibitors of Myelination in Development and Disease

Judith B. Grinspan[1]
Children's Hospital of Philadelphia, Philadelphia, Pennsylvania, USA
Perelman School of Medicine, University of Pennsylvania, Philadelphia, Pennsylvania, USA
[1]Corresponding author: e-mail address: grinspan@email.chop.edu

Contents

Abstract

Myelin, the lipid membrane that surrounds axons, is critical for the propagation of nervous impulses and axonal maintenance. The destruction of myelin or lack of myelin formation due to disease or injury causes severe motor and cognitive disability. Regeneration of myelin is theoretically possible but rarely happens. Myelin is synthesized as the plasma membrane of the oligodendrocyte in the central nervous system. During development, myelin and oligodendrocytes are generated from oligodendrocyte progenitors through a process modulated by extrinsic growth factors signaling to cell-intrinsic proteins. Among the key extrinsic factors are the bone morphogenetic proteins (BMPs), potent inhibitors of oligodendrocyte differentiation and myelin protein expression, likely serving to regulate myelination temporally and spatially. BMPs also promote astrocyte generation. Given the inhibitory role of BMP in oligodendrogliogenesis during

Vitamins and Hormones, Volume 99
ISSN 0083-6729
http://dx.doi.org/10.1016/bs.vh.2015.05.005

development, the expression of BMP during demyelinating disease or injury was investigated, as was whether BMP upregulation could serve to prevent regeneration by both direct inhibition of myelination and increases in astrogliosis. BMPs, predominantly BMP4, were increased in animal models of spinal cord injury, stroke, multiple sclerosis, and perinatal white matter injury. A number of studies inhibited BMP signaling by infusing the injury site with the BMP-specific inhibitor noggin or transplanting stem cells engineered to secrete noggin. In most cases, noggin increased the numbers of mature oligodendrocytes and decreased numbers of astrocytes. Some studies also showed functional improvement. BMP is one of several inhibitory growth factors that now appear to inhibit myelin regeneration. Common downstream mechanisms among these factors are likely to be identified.

1. INTRODUCTION

Bone morphogenetic proteins (BMPs), although first identified for their important roles in bone formation, have been recognized as key regulators of development in many organ systems of the body, especially the central nervous system. BMPs modulate the development of neural stem cells capable of becoming both neurons and glia. Oligodendrocytes, glial cells that synthesize the myelin as part of their plasma membrane, are particularly sensitive to BMP regulation during embryonic development. In this chapter, we demonstrate that BMPs are also powerful inhibitors of remyelination following demyelinating disease and injury. However, to appreciate the significance of the role of BMP in myelination and remyelination, one needs to first understand the importance of myelination to the nervous system.

2. MYELIN IS REQUIRED FOR SALUTATORY CONDUCTION OF NERVOUS IMPULSES AND AXONAL MAINTENANCE

Myelin, the lipid-rich membrane that surrounds axons, ensures conduction of nervous impulses in a salutatory manner and also maintains axonal integrity. Lack of myelin results in physical deficits, cognitive and behavioral deficits and can be life-threatening. Spontaneous regeneration of myelin following a pathological event can occur but usually only from a single acute event (Lasiene, Shupe, Perlmutter, & Horner, 2008). Most diseases of myelin are more chronic in nature. The most prevalent demyelinating disease in the adult, multiple sclerosis (MS), features episodic autoimmune damage to myelin. Although remyelination is sometimes achieved early in the disease, as the disease progresses, this occurs less and less frequently (Goldschmidt, Antel, Konig, Bruck, & Kuhlmann, 2009). Thus, axons remain chronically demyelinated and vulnerable to degeneration as myelin provides axonal maintenance and support. In the newborn, perinatal white matter injury also features a lack of myelin (Volpe, 2001). Infants born severely preterm or following intrauterine growth retardation have diffuse or focal lack of myelin, resulting in cognitive and behavioral deficits in as much as 50% of babies below 1500 g birth weight (Back et al., 2002; Haynes et al., 2003; Kinney, 2006). In its most severe form, perinatal white matter injury results in cerebral palsy. Spinal cord injury also features disruption of myelin. The unifying feature of these diseases is the lack of myelin for salutatory conduction and the inability of the CNS to regenerate new myelin. Myelin is synthesized as the plasma membrane of oligodendrocytes, specialized glial cells in the CNS that arise from oligodendrocyte progenitor cells (OPCs)(Raff, Miller, & Noble, 1983). Although myelination is a developmental phenomenon starting around birth, populations of OPCs remain in the CNS throughout life. The challenge of remyelination is how to facilitate the maturation of these precursors and their synthesis of myelin. Accordingly, much research has been directed at how oligodendrocytes mature from precursors during development with the hope that these lessons can be applied to regeneration.

3. OLIGODENDROCYTES DEVELOP FROM PROGENITORS THROUGH AN ORDERLY PROCESS CONTROLLED BY EXTRINSIC AND INTRINSIC SIGNALING FACTORS

OPCs progress to mature myelinating oligodendrocytes through an orderly process involving specification, proliferation, migration, and differentiation. Several stages of the oligodendrocyte lineage have been well described and characterized. OPCs, expressing the antigen identified by the A2B5 antibody as well as PDGF receptor alpha and the proteoglycan NG2 (Nishiyama, Lin, Giese, Heldin, & Stallcup, 1996; Ranscht, Clapschaw, Price, Noble, & Seifert, 1982), give rise to pro-oligodendrocytes that express the POA antigen identified by the O4 antibody, marking the end stages of the precursor part of the lineage (Bansal, Stefansson, & Pfeiffer, 1992). The first two proteins expressed that are involved in myelin synthesis are 3', 5' cyclic nucelotide phosphodiesterase (CNP) and galactocerebroside (GalC) and these mark the beginning of differentiation (Raff et al., 1978; Scherer et al., 1994). Several days after GalC expression, oligodendrocytes express the myelin proteins proteolipid protein (PLP), which forms 50% of CNS myelin, myelin basic protein (MBP), myelin-associated glycoprotein (MAG, for review see Nave, 2010). Myelin oligodendrocyte glycoprotein (MOG) is generally thought to be the myelin protein expressed last and often used to mark very mature cells (Piddlesden, Lassman, Laffafian, Morgan, & Linington, 1991).

This progression is promoted by several external signaling factors such as sonic hedgehog for specification (Alberta et al., 2001; Orentas, Hayes, Dyer, & Miller, 1999; Pringle et al., 1996), fibroblast growth factor for proliferation, platelet-derived growth factor for migration and survival (McKinnon, Smith, Behar, Smith, & Dubois-Dalcq, 1993; Milner et al., 1997), thyroid hormone for differentiation (Barres, Lazar, & Raff, 1994). These factors transmit or activate internal transcription factors that include Olig1, Olig2, Nkx2.2, Sox 10, and Sox 17, which then promote transcription of genes necessary for myelin formation (Lu, Cai, Rowitch, Cepko, & Stiles, 2001; Lu et al., 2002; Nicolay, Doucette, & Nazarali, 2007; Stolt et al., 2002; Zhou, Choi, & Anderson, 2001). But external signaling factors may also serve to inhibit these processes through interaction with these and other transcription factors and elements such as histones (Marin-Husstege, Muggironi, Liu, & Casaccia-Bonnefil, 2002), thus regulating the extent, location, and timing of oligodendrogliogenesis and myelination. BMP is one of these inhibitory signaling factors.

4. BMPs IN NERVOUS SYSTEM DEVELOPMENT

BMPs are a family of secreted signaling factors in the transforming growth factor beta family that were identified first for their role in bone formation. They direct functions as diverse as proliferation, apoptosis, maturation, and migration. There are 20 structurally distinct forms of BMP but the number that appears to be involved in oligodendrocyte function and myelination is small. Although BMP2, 4, and 7 have been noted to have effects on glia, most studies in the glial field confine themselves to BMP4.

During nervous system development, BMPs are highly expressed dorsally in the roof plate although they are also found in some ventral areas (Liem, Jessell, & Briscoe, 2000; Liem, Tremml, Roelink, & Jessell, 1995; Miller et al., 2004). In these ventral areas, the endogenous BMP-specific inhibitor noggin acts to stop the influence of BMPs. Dorsally, BMPs exhibit a dorsalizing effect on developing neurons such that inhibition of BMP signaling expands ventral domains of the neural tube (Wine-Lee et al., 2004).

BMP is well known to oppose the actions of Shh. Mostly the actions of BMPs are known to be local and BMPs have been found to be bound by heparin sulfate proteoglycan molecules in the extracellular matrix and their diffusion is limited (Ohkawara, Iemura, ten Dijke, & Ueno, 2002). However, the influence of BMPs on patterning can occur at a distance through partnering with other dorsal signaling families such as the Wnts (Feigenson, Reid, See, Crenshaw, & Grinspan, 2011; Wine-Lee et al., 2004).

5. BMPs INHIBIT OLIGODENDROGLIOGENESIS DURING DEVELOPMENT: EVIDENCE *IN VITRO* AND *IN VIVO*

The potential of BMPs to affect the development of oligodendrocytes and myelin was identified by two lines of inquiry. In the first, addition of soluble BMP to cultures of rodent neurospheres directed the development of neural stem cells to astrocytes over oligodendrocytes or neurons (Gross et al., 1996). Later experiments using cultures of rodent OPCs or pre-progenitors, an even earlier stage of the lineage of the oligodendrocyte lineage, showed that treatment with BMP2 or 4 inhibited the formation of mature oligodendrocytes and instead generated astrocyte-like cells in a dose-dependent manner (Fig. 1) (Grinspan et al., 2000; Mabie et al., 1997). At the highest concentrations of BMP4 (50–100 ng/ml), differentiation was completely inhibited such that OPCs extended no processes and had minimal labeling with early differentiation markers such as Gal C (Grinspan et al., 2000). The astrocytes generated from OPCs expressed the astrocyte filament GFAP in addition to markers of OPCs such as the surface antigen identified by the A2B5 antibody and have been historically

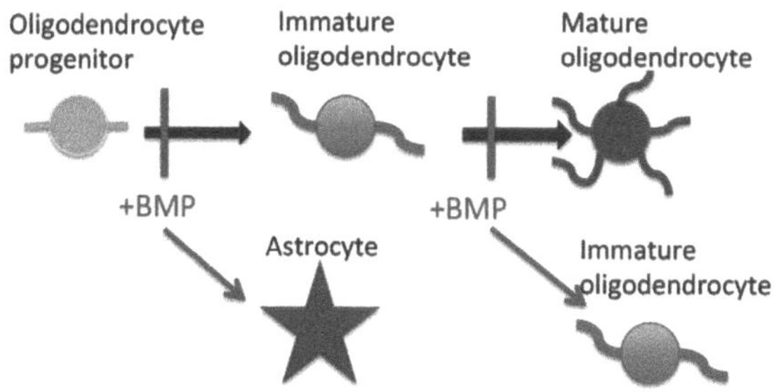

Figure 1 BMP inhibits oligodendrocyte maturation during development. BMPs, predominately BMP4, inhibit the differentiation of OPCs to immature oligodendrocytes and generate astrocyte-like cells *in vitro* (A2B5+/GGAP+). BMP4 treatment of immature oligodendrocytes inhibits myelin protein expression.

called type 2 astrocytes for their hybrid nature (Raff et al., 1983). It is unclear whether they exist *in vivo*; however, the propensity for BMP to favor astrocyte formation over oligodendrocyte formation is clear from other studies (Gross et al., 1996).

The second line of inquiry that led to the identification of the inhibitory properties of BMP came from studies to determine the location of the origin of OPCs. Early in development, most OPCs in the spinal cord are specified from Olig2-expressing cells in the ventral ventricular zone in a domain under the control of sonic hedgehog secreted by the notocord and floor plate (Lu et al., 2000; Zhou, Wang, & Anderson, 2000). Removal of the source of sonic hedgehog in the early development results in almost no OPCs. A more dorsally derived population of OPCs eventually arises but not till later in development and from an area close to the midline (Chandran et al., 2003; Kessaris et al., 2006). The early lack of OPCs in dorsal areas suggested an impediment to oligodendrogliogenesis in the dorsal part of the neural tube. Wada et al. tested this by using spinal cord explant culture and placing pieces of dorsal cord over the ventral sections and inhibiting the generation of oligodendrocytes ventrally (Wada et al., 2000). Removing dorsal areas, conversely, increased the generation of ectopic oligodendrocytes indicating the presence of an inhibitor located dorsally. In this model, flooding the explant cultures with BMP did not mimic the inhibition and the investigators concluded that the inhibitory factor was not BMP. However, later studies showed that beads coated with BMP and placed in the ventral midline could mimic the inhibition locally (Ohkawara et al., 2002; See et al., 2004). One explanation for the effect of BMP when encapsulated in porous beads as opposed to flooding culture medium of an explant was that BMPs are known to bind to heparin sulfate proteoglycans of the extracellular matrix and thus have a potent local effect but not necessarily a global effect (Ohkawara et al., 2002).

The inhibitory effect of BMP in culture is not only limited to lack of initiation of differentiation. BMPs have a role in the regulation of myelin protein expression that is complementary to and independent of their role in morphological differentiation. Following the activation of differentiation machinery in newly differentiating oligodendrocytes, the cells begin to express proteins involved in the formation of the myelin sheath in an orderly manner as previously described (Grinspan, Wrabetz, & Kamholz, 1993). BMP treatment of oligodendrocyte cultures 2 days after differentiation has begun but before expression of most myelin proteins results in a dose-dependent lack of myelin proteins, especially the proteins synthesized later in development like PLP, MBP, and MOG. (See et al., 2004). This late treatment with BMP did not increase the percentage of GFAP + cells in the cultures suggesting that at the immature oligodendrocyte stage, cells were committed to the oligodendrocyte lineage and fate switching to the astrocyte lineage was no longer possible.

The *in vitro* developmental effects of BMP were validated using animal models of development in which either BMP was overexpressed or the BMP inhibitor noggin, which binds BMP ligand before it contacts the BMP receptors, was used to decrease BMP expression. The addition of noggin-soaked beads to chick neural tube resulted in ectopic dorsal generation of oligodendrocytes (Mekki-Dauriac, Agius, Kan, & Cochard, 2002). Using an opposite approach, depletion of noggin from rat optic nerve resulted in decreased numbers of oligodendrocytes and increased numbers of astrocytes (Kondo & Raff, 2004). In *Xenopus*, implantation of BMP-coated beads inhibited oligodendrogliogenesis, whereas anti-BMP-coated beads induced ectopic OPCs to appear in the area of the beads (Miller et al., 2004). Gomes et al. generated a mouse in which the BMP4 was overexpressed under the control of the neuron-specific enolase promoter (Gomes, Mehler, & Kessler, 2003). This resulted in a mild decrease in oligodendrocytes in some areas and a more robust increase in astrocytes. Thus, an inhibitory effect of excess BMP signaling can be demonstrated. This does not inform us about the endogenous role of BMPs but will be important later for the role of BMPs in injury.

6. ENDOGENOUS ROLE OF BMPs IN DEVELOPMENT

Studies of the endogenous role of BMPs in the generation of oligodendrocytes and myelin have produced conflicting results. These studies are based on conditional genetic deletions of BMP receptors. Because BMPs are involved in so many fundamental aspects of development, global genetic deletions of specific BMP ligands or their receptors result in embryonic lethality. Additionally, conditional deletions of BMP ligands would need to target multiple BMPs to avoid compensation. BMP signals through serine–threonine receptor dimer consisting of BMP receptors, BMPR1 and R2, which themselves have multiple subtypes. The BMP dimer binds to the high-affinity type 2 receptors which then binds to and phosphorylates the type 1 receptors (Wrana, Attisano, Wieser, Ventura, & Massague, 1994). These then phosphorylate receptor Smads 1, 5, or 8, which complexes with Smad 4 and enters the nucleus to initiate transcription

(Kretzschmar, Doody, & Massague, 1997; Zhang & Miller, 1996). A conditional knockout of the Bmpr1a receptor, *Bmpr1a,* was created using Cre/LoxP technology based on the proximal *cis*-active transcriptional regulatory elements of the POU-domain gene (Brn4/Pou3f4), which is expressed in the neural tube at approximately E9 (Ahn, Mishina, Hanks, Behringer, & Crenshaw, 2001). This conditional knockout was crossed with traditional *Bmpr1b* knockouts to generate double knockouts (Wine-Lee et al., 2004). The resulting mice have dorsal/ventral neural tube patterning defects and limb deformities and neonatal lethality (Wine-Lee et al., 2004). When analyzed at P0 for the oligodendrocyte lineage effects, the number of OPCs was the same as the controls but the number of PLP+ or MBP + oligodendrocytes was decreased by more than 50%, as was the number of GFAP+ astrocytes (See et al., 2007). When the OPCs from the double knockouts were grown in culture, they were able to differentiate and clearly lacked BMP receptors since they did not respond to exogenous BMP and had no staining for nuclear phospho-Smad. These results suggested that some amount of BMP was necessary for timely myelination, either directly to the OPCs or indirectly through the astrocytes. Given the ability of the OPCs from these mice to differentiate in culture, the latter may be more likely. This study found that both BMP type 1 receptors needed to be deleted to see any effects on oligodendrogliogenesis (See et al., 2007).

A second model disrupted BMPR1a only using an Olig1-Cre, which deletes expression of Olig 1 from the neural tube by E13.5. While this mutant had no changes in OPCs or astrocytes, the numbers of oligodendrocytes and calbindin-positive interneurons increased at P21 (Samanta et al., 2007). Although these data are at odds with the first study, the role of astrocytes in the generation of oligodendrocytes should be underestimated. One common factor in both of these studies is the lack of effect of deleting BMP receptors on the generation of OPCs. Both Cres used targeted cells before the majority of the OPCs were specified. In general, most *in vitro* or *in vivo* studies only assessed oligodendrogliogeneisis using markers that identified the late progenitor stage (O4) through maturity and so did not examine OPC generation. The exception was Miller et al. who found a milder effect on OPCs using the A2B5 antibody than using more mature markers (Miller et al., 2004). We hypothesize from these data that BMPs do not play a major role in OPC specification but do play a major role in differentiation. This may also be true in human stem cell cultures where noggin added late in the culture process was the key to obtaining differentiated oligodendrocytes (Izrael et al., 2007).

7. DOWNSTREAM OF BMPs: PUTATIVE MECHANISM OF BMP ACTION

The mechanism by which OPCs mature and the controls on this mechanism are complex and still incompletely understood. As OPCs begin to mature, proliferation is halted through downregulation of factors such as P27kip1, CDK2, and p53. Inhibition of proliferation is necessary but not sufficient to induce differentiation (Casaccia-Bonnefil et al., 1999; Tang et al., 1999). A number of transcription factors are required for differentiation, including Olig 1 and 2, Sox 10, Nkx2.2, Sox 17 (for review, see Emery, 2010), but a unifying mechanism for external control of these intrinsic factors has not emerged. Insulin-like growth factor 1, thyroid hormone, and EGFR potentiate differentiation (Aguirre, Dupree, Mangin, & Gallo, 2007; Barres et al., 1994; McMorris & Dubois-Dalcq, 1988) and at least four families of growth factors are known to oppose differentiation. These include: BMPs, Wnts, notch, and FGF2 (Bansal, 2002; Feigenson, Reid, See, Crenshaw, & Grinspan, 2009; Grinspan et al., 2000 John et al., 2002). Most likely, BMP signaling is mediated through a family of four proteins known as inhibitors of DNA binding or inhibitors of differentiation (Id). These bind to bHLH transcription factors and inhibit DNA binding and are upregulated in response to BMP signaling. (Wang, Sdrulla, Johnson, Yokota, & Barres, 2001; Wine-Lee et al., 2004). Overexpression of Ids2 and 4 in cultured oligodendrocytes promotes OPCs to adopt an astrocyte-like phenotype, mimicking the effects of BMP4 (Kondo & Raff, 2000; Wang et al., 2001). Samanta et al. found that Id proteins bound to the critical oligodendrocyte transcription factors Olig1 and Olig2 in the cytoplasm of neural precursor cells and prevented them from entering the nucleus and initiating transcription (Samanta & Kessler, 2004). Conversely, overexpression of Olig1 and Olig2 in adult OPCs rescued oligodendrocyte differentiation and blocked the generation of astrocytes during treatment with BMP2 or 4(Cheng et al., 2007). Olig2 may also interact with the Smad protein downstream of BMP (Bilican, Fiore-Heriche, Compston, Allen, & Chandran, 2008). In transgenic mice, overexpression of noggin increases Olig2 progenitors, as does a genetic deletion of Smad 4, thus providing evidence for an *in vivo* relationship between Olig 2 and BMPs (Colak et al., 2008). An additional layer of regulation of OPC differentiation comes from studies of the relationship between BMPs and histone acetylation necessary for chromosomal relaxation and gene transcription (Marin-Husstege et al., 2002).

BMP treatment decreases the activity of Hdacs that then permit transcription of genes that promote astrocyte generation and inhibition of oligodendrocyte differentiation (Wu et al., 2012).

Given the role of BMPs in the inhibition of oligodendrocyte differentiation during development, how do OPCs then manage to mature and wrap axons? OPCs are specified in humans, rats, and mice in the later third of embryonic development, and mature oligodendrocytes do not appear to any extent until after birth. In rats and mice, actual myelination starts soon after birth in the cervical spinal cord and proceeds rostrally and caudally simultaneously. Myelination in these animals is not complete until postnatal day 21. Coincidently, studies have showed that BMP levels decrease significantly at birth (Miller et al., 2004). One can then speculate that the decrease in BMP below a critical threshold permits maturation of OPCs and myelination to begin.

8. EXPRESSION OF BMPs IS INCREASED IN DEMYELINATION PATHOLOGIES

The identification of BMPs as inhibitors of oligodendrogliogenesis during development suggested potential roles in demyelination/remyelination. Initially, a variety of models of neuropathologies were examined to determine the levels of BMP signaling, which BMPs and which cells are making the BMP on both the message and protein level. These pathologies fall into three categories: traumatic spinal cord injury, hypoxia/ischemia in adult or newborn, and demyelinating injury caused by chemical exposure or immune-mediated, the latter two categories modeling stroke, perinatal white matter injury, and multiple sclerosis (Table 1).

An increase in BMP4 and BMP7 was seen in compression injury of the spinal cord. BMP2 was not elevated (Chen et al., 2005; Setoguchi et al., 2001). The BMPs in these studies double labeled with astrocytes and neurons and the authors speculated that the BMP increase would be beneficial by promoting neuronal survival and general gliogenesis. Carotid artery occlusion causing ischemic damage to the brain upregulated BMP6 (Martinez et al., 2001). The BMP6 was thought to be released from neurons (Martinez et al., 2001).

Several models of multiple sclerosis were shown to have elevated BMPs. In an experimental autoimmune encephalomyelitis model in which the animal is immunized with the myelin protein MOG and mounts an immune response characterized by an inflammatory infiltrate, focal demyelination,

Table 1 Summary of Studies on Expression and Role of BMPs in Demyelinating Diseases and Injury in the CNS

Injury or Disease Model	BMP Family Member Elevated	Cell Type Association	Evidence of Role in Inhibition of Remyelination	References
Studies examining BMP expression in demyelinating disease or injury				
Spinal cord compression injury	BMP4, BMP7	Astrocytes and neurons		Chen, Leong, and Schachner (2005)
Spinal cord	BMP4, BMP7	BMP7 in oligodendrocytes		Setoguchi et al. (2001)
Ischemia after carotid artery occlusion	BMP6	Neurons		Martinez, Carnazza, Dii Giacomo, Sorrenti, and Vanella (2001)
Experimental autoimmune encephalomyelitis (EAE)	BMP4, 6, and 7	Microglia, some oligodendrocytes, and astrocytes		Ara et al. (2008)
Lysolecithin demyelination in spinal cord	BMP4, 7	Astrocytes		Fuller et al. (2007)
Ethidium bromide induced myelination in brain	BMP4	OPCs		Zhao, Fancy, Magy, Urwin, and Franklin (2005)

Continued

Table 1 Summary of Studies on Expression and Role of BMPs in Demyelinating Diseases and Injury in the CNS—cont'd

Injury or Disease Model	BMP Family Member Elevated	Cell Type Association	Evidence of Role in Inhibition of Remyelination	References
Studies examining inhibition of BMP signaling in demyelinating conditions				
Cuprizone model of demyelination	BMP4, PhoshoSmad, 1, 5, 8	PhosphoSmad in oligodendrocytes and astrocytes	Noggin infusion decreased astrocytes and increased mature oligodendrocytes	Cate et al. (2010)
Cuprizone model of demyelination	BMP4, PhosphoSmad increased	PhosphoSmad in oligodendrocytes, some astrocytes	BMP infusion increased OPCs, noggin increased oligodendrocytes and myelin	Sabo, Aumann, Merlo, Kilpatrick, and Cate (2011)
Lysolecithin induced myelination	Increase in BMP inhibitor chordin		Chordin infusion increased numbers of oligodendrocytes	Jablonska et al. (2010)
Hemorrhage of the newborn, hypoxia/ischemia	BMP4 markedly increased	Neurons, OPCs, and mature OIs	Noggin infusion improved neurobehavior, increased myelin proteins, reduced gliosis	Dummula et al. (2011)
Intraventricular hemorrhage of newborn—human	BMP4 elevated			Dummula et al. (2011)
Stroke model of neonatal hypoxia	BMP4 elevated		Smaller infarct sizes, mice more likely to survive, more MBP+ cells	Dizon, Maa, and Kessler. (2011)
Adult stroke model			Reduced infarct model, motor function protected	Samanta, Alden, Gobeske, Kan, and Kessler (2010)

Intrauterine growth retardation	BMP4		OPCs from model had restricted differentiation, rescued with noggin *in vitro*	Reid et al. (2012)
Lesioned spinal cord	BMP2		Neural precursor cells expressing noggin increased functional recovery	Setoguchi et al. (2004)
Spinal cord compression	BMP4	Astrocytes and oligodendrocytes	Agmatine decreased BMP4, promoted remyelination	Park et al. (2013)
Spinal cord contusion	BMP2, 4, & 7, PhosphoSmad increased	Neurons, oligodendrocytes, and microglia	Noggin infusion decreased phosphoSmad signaling	Xiao et al. (2010)
Spinal cord transections	BMP2, 4		Stromal cells engineered to produce noggin increased MBP expressing cells	Izrael et al. (2007)
Contused spinal cord			Neural stem cell overexpressing noggin exacerbated lesion	Enzmann et al. (2005)

The first section summarizes studies in which only expression of BMPs and the neural cell types associated were investigated. Upregulation of BMPs 4, 6, and 7 was noted although increases in BMP4 were the most common and the largest. BMP expression was associated with neurons, astrocytes, microglia and oligodendrocytes. The second section additionally summarizes studies in which BMP signaling was inhibited. In most cases, decreases in BMP signaling by the BMP inhibitors noggin, chordin or agmatine resulted in increases in oligodendrocytes, decreases in astrocytes, and functional recovery depending on the model.

and paralysis, BMPs 4, 6, and 7 were significantly elevated by 14 days after injection (Ara et al., 2008). BMP2 was not elevated. BMP4mRNA was increased at a 50– to 100-fold higher level than BMPs 6, and 7. Immunostaining showed BMP4 labeling in macrophages and in some astrocytes and oligodendrocytes. A model in which lysolecithin was injected into the spinal cord to cause a demyelinated lesion showed elevations in BMP 4 and 7 (Fuller et al., 2007). Astrocytes showed expression of phospho-Smad and released chondroitin sulfate proteoglycan that is an essential component of the glial scar that can inhibit remyelination.

A model in which ethidium bromide was injected into the caudal cerebral peduncle of a rat causing demyelination was examined by *in situ* hybridization for expression of BMP2, noggin, and BMP7, which were either undetectable or weak; however, BMP4 expression was easily detectable (Zhao et al., 2005). Double labeling confirmed the BMP4 to be in OPCs that expressed BMP4 only during remyelination. This is surprising result since it suggests that BMP4 in this situation favors remyelination and does not impede it. The authors speculate that perhaps antagonists of BMPs such as chordin or follistatin were in abundance to counter the effect.

Increases in BMP family members have also been demonstrated in multiple sclerosis patients (Deininger, Meyermann, & Schluesener, 1995).

The common factor in all of these studies is the significant increase in BMPs, predominantly BMP4, in demyelinating situations. All four major types of cells in the brain, neurons, astrocytes, oligodendrocytes and microglia, have been identified as associated with BMP expression. This has been performed by immunohistochemistry, except in the last study which employed *in situ* hybridization (Zhao et al., 2005). Given that BMPs are diffusible growth factors but has been known to adhere to the extracellular membrane of cells, BMP immunoreactivity could reflect the cells that synthesized the BMP or the cells that it adhered to. BMPs can also work in an autocrine manner. Neurons and astrocytes are known to make BMP as well as OPCs and oligodendrocytes (Kondo & Raff, 2004; See et al., 2004). Thus who is making the BMP in these situations is still an open question.

9. EVIDENCE THAT BMPs RESTRICT REMYELINATION

The identification of a marked increase in BMP signaling in demyelinating disease and injury plus the function of BMP to block oligodendrocyte differentiation during development suggests that it may serve to inhibit remyelination in these varied pathologies but does not prove it. For this, a number of studies have attempted to block BMP signaling to improve remyelination (Table 1). One of the challenges of this area of study is dissecting whether BMPs are inhibiting remyelination directly or by increasing astrogliosis or some of both. Astrocytes secrete damaging cytokines like TNF-alpha (Back et al., 2005; Bannerman, Hahn, Soulika, Gallo, & Pleasure, 2007; Raine, Bonetti, & Cannella, 1998; Selmaj, Raine, Cannella, & Brosnan, 1991) and can create a physical barrier to recruitment of OPCs to demyelinated lesions by the secretion of chondroitin sulfate proteoglycan and glycosaminoglycan hyaluronan (Back et al., 2005; Bannerman et al., 2007) although reactive astrocytes are also known to be beneficial to remyelination, producing chemoattractants for OPCs, and maturation factors such as leukemia in Inhibitory factor (LIF) (Ishibashi et al., 2006; Tsai & Miller, 2002).

Two studies using MS models of demyelination have shown that BMP inhibition increases oligodendrogliogenesis. In the cuprizone model of multiple sclerosis, animals are fed the copper chelator cuprizone for 5–6 weeks and develop a very specific demyelinating lesion in the posterior corpus callosum accompanied by astrogliosis and inflammation but recover over a 3-week period following the removal of the drug (Matsushima & Morell, 2001). Using this model, Cate et al. first demonstrated increases in BMP4, its receptors and phosphor-SMAD 1, 5, 8 in the subventricular zone (SVZ) of cuprizone-treated mice (Cate et al., 2010). Astrocytes were increased in the SVZ. Infusion of the specific BMP-inhibitor noggin decreased astrocytes and increased numbers of mature oligodendrocytes in the SVZ suggesting that BMP is affecting the lineage commitment of neural precursors that could modulate recovery. This group then went on to show active BMP signaling by expression of phosphor-SMAD 1, 5, 8 in the *corpus callosum* of cuprizone-treated mice. Infusion of additional BMP4 during demyelination increased numbers of OPCs (Sabo et al., 2011). These OPCs did not appear to go on to differentiate to mature oligodendrocytes but rather demonstrated increased caspase 3 staining, suggesting that these new OPCs were unable to survive to maturation and myelination. Numerous studies have shown increases in OPCs upon demyelinating disease but these cells are unable to differentiate (Nait-Oumesmar et al., 2007). Infusion of noggin for the final 7 days of a 6-week course of cuprizone increased the numbers of mature oligodendrocytes as assessed by counting cells stained with several different myelin proteins. Electron microscopy on noggin-infused brains showed increased myelin G ratios, indicating the thin myelin sheaths characteristic of remyelinated axons.

Although these and most studies attempting to inhibit BMP signaling have used exogenous administration of noggin, several other endogenous inhibitors of BMP have been identified. *Corpus callosum* tissue from a lysolecithin-induced demyelination of the *corpus callosum* showed upregulation of the BMP-inhibitor chordin (Jablonska et al., 2010). Exogenous chordin infused into mice with lysolecithin-induced demyelination of the *corpus callosum* increased numbers of oligodendrocytes. Cell number and proliferation analysis showed that this was from enhanced differentiation not expansion of progenitors. These studies suggest that neural stem cell populations may have endogenous repair mechanisms that serve to promote lineage plasticity for repair, although perhaps these are overwhelmed as demyelinating disease like MS progress.

Inhibition of BMP was also shown to be beneficial to myelination in two ischemic injury models of the newborn. One of the causes of dysmyelination and white matter injury in the newborn is due to intraventricular hemorrhage. This hemorrhage typically begins in the germinal matrix, thus affecting nascent oligodendrocytes (Dummula et al., 2011). Many studies have shown that the stage of the oligodendrocyte lineage most vulnerable to hypoxic-ischemic insults and oxidative stress are "pre-oligodendrocytes," which are at the end of their OPC stage but have not yet contacted axons and begun to make myelin proteins and myelinate (Back et al., 2001; Segovia et al., 2008). Dummula et al. administered glycerol to newborn rabbits to induce hemorrhage and found increased apoptosis, reduced proliferation, and reduced differentiation of oligodendrocyte lineage cells (Dummula et al., 2011). BMP4, but not BMP2, was markedly increased on the message and protein levels and found in neurons and oligodendrocytes. Infusion with noggin by *cannulae* into the cerebral ventricles improved neurobehavioral assessments including muscle tone, gait, and righting reflex. Noggin-treated pups showed higher amounts of MBP and MAG by Western blot and immunostaining as well as increased myelin density. As seen before, the OPC proliferation was unaffected. Astrogliosis was also reduced.

Dummula et al. uniquely evaluated human postmortem samples from premature infants with and without intraventricular hemorrhage and found BMP4 immunoreactivity in the VZ, SVZ, and adjacent white matter of the infants with hemorrhage (Dummula et al., 2011). The BMP staining was associated with neurons, OPC, and mature oligodendrocytes. BMP4 levels were higher by Western blot in the VZ, SVZ, and white matter but not other brain regions.

An interesting alternative to infusing BMP inhibitors to determine the role of BMPs in white matter injury is to transgenically overexpress them in the brain. The Kessler lab has generated a mouse in which noggin expression is driven by the neuron-specific enolase promoter (Gomes et al., 2003). The mice are phenotypically normal but do have increased expression of noggin in the brain. This mouse was utilized to determine the role of BMP also in neonatal hypoxia ischemia using another stroke-like model in which the carotid artery was ligated at P7 and the mice were then placed in a hypoxic chamber for an hour (Dizon et al., 2011). An infarct occurs and the extent can be measured. BMP expression was significantly increased on the protein but not the message level within 24 h of the lesion. Noggin-overexpressing mice had smaller infarct sizes and were more likely to survive the injury. The number of cells staining with MBP was increased in lesioned noggin mice compared to lesioned wildtype. Most importantly, using a digital system to assess gait, the lesioned wildtype mice showed significant disability compared to nonlesioned mice but the lesioned noggin mice were no different than their nonlesioned counterparts. One difference between this study and the previous infarct study (Dummula et al., 2011) was that the number of Olig2+ oligodendrocytes was increased in the noggin-overexpressing mice. Since proliferation was unchanged, the authors speculate that noggin may remove any inhibition of neural precursor specification to the oligodendroglial lineage (Dizon et al., 2011). Although both of these studies feature neonatal stroke models, the models, species tested, and methods of noggin presentation differ which may explain the differing results.

A similar study in adult noggin-overexpressing mice in which a stroke model was also created showed that noggin reduced infarct volume protected motor function. In this model, the number of OPCs surrounding the infarct increased, whereas the number of MBP and CNPase cells decreased (Samanta et al., 2010).

The white matter loss seen in neonates corresponding to behavioral and cognitive defects and cerebral palsy can appear as a lesion as generated by the stroke models just described, but more often, perinatal white matter injury appears as a diffuse lack of myelin. A useful animal model for this is intrauterine growth retardation (IUGR) generated by ligation of the intrauterine artery during the last trimester of pregnancy. This model decreases circulation by 50% and generates hypoxia, ischemia, and oxidative stress (Peterside, Selak, & Simmons, 2003; Selak, Storey, Peterside, & Simmons, 2003). Rat pups born to IUGR dams are smaller than normal but viable. These animals have a diffuse lack of mature oligodendrocytes and myelinating axons but not OPCs in the *corpus callosum* during development (up to 21 days postbirth) but the oligodendrocyte number approaches control levels by 8 weeks (Reid et al., 2012). However, behavioral and cognitive deficits persist. BMP4 levels are markedly increased on the message and protein level during the developmental period, returning to normal by 8 weeks. Remarkably, OPCs cultured from the IUGR pups at P2 retain excess BMP signaling as shown by phospho-Smad expression. Differentiation to mature oligodendrocyte is decreased 50% compared to controls but is completely rescued by added recombinant noggin at the time of differentiation.

Spinal cord injury models have also been employed to demonstrate the inhibition of remyelination by BMPs. In the simplest and earliest of these models, rats with lesioned spinal cords were transplanted with neuronal precursor cells engineered to overexpress noggin. A significant increase in functional recovery was noted as well as an increase in neurons, astrocytes and oligodendrocytes (Setoguchi et al., 2004).

Park et al. used the drug agmatine, an NMDA receptor antagonist and nitric oxide inhibitor, to treat mice with compression spinal cord injuries. Agmatine promoted remyelination, decreased neuronal loss, and the glial scar (Park et al., 2013). Interestingly, agmatine increased amounts of BMP2 and BMP7 in neurons and oligodendrocytes but decreased BMP4 in oligodendrocytes and astrocytes at the lesion site. This study highlights the importance specifically of BMP4 in the inhibition of remyelination and suggests alternative roles for BMP2 and BMP7. It was not clear whether agmatine was affecting BMP signaling through its antagonism of NMDA receptors or nitric oxide.

Xaio et al. used a contusive injury to the spinal cord and found increases in BMP2, 4, and 7, as well as phosphor-Smad, within hours after injury (Xiao, Du, Wu, & Yip, 2010). BMPs were coexpressed with markers of neurons oligodendrocytes and microglia. Noggin infusion decreased phosphor-Smad signaling but did not affect GFAP expression.

Manipulation of BMP signaling has turned out to be a key to the success of generating oligodendrocytes from transplanted stem cells. *In vitro*, bone marrow stromal cells have been shown to enhance the differentiation of adult neural progenitor cells in to oligodendrocytes (Sander et al., 2013). Rats with spinal cord transections transplanted with the bone marrow stromal cells did not have enhanced myelination but did have upregulation of BMPs 2 and 4. Stromal cells engineered to produce noggin were able to block the BMP effect and increased significantly the number of MBP expressing cells *in vitro*. Izrael et al. pretreated human OPCs generated from stem cells with noggin and then transplanted into Shiverer mice which have a mutation in MBP and show no MBP immunoreactivity (Izrael et al., 2007). Thus, all MBP seen after transplant comes from the donor cells. Myelination was highly enhanced by this method. However, in another study, OPCs engineered to express noggin were transplanted into contused spinal cord and failed to differentiate (Enzmann et al., 2005). Neural stem cells engineered to express noggin actually exacerbated the lesions. It is unclear why this study radically differed from the previous ones.

The studies reviewed here show, for the most part, that BMP4 is increased in pathological situations that also cause demyelination in the CNS and BMPs serve to inhibit regeneration (Fig. 2). One question that has scarcely been addressed is what aspects of the pathology cause the upregulation of BMP and through what mechanism? There are common processes occurring at the axon in MS, perinatal white matter injury, and spinal cord injury and these include inflammation (Raine, 1990), hypoxia/ischemia (Hertz, 2008; Khwaja & Volpe, 2008), oxidative stress

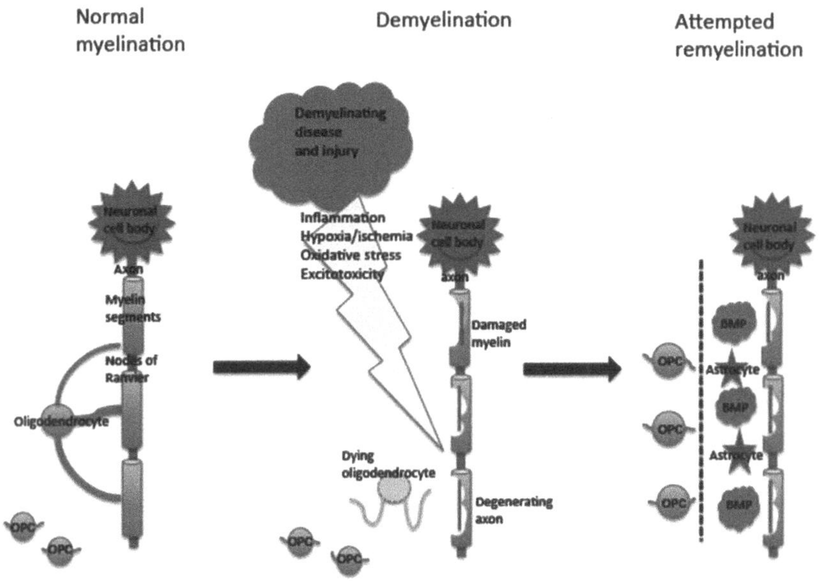

Figure 2 Demyelinating disease and injury create processes such as inflammation, hypoxia/ischemia, oxidative stress, and excitotoxicity that damage myelin resulting, eventually, in axonal degeneration. OPCs are recruited to attempt to remyelinate but are hampered by upregulation of BMPs that both directly and indirectly through astrogliosis inhibits available OPCs from generating new myelin.

(Alberdi et al., 2006; Connor & Menzies, 1996; Thorburne & Juurlink, 1996), and excitotoxicity (Alberdi et al., 2006) but it is not known which of these upregulates BMP postbirth and through what pathway. Reid et al. presented *in vitro* evidence that oxidative stress increases BMP4 by treating cultures of OPC with oxidants at the time of differentiation that inhibited differentiation and increased BMP and phospho-Smad (Reid et al., 2012). Noggin added to cultures with the oxidants rescued the differentiation block. Additionally, cultures of OPCs from genetically altered mice lacking BMP receptors differentiated despite the addition of oxidants. But the signaling pathway from the oxidative stress to BMP is unknown, as are the contributions of other damaging processes in the regulation of BMP signaling (Fig. 2).

It is clear from the studies reviewed above that an increase in astrogliosis accompanies demyelinating pathology, regardless of the etiology. Although studies in which BMP expression is decreased by inhibitors generally demonstrate decreased astrogliosis as well as increased numbers of mature oligodendrocytes; it is not yet possible to dissect out what portion of the effects of BMP on the inhibition of myelination are through the astrocytes and what are direct. One would need to design a mouse in which both astrocyte proliferation and hypertrophy could be conditional deleted at time of injury.

10. BMPs INTERACT WITH OTHER INHIBITORS OF MYELINATION AND REMYELINATION

From the perspective of the oligodendrocyte biologist, BMP4 is a potent inhibitor of myelination and remyelination and seems to be a prime candidate for intervention to promote regeneration. However, several other external signaling factors have been identified, which also inhibit myelination during development and are upregulated following demyelination. These include Wnt, FGF2, notch, LINGO, and GPR17. The Wnts are also a dorsally derived signaling factors that are often associated with BMPs. Treatment of OPCs in culture with soluble Wnt proteins or overexpression of beta catenin, the downstream effector of Wnt decreases oligodendrocyte differentiation both *in vivo* and *in vitro* (Fancy et al., 2009; Feigenson et al., 2009; Ye et al., 2009). Expression of Wnt pathway elements such as the transcription factor TCF/LEF downstream of Wnt and beta catenin are increased in multiple sclerosis and perinatal white matter injury lesions (Fancy et al., 2009). In the early development of the nervous system, BMPs and Wnts interact since their expression is temporally and spatially similar and they are involved in many of the same functions. However, depending on the context, they can be inductive, antagonistic, or synergistic (Soshnikova et al., 2003). In the development of oligodendrocytes, evidence suggests that Wnts require elements of the BMP pathway to inhibit differentiation (Feigenson et al., 2011; Kasai, Satoh, & Akiyama, 2005). In culture, Wnt-treated OPC are rescued from their differentiation block by cotreatment with noggin and Wnt treatment does not decrease differentiation in OPCs cultured from mice lacking BMP type 1 receptors (Feigenson et al., 2011). Conversely however, inhibition of Wnt signaling using Dkk inhibitor or OPCs from mice in which beta catenin is deleted from oligodendrocyte lineage cells does not affect BMP signaling, suggesting that BMP signaling is downstream of the Wnt pathway and not synergistic in this case. Weng et al. showed that the BMP and Wnt pathways were functionally linked by a common protein, Smad interacting protein 1 (Sip1), in conjunction with Smad 7 (Weng et al., 2012). Sip1 also interacts with Hes 1 downstream of the Notch pathway, another inhibitory pathway shown to be upregulated during demyelination (John et al., 2002; Jurynczyk, Jurewicz, Bielecki, Raine, & Selmaj, 2005). A transcriptome study, later validated in cultured OPCs, demonstrated that BMP4 promoted expression of Notch target genes (Wu et al., 2012). BMP4 was thus shown to connecting three of the five inhibitory pathways identified so far. It is likely that more interactions between these inhibitory signaling factors will be identified in the future, possibly leading us to common downstream regulatory elements that could be modulated to promote regeneration therapy in a wide variety of de- and dysmyelinating conditions.

11. CONCLUSIONS AND FUTURE DIRECTIONS

The study of the role of BMPs in oligodendrocyte development and myelination has demonstrated that this family of signaling factors are powerful inhibitors of oligodendrogliogenesis and myelination in development and in disease. The role of BMPs in oligodendrocyte development was identified first leading to speculation that BMPs might be involved in pathologies, which has now been verified. Oligodendrocyte biologists have long claimed that the study of oligodendroglial development could identify factors important in remyelination and this is clearly proof of that approach. Multiple studies showing improvement in myelination and motor function following the inhibition of BMP signaling are encouraging and suggest directions for therapeutic interventions. However, a number of challenges exist such as possibly beneficial effects of BMP on other cell types in the brain in the neonatal or adult CNS. Also, the identification of other inhibitory factors for remyelination that interact with BMP suggests that a more global approach to trying to modulate multiple factors might be more powerful. In this regard, more research into the common downstream signaling pathways from these inhibitors is necessary. Interestingly, the role of BMP unites pathologies with different etiologies such as inflammation, hypoxia/ischemia, and injury that have common paths of destruction of the nervous system. Future research will hopefully begin to understand how the common damaging factors unleashed in all of these demyelinating diseases can increase BMPs, which will undoubtedly benefit basic research on myelin in a number of areas.

ACKNOWLEDGMENTS

This work is supported by National MS Society RG4558A8/2 (J.B.G.), RO1 MH098742, and the Cellular Neuroscience Core of the Institutional Intellectual and Developmental Disabilities Research Core of the Children's Hospital of Philadelphia (HD26979).

REFERENCES

Aguirre, A., Dupree, J. L., Mangin, J. M., & Gallo, V. (2007). A functional role for EGFR signaling in myelination and remyelination. *Nature Neuroscience*, *10*(8), 990–1002.

Ahn, K., Mishina, Y., Hanks, M. C., Behringer, R. R., & Crenshaw, E. B., 3rd. (2001). BMPR-IA signaling is required for the formation of the apical ectodermal ridge and dorsal-ventral patterning of the limb. *Development*, *128*(22), 4449–4461.

Alberdi, E., Sanchez-Gomez, M. V., Torre, I., Domercq, M., Perez-Samartin, A., Perez-Cerda, F., et al. (2006). Activation of kainate receptors sensitizes oligodendrocytes to complement attack. *Journal of Neuroscience: The Official Journal of the Society for Neuroscience*, *26*(12), 3220–3228.

Alberta, J. A., Park, S. K., Mora, J., Yuk, D. I., Pawlitzky, I., Iannarelli, P., et al. (2001). Sonic hedgehog is required during an early phase of oligodendrocyte development in mammalian brain. *Molecular and Cellular Neurosciences*, *18*(4), 434–441.

Ara, J., See, J., Mamontov, P., Hahn, A., Bannerman, P., Pleasure, D., et al. (2008). Bone morphogenetic proteins 4, 6, and 7 are up-regulated in mouse spinal cord during experimental autoimmune encephalomyelitis. *Journal of Neuroscience Research*, *86*(1), 125–135.

Back, S. A., Han, B. H., Luo, N. L., Chricton, C. A., Xanthoudakis, S., Tam, J., et al. (2002). Selective vulnerability of late oligodendrocyte progenitors to hypoxia-ischemia. *The Journal of Neuroscience*, *22*(2), 455–463.

Back, S. A., Luo, N. L., Borenstein, N. S., Levine, J. M., Volpe, J. J., & Kinney, H. C. (2001). Late oligodendrocyte progenitors coincide with the developmental window of vulnerability for human perinatal white matter injury. *The Journal of Neuroscience*, *21*(4), 1302–1312.

Back, S. A., Tuohy, T. M., Chen, H., Wallingford, N., Craig, A., Struve, J., et al. (2005). Hyaluronan accumulates in demyelinated lesions and inhibits oligodendrocyte progenitor maturation. *Nature Medicine*, *11*(9), 966–972.

Bannerman, P., Hahn, A., Soulika, A., Gallo, V., & Pleasure, D. (2007). Astrogliosis in EAE spinal cord: Derivation from radial glia, and relationships to oligodendroglia. *Glia, 55*(1), 57–64.
Bansal, R. (2002). Fibroblast growth factors and their receptors in oligodendrocyte development: Implications for demyelination and remyelination. *Developmental Neuroscience, 24*(1), 35–46.
Bansal, R., Stefansson, K., & Pfeiffer, S. E. (1992). Proligodendroblast antigen (POA), a developmental antigen expressed by A007/O4-positive oligodendrocyte progenitors prior to the appearance of sulfatide and galactocerebroside. *Journal of Neurochemistry, 58*, 2221–2229.
Barres, B. A., Lazar, M. A., & Raff, M. C. (1994). A novel role for thyroid hormone, glucocorticoids and retinoic acid in timing oligodendrocyte development. *Development, 120*(5), 1097–1108.
Bilican, B., Fiore-Heriche, C., Compston, A., Allen, N. D., & Chandran, S. (2008). Induction of Olig2 precursors by FGF involves BMP signalling blockade at the Smad level. *PLoS One, 3*(8), e2863.
Casaccia-Bonnefil, P., Hardy, R. J., Teng, K. K., Levine, J. M., Koff, A., & Chao, M. V. (1999). Loss of p27Kip1 function results in increased proliferative capacity of oligodendrocyte progenitors but unaltered timing of differentiation. *Development, 126*(18), 4027–4037.
Cate, H. S., Sabo, J. K., Merlo, D., Kemper, D., Aumann, T. D., Robinson, J., et al. (2010). Modulation of bone morphogenic protein signalling alters numbers of astrocytes and oligodendroglia in the subventricular zone during cuprizone-induced demyelination. *Journal of Neurochemistry, 115*(1), 11–22.
Chandran, S., Kato, H., Gerreli, D., Compston, A., Svendsen, C. N., & Allen, N. D. (2003). FGF-dependent generation of oligodendrocytes by a hedgehog-independent pathway. *Development, 130*(26), 6599–6609.
Chen, J., Leong, S. Y., & Schachner, M. (2005). Differential expression of cell fate determinants in neurons and glial cells of adult mouse spinal cord after compression injury. *The European Journal of Neuroscience, 22*(8), 1895–1906.
Cheng, X., Wang, Y., He, Q., Qiu, M., Whittemore, S. R., & Cao, Q. (2007). Bone morphogenetic protein signaling and olig1/2 interact to regulate the differentiation and maturation of adult oligodendrocyte precursor cells. *Stem Cells, 25*(12), 3204–3214.
Colak, D., Mori, T., Brill, M. S., Pfeifer, A., Falk, S., Deng, C., et al. (2008). Adult neurogenesis requires Smad4-mediated bone morphogenic protein signaling in stem cells. *The Journal of Neuroscience, 28*(2), 434–446.
Connor, J. R., & Menzies, S. L. (1996). Relationship of iron to oligodendrocytes and myelination. *Glia, 17*(2), 83–93.
Deininger, M., Meyermann, R., & Schluesener, H. (1995). Detection of two transforming-growth-factor-beta-related morphogens, bone morphogenetic proteins-4 and −5, in RNA of multiple sclerosis and Creutzfeld-Jakob disease lesion. *Acta Neuropathologica, 90*, 76–79.
Dizon, M. L., Maa, T., & Kessler, J. A. (2011). The bone morphogenetic protein antagonist noggin protects white matter after perinatal hypoxia-ischemia. *Neurobiology of Disease, 42*(3), 318–326.
Dummula, K., Vinukonda, G., Chu, P., Xing, Y., Hu, F., Mailk, S., et al. (2011). Bone morphogenetic protein inhibition promotes neurological recovery after intraventricular hemorrhage. *The Journal of Neuroscience, 31*(34), 12068–12082.
Emery, B. (2010). Regulation of oligodendrocyte differentiation and myelination. *Science, 330*(6005), 779–782.
Enzmann, G. U., Benton, R. L., Woock, J. P., Howard, R. M., Tsoulfas, P., & Whittemore, S. R. (2005). Consequences of noggin expression by neural stem, glial,

and neuronal precursor cells engrafted into the injured spinal cord. *Experimental Neurology, 195*(2), 293–304.

Fancy, S. P., Barazini, S. E., Zhao, C., Yuk, D. I., Irvine, K. A., Kaing, S., et al. (2009). Dysregulation of the Wnt pathway inhibits timely myelination and remyelination in the mammalian CNS. *Genes and Development, 23*, 1571–1585.

Feigenson, K., Reid, M., See, J., Crenshaw, E. B., III, & Grinspan, J. B. (2011). Canonical Wnt signaling requires the BMP pathway to inhibit oligodendrocyte maturation. *ASN Neuro, 3*(3), e00061. http://dx.doi.org/10.1042/AN20110004.

Feigenson, K., Reid, M., See, J., Crenshaw, E. B., II, & Grinspan, J. B. (2009). Wnt signaling is sufficient to perturb oligodendrocyte maturation. *Molecular and Cellular Neuroscience, 42*, 255–265.

Fuller, M. L., Dechant, A. K., Rothstein, B., Caprariello, A., Wang, R., Hall, A. K., et al. (2007). Bone morphogenetic proteins promote gliosis in demyelinating spinal cord lesions. *Annals of Neurology, 62*(3), 288–300.

Goldschmidt, T., Antel, J., Konig, F. B., Bruck, W., & Kuhlmann, T. (2009). Remyelination capacity of the MS brain decreases with disease chronicity. *Neurology, 72*(22), 1914–1921.

Gomes, W. A., Mehler, M. F., & Kessler, J. A. (2003). Transgenic overexpression of BMP4 increases astroglial and decreases oligodendroglial lineage commitment. *Developmental Biology, 255*(1), 164–177.

Grinspan, J. B., Edell, E., Carpio, D. F., Beesley, J. S., Lavy, L., Pleasure, D., et al. (2000). Stage-specific effects of bone morphogenetic proteins on the oligodendrocyte lineage. *Journal of Neurobiology, 43*(1), 1–17.

Grinspan, J., Wrabetz, L., & Kamholz, J. (1993). Oligodendrocyte maturation and myelin gene expression in PDGF-treated cultures from rat cerebral white matter. *Journal of Neurocytology, 22*, 322–333.

Gross, R. E., Mehler, M. F., Mabie, P. C., Zang, Z., Santschi, L., & Kessler, J. A. (1996). Bone morphogenetic proteins promote astroglial lineage commitment by mammalian subventricular zone progenitor cells. *Neuron, 17*(4), 595–606.

Haynes, R. L., Folkerth, R. D., Keefe, R. J., Sung, I., Swzeda, L. I., Rosenberg, P. A., et al. (2003). Nitrosative and oxidative injury to premyelinating oligodendrocytes in periventricular leukomalacia. *Journal of Neuropathology and Experimental Neurology, 62*(5), 441–450.

Hertz, L. (2008). Bioenergetics of cerebral ischemia: A cellular perspective. *Neuropharmacology, 55*(3), 289–309 (Review).

Ishibashi, T., Dakin, K. A., Stevens, B., Lee, P. R., Kozlov, S. V., Stewart, C. L., et al. (2006). Astrocytes promote myelination in response to electrical impulses. *Neuron, 49*(6), 823–832.

Izrael, M., Zhang, P., Kaufman, R., Shinder, V., Ella, R., Amit, M., et al. (2007). Human oligodendrocytes derived from embryonic stem cells: Effect of noggin on phenotypic differentiation in vitro and on myelination in vivo. *Molecular and Cellular Neurosciences, 34*, 310–323.

Jablonska, B., Aguirre, A., Raymond, M., Szabo, G., Kitabatake, Y., Sailor, K. A., et al. (2010). Chordin-induced lineage plasticity of adult SVZ neuroblasts after demyelination. *Nature Neuroscience, 13*(5), 541–550.

John, G. R., Shankar, S. L., Shafit-Zagardo, B., Massimi, A., Lee, S. C., Raine, C. S., et al. (2002). Multiple sclerosis: Re-expression of a developmental pathway that restricts oligodendrocyte maturation. *Nature Medicine, 8*(10), 1115–1121 (see comment).

Jurynczyk, M., Jurewicz, A., Bielecki, B., Raine, C. S., & Selmaj, K. (2005). Inhibition of Notch signaling enhances tissue repair in an animal model of multiple sclerosis. *Journal of Neuroimmunology, 170*(1–2), 3–10.

Kasai, M., Satoh, K., & Akiyama, T. (2005). Wnt signaling regulates the sequential onset of neurogenesis and gliogenesis via induction of BMPs. *Genes to Cells*, *10*, 777–783.

Kessaris, N., Fogarty, M., Iannarelli, P., Grist, M., Wegner, M., & Richardson, W. D. (2006). Competing waves of oligodendrocytes in the forebrain and postnatal elimination of an embryonic lineage. *Nature Neuroscience*, *9*, 173–179.

Khwaja, O., & Volpe, J. J. (2008). Pathogenesis of cerebral white matter injury of prematurity. *Archives of Disease in Childhood. Fetal and Neonatal Edition*, *93*(2), F153–F161.

Kinney, H. C. (2006). The near-term (late preterm) human brain and risk for periventricular leukomalacia: A review. *Seminars in Perinatology*, *30*(2), 81–88.

Kondo, T., & Raff, M. (2000). The Id4 HLH protein and the timing of oligodendrocyte differentiation. *The EMBO Journal*, *19*, 1998–2007.

Kondo, T., & Raff, M. C. (2004). A role for Noggin in the development of oligodendrocyte precursor cells. *Developmental Biology*, *267*(1), 242–251.

Kretzschmar, M., Doody, J., & Massague, J. (1997). Opposing BMP and EGF signalling pathways converge on the TGF-beta family mediator Smad1. *Nature*, *389*(6651), 618–622.

Lasiene, J., Shupe, L., Perlmutter, S., & Horner, P. (2008). No evidence for chronic demyelination in spared axons after spinal cord injury in a mouse. *The Journal of Neuroscience: The Official Journal of the Society for Neuroscience*, *28*(15), 3887–3896.

Liem, K. F., Jr., Jessell, T. M., & Briscoe, J. (2000). Regulation of the neural patterning activity of sonic hedgehog by secreted BMP inhibitors expressed by notochord and somites. *Development*, *127*(22), 4855–4866.

Liem, K. F., Jr., Tremml, G., Roelink, H., & Jessell, T. M. (1995). Dorsal differentiation of neural plate cells induced by BMP-mediated signals from epidermal ectoderm. *Cell*, *82*(6), 969–979.

Lu, Q. R., Cai, L., Rowitch, D., Cepko, C. L., & Stiles, C. D. (2001). Ectopic expression of Olig1 promotes oligodendrocyte formation and reduces neuronal survival in developing mouse cortex. *Nature Neuroscience*, *4*(10), 973–974.

Lu, Q. R., Sun, T., Zhu, Z., Ma, N., Garcia, M., Stiles, C. D., et al. (2002). Common developmental requirement for Olig function indicates a motor neuron/oligodendrocyte connection. *Cell*, *109*(1), 75–86.

Lu, Q. R., Yuk, D.-I., Alberta, J. A., Zhu, Z., Pawlitzky, I., Chan, J., et al. (2000). Sonic hedgehog-regulated oligodendrocyte lineage genes encoding bHLH proteins in the mamalian central nervous system. *Neuron*, *25*, 317–329.

Mabie, P. C., Mehler, M. F., Marmur, R., Papavasiliou, A., Song, Q., & Kessler, J. A. (1997). Bone morphogenetic proteins induce astroglial differentiation of oligodendroglial-astroglial progenitor cells. *The Journal of Neuroscience*, *17*(11), 4112–4120.

Marin-Husstege, M., Muggironi, M., Liu, A., & Casaccia-Bonnefil, P. (2002). Histone deacetylase activity is necessary for oligodendrocyte lineage progression. *The Journal of Neuroscience*, *22*(23), 10333–10345.

Martinez, G., Carnazza, M., Dii Giacomo, C., Sorrenti, V., & Vanella, A. (2001). Expression of bone morphogenetic protein 6 and transforming growth factor b1 in the rat brain after a mild and reversible ischemic damage. *Brain Research*, *894*, 1–11.

Matsushima, G. K., & Morell, P. (2001). The neurotoxicant, cuprizone, as a model to study demyelination and remyelination in the central nervous system. *Brain Pathology*, *11*(1), 107–116.

McKinnon, R. D., Smith, C., Behar, T., Smith, T., & Dubois-Dalcq, M. (1993). Distinct effects of bFGF and PDGF on oligodendrocyte progenitor cells. *Glia*, *7*, 245–254.

McMorris, F. A., & Dubois-Dalcq, M. (1988). Insulin-like growth factor 1 promotes cell proliferation and oligodendroglial commitment in rat glial progenitor cells developing in vitro. *Journal of Neuroscience Research*, *21*, 199–209.

Mekki-Dauriac, S., Agius, E., Kan, P., & Cochard, P. (2002). Bone morphogenetic proteins negatively control oligodendrocyte precursor specification in the chick spinal cord. *Development*, *129*(22), 5117–5130.

Miller, R. H., Dinsio, K., Wang, R., Geertman, R., Maier, C. E., & Hall, A. K. (2004). Patterning of spinal cord oligodendrocyte development by dorsally derived BMP4. *Journal of Neuroscience Research*, *76*(1), 9–19.

Milner, R., Anderson, H. J., Rippon, R. F., McKay, J. S., Franklin, R. J., Marchionni, M. A., et al. (1997). Contrasting effects of mitogenic growth factors on oligodendrocyte precursor cell migration. *Glia*, *19*(1), 85–90.

Nait-Oumesmar, B., Picard-Riera, N., Kerninon, C., Decker, L., Seilhean, D., Hoglinger, G. U., et al. (2007). Activation of the subventricular zone in multiple sclerosis: Evidence for early glial progenitors. *Proceedings of the National Academy of Sciences of the United States of America*, *104*(11), 4694–4699.

Nave, K. A. (2010). Myelination and support of axonal integrity by glia. *Nature*, *468*(7321), 244–252.

Nicolay, D. J., Doucette, J. R., & Nazarali, A. J. (2007). Transcriptional control of oligodendrogenesis. *Glia*, *55*(13), 1287–1299.

Nishiyama, A., Lin, X.-H., Giese, N., Heldin, C.-H., & Stallcup, W. B. (1996). Interaction between NG2 proteonglycan and PDGF α-receptor on O2A progenitor cells is required for optimal response to PDGF. *Journal of Neuroscience Research*, *43*, 315–330.

Ohkawara, B., Iemura, S., ten Dijke, P., & Ueno, N. (2002). Action range of BMP is defined by its N-terminal basic amino acid core. *Current Biology*, *12*(3), 205–209.

Orentas, D. M., Hayes, J. E., Dyer, K. L., & Miller, R. H. (1999). Sonic hedgehog signaling is required during the appearance of spinal cord oligodendrocyte precursors. *Development*, *126*, 2419–2429.

Park, Y. M., Lee, W. T., Bokara, K. K., Seo, S. K., Park, S. H., Kim, J. H., et al. (2013). The multifaceted of effects of agmatine on functional recovery after spinal cord injury through modulations of BMP-2/4/7 expressions in neurons and glial cells. *PLoS One*, *8*, e53911.

Peterside, I. E., Selak, M. A., & Simmons, R. A. (2003). Impaired oxidative phosphorylation in hepatic mitochondria in growth-retarded rats. *American Journal of Physiology. Endocrinology and Metabolism*, *285*(6), E1258–E1266.

Piddlesden, S., Lassman, H., Laffafian, I., Morgan, B. P., & Linington, C. (1991). Antibody-mediated demyelination in experimental allergic encephalomyelitis is independent of complement membrane attack complex formation. *Clinical and Experimental Immunology*, *83*, 245–250.

Pringle, N. P., Yu, W. P., Guthrie, S., Roelink, H., Lumsden, A., Peterson, A. C., et al. (1996). Determination of neuroepithelial cell fate: Induction of the oligodendrocyte lineage by ventral midline cells and sonic hedgehog. *Developmental Biology*, *177*(1), 30–42.

Raff, M. C., Miller, R. H., & Noble, M. (1983). A glial progenitor cell that develops in vitro into an astrocyte or an oligodendrocyte depending on the culture medium. *Nature*, *303*, 390–396.

Raff, M. C., Mirsky, R., Fields, K. L., Lisak, R. P., Dorfman, S. H., Silberberg, D. H., et al. (1978). Galactocerebroside: A cell specific surface antigen marker for oligodendrocytes in culture. *Nature*, *274*, 813–816.

Raine, C. S. (1990). Multiple sclerosis: Immunopathologic mechanisms in the progression and resolution of inflammatory demyelination. *Research Publications - Association for Research in Nervous and Mental Disease*, *68*, 37–54.

Raine, C. S., Bonetti, B., & Cannella, B. (1998). Multiple sclerosis: Expression of molecules of the tumor necrosis factor ligand and receptor families in relationship to the demyelinated plaque. *Revue Neurologique*, *154*(8–9), 577–585.

Ranscht, B., Clapschaw, P. A., Price, J., Noble, M., & Seifert, W. (1982). Development of oligodendrocytes and Schwann cells studied with a monoclonal antibody against galactocerebroside. *Proceedings of the National Academy of Sciences of the United States of America, 79*, 2709–2713.

Reid, M. V., Murray, K. A., Marsh, E. D., Golden, J. A., Simmons, R. A., & Grinspan, J. B. (2012). Delayed myelination in an intrauterine growth retardation model is mediated by oxidative stress upregulating bone morphogenetic protein 4. *Journal of Neuropathology and Experimental Neurology, 71*(7), 640–653.

Sabo, J. K., Aumann, T. D., Merlo, D., Kilpatrick, T. J., & Cate, H. S. (2011). Remyelination is altered by bone morphogenic protein signaling in demyelinated lesions. *The Journal of Neuroscience, 31*(12), 4504–4510.

Samanta, J., Alden, T., Gobeske, K., Kan, L., & Kessler, J. A. (2010). Noggin protects against ischemic brain injury in rodents. *Stroke, 41*, 357–362.

Samanta, J., Burke, G. M., McGuire, T., Pisarek, A. J., Mukhopadhyay, A., Mishina, Y., et al. (2007). BMPR1a signaling determines numbers of oligodendrocytes and calbindin-expressing interneurons in the cortex. *The Journal of Neuroscience, 27*(28), 7397–7407.

Samanta, J., & Kessler, J. A. (2004). Interactions between ID and OLIG proteins mediate the inhibitory effects of BMP4 on oligodendroglial differentiation. *Development, 131*, 4131–4142.

Sander, B., Rivera, F. J., Caioni, M., Nicholson, L., Eckstein, V., Bogdahn, U., et al. (2013). Bone morphogenetic proteins prevent bone marrow stromal cell-mediated oligodendroglial differentiation of transplanted adult neural progenitor cells in the injured spinal cord. *Stem Cell Research, 11*, 758–771.

Scherer, S., Braun, P., Grinspan, J., Collarini, E., Wang, D.-Y., & Kamholz, J. (1994). Differential regulation of the 2'3'-cyclic nucleotide 3'-phosphodiesterase gene in oligodendrocyte development. *Neuron, 12*, 1–20.

See, J., Mamontov, P., Ahn, K., Wine-Lee, L., Crenshaw, E. B., 3rd, & Grinspan, J. B. (2007). BMP signaling mutant mice exhibit glial cell maturation defects. *Molecular and Cellular Neurosciences, 35*(1), 171–182.

See, J., Zhang, X., Eraydin, N., Mun, S. B., Mamontov, P., Golden, J. A., et al. (2004). Oligodendrocyte maturation is inhibited by bone morphogenetic protein. *Molecular and Cellular Neurosciences, 26*(4), 481–492.

Segovia, K. N., McClure, M., Moravec, M., Luo, N. L., Wan, Y., Gong, X., et al. (2008). Arrested oligodendrocyte lineage maturation in chornic perinatal whilte matter injury. *Annals of Neurology, 63*, 520–530.

Selak, M. A., Storey, B. T., Peterside, I., & Simmons, R. A. (2003). Impaired oxidative phosphorylation in skeletal muscle of intrauterine growth-retarded rats. *American Journal of Physiology. Endocrinology and Metabolism, 285*(1), E130–E137.

Selmaj, K., Raine, C. S., Cannella, B., & Brosnan, C. F. (1991). Identification of lymphotoxin and tumor necrosis factor in multiple sclerosis lesions. *The Journal of Clinical Investigation, 87*(3), 949–954.

Setoguchi, T., Nakashima, K., Takizawa, T., Yanagisawa, M., Ochiai, W., Okabe, M., et al. (2004). Treatment of spinal cord injury by transplantation of fetal neural precursors cells engineered to express BMP inhibitor. *Experimental Neurology, 189*, 33–44.

Setoguchi, T., Yone, K., Matsuoka, E., Takenouchi, H., Nakashima, K., Sakou, T., et al. (2001). Traumatic injury-induced BMP 7 expression in the adult rat spinal cord. *Brain Research, 921*, 219–225.

Soshnikova, N., Zechner, D., Huelsken, J., Mishina, Y., Behringer, R. R., Taketo, M. M., et al. (2003). Genetic interaction between Wnt/beta-catenin and BMP receptor signaling during formation of the AER and the dorsal-ventral axis in the limb. *Genes & Development, 17*(16), 1963–1968.

Stolt, C. C., Rehberg, S., Ader, M., Lommes, P., Riethmacher, D., Schachner, M., et al. (2002). Terminal differentiation of mylein-forming oligodendroyctes depends on the transcription factor Sox 10. *Genes and Development*, *16*, 165–170.

Tang, X.-M., Beesley, J. S., Grinspan, J. B., Seth, P., Kamholz, J., & Cambi, F. (1999). Cell cycle arrest induced by ectopic expression of p27 is not sufficient to promote oligodendorcyte differentiation. *Journal of Cellular Biochemistry*, *76*, 270–279.

Thorburne, S., & Juurlink, B. H. (1996). Low glutathione and high iron govern the suceptibility of oligodendroglial precursors to oxidative stress. *Journal of Neurochemistry*, *67*, 1014–1022.

Tsai, H. H., & Miller, R. H. (2002). Glial cell migration directed by axon guidance cues. *Trends in Neurosciences*, *25*(4), 173–175 (discussion 175–176).

Volpe, J. J. (2001). Neurobiology of periventricular leukomalacia in the premature infant. *Pediatric Research*, *50*(5), 553–562.

Wada, T., Kagawa, T., Ivanova, A., Zalc, B., Shirasaki, R., Murakami, F., et al. (2000). Dorsal spinal cord inhibits oligodendrocyte development. *Developmental Biology*, *227*(1), 42–55.

Wang, S., Sdrulla, A., Johnson, J. E., Yokota, Y., & Barres, B. A. (2001). A role for the helix-loop-helix protein Id2 in the control of oligodendrocyte development. *Neuron*, *29*(3), 603–614.

Weng, Q., Chen, Y., Wang, H., Xu, X., Yang, B., He, Q., et al. (2012). Dual-mode modulation of Smad signaling by Smad-interacting protein Sip1 is required for myelination in the central nervous system. *Neuron*, *73*(4), 713–728.

Wine-Lee, L., Ahn, K. J., Richardson, R. D., Mishina, Y., Lyons, K. M., & Crenshaw, E. B., 3rd. (2004). Signaling through BMP type 1 receptors is required for development of interneuron cell types in the dorsal spinal cord. *Development*, *131*(21), 5393–5403.

Wrana, J. L., Attisano, L., Wieser, R., Ventura, F., & Massague, J. (1994). Mechanism of activation of the TGF-beta receptor. *Nature*, *370*(6488), 341–347.

Wu, M., Hernandez, M., Shen, S., Sabo, J. K., Kelkar, D., Wang, J., et al. (2012). Differential modulation of the oligodendrocyte transcriptome by sonic hedgehog and bone morphogenetic protein 4 via opposing effects on histone acetylation. *Journal of Neuroscience: The Official Journal of the Society for Neuroscience*, *32*(19), 6651–6664.

Xiao, Q., Du, Y., Wu, W., & Yip, H. K. (2010). Bone morphogenetic proteins mediate cellular response, and together with Noggin, regulate astrocyte differentiation after spinal cord injury. *Experimental Neurology*, *221*(2), 353–366.

Ye, F., Chen, Y., Hoang, T., Montgomery, R. L., Zhao, X. H., Bu, H., et al. (2009). HDAC1 asnd HDAC2 regulate oligodendrocyte differentiation by disrupting the beta-catenin-TCF interaction. *Nature Neuroscience*, *12*, 829–838.

Zhang, H., & Miller, R. H. (1996). Density-dependent feedback inhibition of oligodendrocyte precursor expansion. *The Journal of Neuroscience*, *16*(21), 6886–6895.

Zhao, C., Fancy, S. P. J., Magy, L., Urwin, J. E., & Franklin, R. J. M. (2005). Stem cells, progenitors and myelin repair. *Journal of Anatomy*, *207*, 251–258.

Zhou, Q., Choi, G., & Anderson, D. J. (2001). The bHLH transcription factor Olig2 promotes oligodendrocyte differentiation in collaboration with Nkx2.2. *Neuron*, *31*(5), 791–807.

Zhou, Q., Wang, S., & Anderson, D. J. (2000). Identification of a novel family of oligodendrocyte lineage-specific basic helix-loop-helix transcription factors. *Neuron*, *25*, 331–343.

CHAPTER SEVEN

Bone Morphogenetic Proteins in Preimplantation Embryos

Isabel La Rosa[1]
Laboratory of Animal Biotechnology, Agriculture Faculty, University of Buenos Aires (UBA), Buenos Aires, Argentina
[1]Corresponding author. e-mail address: ilarosa@agro.uba.ar

Contents

Abstract

The bone morphogenetic proteins' (BMPs) pathway is one of the evolutionarily oldest used by animal embryos and is required for dorsal–ventral patterning of the early embryo of both vertebrates and invertebrates. Nevertheless, the role of this system in preimplantation embryo development has not been extensively studied yet. Taking into account that the preimplantation period is different among species though the BMP system is conserved, information regarding comparative embryo development and the role of BMPs in different mammalian models is revised and discussed in this chapter. BMP system is expressed by maternal tissues (the ovary, the oviduct, and the uterus) as well as by the embryo and extraembryonic tissues. The reviewed information demonstrates a very important role for BMP signaling system at different stages of embryo preimplantation development from acquisition of gamete competence to regulation of trophoblast development and differentiation in mice as well as in ungulates.

Vitamins and Hormones, Volume 99
ISSN 0083-6729
http://dx.doi.org/10.1016/bs.vh.2015.04.001

1. BONE MORPHOGENETIC PROTEINS AND NOGGIN

The critical step of transformation to multicellularity must have been accompanied by the development of adhesion molecules, extracellular matrix proteins (such as collagen), and cell-to-cell communication. It is essential to identify the critical signaling pathways, in particular those involved in control of development, cellular differentiation, and body plan formation. Interesting, the emergence of the TGFβ pathway coincided with appearance of the first known animal species (Huminiecki et al., 2009).

Bone morphogenetic proteins (BMPs) belong to the TGFβ superfamily of growth factors. There are around 20 described members of this family (Chen, Zhao, & Mundy, 2004). BMPs are so-called cysteine-knot proteins for they contain many cysteine residues and form a ring. They are secreted as monomers and form homo- as well as heterodimers in the extracellular matrix (Vitt, 2001).

Based on their sequence homology, BMPs are grouped into several classes. The DPP class (BMP2, BMP4, and DPP) is at one end of the spectrum followed by the 60A class (BMP5–8 and 60A) and with a continuous sequence drift toward the Activins/Inhibins and the TGFβ at the other. Members of this superfamily are structurally related but functionally diverse polypeptides (Hu et al., 2004).

BMPs exert their functions by binding their membrane receptors BMPRI and BMPRII. Type I receptors propagate the signal transduction by phosphorylating Smad proteins specifically Smad 1, 5, and 8. Once these intracellular mediators are phosphorylated, they form a complex with Smad 4, and this complex is then translocated to the nucleus where it regulates activation or repression of transcription in numerous genes (Massagué, Blain, & Lo, 2000; Miyazono, Kusanagi, & Inoue, 2001). Also there is a "smad-independent" way which implies phosphorylation of various kinases (i.e., MAPK, JUNK) that mediates many BMP effects (Sieber, Kopf, Hiepen, & Knaus, 2009). Among BMP's target genes are the "inhibition of differentiation" (*ID*) gene family members (Hollnagel et al., 1999; Kowanetz, Valcourt, Bergstro, & Heldin, 2004).

Regulation of the extracellular levels of these factors is critical for normal development in any of the system in which these proteins are implicated. BMP signaling systems is negatively regulated at different cellular levels, with various types of inhibitors (Chen et al., 2004). Extracellular regulation includes the secretion of "ligand traps," a group of inhibitors that bind directly to their targets and prevent interaction with its receptors. Noggin is a ligand trap inhibitor that binds BMP2, 4, and 7 (Groppe et al., 2002). Homologues of this protein have been indentified in every studied vertebrate. Actually, its structure is closely related to that of BMP ligands which strongly suggests they have evolved from a common ancestor (Groppe et al., 2002). Noggin extracellular localization is determined by its union with heparan-sulfate proteoglycans in the extracellular matrix that keeps it restricted to cell surface. This particular localization permits the regulation of ligand gradients (Paine-Saunders, Viviano, Economides, & Saunders, 2002), fundamental for morphogens. Noggin is a critical regulator of BMP signaling, and its function is remarkable in many systems such as limb, cartilage, and lens formation (Brunet, 1998; Wijgerde, Karp, McMahon, & McMahon, 2005). Intracellular regulation of TGFβ includes inhibitory Smads, Smad 6 and 7. Both are produced in response to ligands so they are implicated in the negative feedback of this signaling system (Von Bubnoff & Cho, 2001). Other examples of intracellular regulation are Smad Ubiquitin Regulatory Factor 1 (Chen et al., 2004) that, besides ubiquitinization, affects interaction with the nuclear translocation factor (Sapkota, Alarcón, Spagnoli, Brivanlou, & Massagué, 2007) and a Smad 1-specific phosphatase that contributes to diminish BMP response (Knockaert, Sapkota, Alarcón, Massagué, & Brivanlou, 2006).

BMPs have roles in brain, eye, hair follicles, kidney, lung, liver, skin, and tooth development. Thus, BMPs have actions beyond bone, and hence, they may be considered as body morphogenetic proteins (Reddi, 2005; Wagner et al., 2010). Indeed, this pathway is one of the evolutionarily oldest used by animal embryos and is required for dorsal–ventral patterning of the early embryo of both vertebrates and invertebrates (Groppe et al., 2002).

2. EMBRYO DEVELOPMENT

Fertilization is the process by which haploid gametes, sperm, and egg unite to produce a genetically distinct individual (Florman & Ducibella, 2006). This marks the start of a new life and is consequence of a precisely ordered sequence of cellular interactions (Florman & Ducibella, 2006). Fusion of maternal and paternal pronuclei (singamy) results in the formation of a zygote, the one cell embryo. For successful of this process, both gametes should be competent. Acquisition of competence by eggs occurs within the ovary and by sperm within the epididymis and the female tract

(Florman & Ducibella, 2006). The oviduct and its secretions provide the optimal microenvironment necessary for the reproductive events that precede implantation including gamete transport, sperm capacitation, fertilization, and early embryonic development (García, Rizo, Valdecantos, Miceli, & Barrera, 2014). In mammalian embryos, the first cell cycle is characterized by a length of about 1 day in which parental genomes remain separated within the same cytoplasmic environment during the first DNA replication (S phase; Comizzoli, Marquant-Le Guienne, Heyman, & Renard, 2000). During the first days of development, the zygote undergoes several rounds of cell division without any increase in cell mass or cellular growth known as cleavage divisions and strictly depends on the large cytosolic compartment of the fertilized egg (Boroviak & Nichols, 2014).

As cleavage continues, embryos form a mass of cells (or blastomeres) packed tightly together into a morula. During compaction, the blastomeres of the morula increase their intercellular interactions, thereby providing the essential spatial queues for the first lineage decision in the mammalian embryo. This allows the establishment of differential compartments initially formulated as the "inside–outside hypothesis" (Boroviak & Nichols, 2014) which stands that, based on their position, the outer cells of the morula become the trophoblast (TE) and the inner cells give rise to the inner cell mass (ICM) of the blastocyst (Boroviak & Nichols, 2014; Lee, Choi, et al., 2014; Simmons & Cross, 2005; Yamanaka, Ralston, Stephenson, & Rossant, 2006). Nevertheless, another hypothesis proposes that cell fate is established earlier, at the eight-cell stage, by the establishment of cell polarity along the radius of the early morula. Subsequent cell divisions lead to symmetric or asymmetric distribution of polarity information, based on the angle of cell division. Polar and apolar cells would then differ in their developmental potential (Lee, Choi, et al., 2014; Madeja et al., 2013; Yamanaka et al., 2006). In farm animals, i.e., cows and pigs, the "inside–outside" hypothesis is not fulfilled as morula compaction, formation of tight junctions in the outer cells and ICM allocation occur independently from each other (Kirchhof et al., 2000; Van Soom et al., 1997). New research has been performed to abort this issue in the bovine species, and the results suggest that cell allocation in bovine is related to the subcellular localization of various transcription factors that determine ICM or TE phenotypes and seems to be started as early as the eight-cell stage (Madeja et al., 2013). Nevertheless, the exact mechanisms that trigger this protein redistribution which in term causes lineage determination in the bovine are still to be clarified.

Independently of the mechanism of segregation, blastocyst formation is the first irreversible event of morphological differentiation, that is, formation of ICM and TE (Babaie et al., 2007; Simmons & Cross, 2005); both tissues have different developmental potential and their derivatives are indeed different. As stated below, cell identities are regulated by the expression of different transcription factors which, in turn, regulate gene expression leading to differentiation. Several factors have been established as markers for ICM and TE cells. In mice, factors important for ICM specification, such as Oct4, Nanog, and Sox2, are restricted to the ICM. By contrast, the transcription factors important for TE specification are Cdx2, Eomes, and Gata3 and are restricted to the TE. The ICM and TE identities are incompatible with each other. In fact, Cdx2 and Oct4 reciprocally inhibit each other in this species (Bedzhov, Graham, Leung, & Zernicka-Goetz, 2014). In cattle, this inhibition is observed but later in development and is related to the ratio of protein concentrations (Berg et al., 2011). It is noticeable that although the transcription factors are the same, the signaling networks that regulate its expression are species-specific.

TE will form the ectodermic part of the placenta while ICM will, in turn, differentiate into the hypoblast (that forms the extraembryonic endoderm) and the epiblast (that gives rise to the embryo itself and to the extraembryonic mesoderm; Beddington & Robertson, 1999). The epiblast cells are considered as "naïve" pluripotent cells as the three embryonic layers and the germ line are derived from this tissue during gastrulation (Bedzhov et al., 2014).

The ICM marker Oct-4 is a POU domain transcription factor critical regulator of pluripotency in the mammalian embryo and is expressed in unfertilized oocytes, in the ICM and epiblasts of pregastrulation embryos, and in primordial germ cells (Babaie et al., 2007; Pesce, Gross, & Schöler, 1998). Localization of the protein is not restricted to the ICM in ungulates' embryos so it could not be considered as an ICM marker; still it could be considered a marker of undifferentiated state (Degrelle et al., 2005; Kirchhof et al., 2000; Pesce et al., 1998). As development continues, the blastocyst expands and hatches from the pellucid zone. From that time, trophectodermic cells could be directly in contact with maternal endometrium and so implantation could take place (Armant, 2005; Spencer, Johnson, Bazer, & Burghardt, 2004).

Starting from fertilization and ending with implantation, preimplantation embryo development can be divided into several distinct stages: fertilization, cell cleavage, morula, and blastocyst (Lee, Folger, Rajput, & Smith, 2014). These stages imply major morphogenetic transitions of the embryos: compaction, cavitation, expansion, and hatching (Kirchhof et al., 2000). All these spatiotemporally regulated processes often appear conserved, even though their timing can dramatically vary between species. To rely on conserved morphological traits rather than on time schedules (i.e., days postfertilization, p.i.), the use of developmental staging has provided worldwide standards for common classifications (Degrelle et al., 2011). For example, differences are evident when comparing rodents' (very early) and ruminants' (more delayed) timing of major embryonic genome activation (Memili & First, 1999; Watson et al., 1994), of entering the uterus, and of implantation (Degrelle et al., 2005; Kirchhof et al., 2000). It is remarkable that these timing differences of genetic and morphological transitions among species are not related to the duration of gestation but seem to be related to the length of the different implantation phases (hours in rodents to days in humans and domestic animals), the evolution of the cell–cell contacts, and the degree of endometrial invasion by the trophoblast (Spencer et al., 2004).

In fact, as ungulates, ruminants are considered "late implanting species". In these species, the trophoblast elongates exponentially to filamentous conceptuses that occupy the entire length of the uterine horn. The elongated or filamentous conceptus is composed mainly of trophectoderm lined with extraembryonic endoderm (Degrelle et al., 2005; Guillomot, Turbe, Hue, & Renard, 2004). Hatched blastocysts do not elongate *in vitro*, but do so when transferred into the uterus of either sheep or cows (Spencer, Sandra, & Wolf, 2008). All this points out that maternal–embryonic cross talk at the uterus is indispensable for development as the blastocyst is the maximum stage reached *in vitro*.

3. BMPs IN EMBRYO DEVELOPMENT

BMPs are implicated in reproduction at many phases. The roles of BMPs in every stage of preimplantation development will be treated in this chapter comparing a variety of mammalian models.

3.1 BMPs in the Gonads: Gamete Formation

In males, BMPs regulate spermatogenesis, germ cell differentiation, and epididymal integrity with different members expressed in different cell types during development as well as in the adulthood (Hu et al., 2004; Rossi & Dolci, 2013), i.e., during postnatal development, BMP4 is important for corpus region of the epididymis while BMP7 and BMP8 are important for caput and cauda (Hu et al., 2004). In the adult testis, BMP4 has been reported to be produced by spermatogonia, but not by Sertoli cells, suggesting that it might work as a paracrine–autocrine factor modulating the establishment of the cycle of the seminiferous epithelium (Rossi & Dolci, 2013).

In females, BMPs also play a role in follicular differentiation and physiology. Members of the superfamily are found in different cell types within the ovaries exerting different functions in rodents, humans, sheep, and cows (Knight & Glister, 2006). BMP4 is expressed first in stromal cells and later in follicular development, in theca cells. Its receptors are found in granulosa cells as well as in the oocyte itself (Fatehi et al., 2005; Glister, Fred Kemp, & Knight, 2004), and it is in charge of primordial to primary follicular transition and regulation of steroidogenesis in granulosa cells (Bristol-Gould & Woodruff, 2006). Other TGFβ members, such as BMP15 and GDF9, are expressed by the oocyte and are essential for follicular growth and oocyte maturation also in many species such as cats (Bristol-Gould & Woodruff, 2006), mice (Elvin, Yan, & Matzuk, 2000), rats (Erickson & Shimasaki, 2003), hens (Elis et al., 2007), goats (Silva, van den Hurk, van Tol, Roelen, & Figueiredo, 2005), and cattle (Juengel et al., 2009). The importance of BMPs for the female reproductive system has been evidenced by spontaneous mutations of these genes in sheep influencing ovulation rates (Montgomery, 2001). Increments as well as decrements of many reproductive traits have been observed in mutant animals, depending on the genes affected (Fabre et al., 2006; McNatty et al., 2004; Mulsant et al., 2001).

The effect of BMP4 during *in vitro* maturation (IVM) on subsequent development of bovine oocytes was assessed by our group, and the results will be discussed here (for details, see La Rosa et al., 2011).

3.1.1 Experimental Design

Cumulus–oocyte complexes (COCs) obtained from slaughtered animal ovaries were *in vitro* matured under three different conditions [BMP4 (mBMP4), Noggin (mNoggin), and Control (mControl)]. COCs were matured in the presence of BMP4 or Noggin. A Control group consisted of COCs matured without BMP4 or Noggin. Nuclear stage was analyzed and transcripts of different genes were quantified by RT-qPCR. Oocytes matured under the three treatments were then submitted to parthenogenic activation (PA) or *in vitro* fertilization (IVF). Oct-4 expression was studied in blastocysts by immunohistochemistry.

3.1.2 Nuclear Maturation and Gene Expression

The addition of BMP4 or Noggin to IVM medium did not affect nuclear maturation under our conditions with most oocytes (around 90%) in all three groups reaching metaphase II and extruding the first polar body. No differences in the relative amount of ZAR1, GDF9, BAX, MATER, and HSP70 transcripts were observed between mBMP4 and mControl groups, indicating no differential degradation or *de novo* synthesis of these messengers in oocytes matured with BMP4. On the other hand, an upregulation of MATER (mean factor: 2.484) and HSP70 (mean factor: 1.755) was observed in mNoggin oocytes (Fig. 1).

We found that whereas BMP4 and Noggin did not affect nuclear maturation, Noggin supplementation upregulated the expression of HSP70 and MATER messengers. The *hsp70* gene encodes a heat-shock protein that binds several proteins, and its overexpression is usually associated with stress (Kregel, 2002). In vascular tissues, Hsp70 protein can enhance BMP activity by binding BMP inhibitors, and it can be a possible link between cellular stress and BMP signaling (Yao, Watson, Ji, & Boström, 2009). From our finding, we speculate that the increase of HSP70 transcripts could be an effort by the cell to keep BMPs at adequate levels or to minimize possible stresses induced by Noggin. Recently, Clementi et al. observed that in two conditional knockout mice models, one for BMP2 and the other for a type I receptor (Alk 2), both phenotypes showed negatively affected immunophilins' expression in uterine cells (Clementi et al., 2013). Immunophilins posses a binding site for Hsp70 and Hsp90 which in turn can bind progesterone receptors, and altogether they form a complex. This complex regulates the intracellular trafficking and function of steroid hormone receptors in mice and humans (Clementi et al., 2013; Nair et al., 1997). It is known that during oocyte IVM, progesterone receptors' protein expression is increased in COCs with effects on subsequent IVF embryo

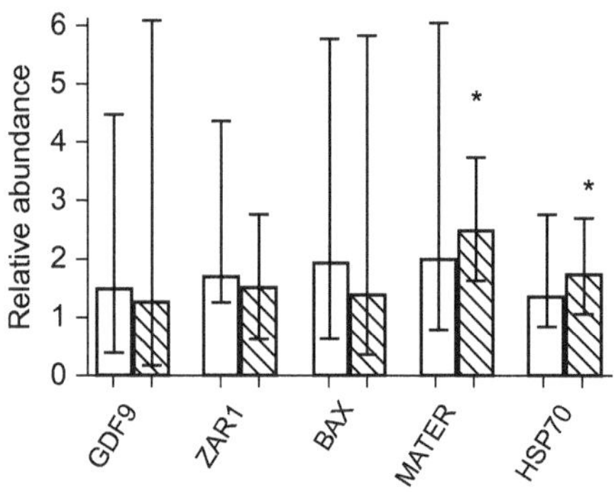

Figure 1 Gene expression in treated oocytes. Quantification of transcripts by RT-qPCR in BMP4- (□) and Noggin (■)-treated oocytes relative to Control oocytes (mean ± extreme ratios). Statistical differences with respect to Control are indicated by *(REST software). *MATER* (mean factor: 2.484) and *HSP70* (mean factor: 1.755).

development (Aparicio et al., 2011). Although not evaluated by our group, this new observation that BMP system inhibition affects progesterone actions in the uterus via regulation of the intracellular receptor complexes permits as to speculate another role for this increased HSP70 expression related to regulation of progesterone receptor functionality. Nevertheless, there are only few reports that link BMP signaling to heat-shock proteins, and so more information is needed to elucidate the exact roles of this important linkage in each system where they participate.

Mater (maternal antigen that embryo requires) is a maternal-effect protein that plays an essential role on early embryo development in the mouse (Tong et al., 2000), but its role in other species is not well known. Mota et al. reported no variation in *mater* gene expression between bovine oocytes with low and high competences (Mota et al., 2010). Whereas Pennetier et al. found that MATER mRNA amount decreases strongly during maturation (Pennetier et al., 2006), and Wood et al. found overexpression in oocytes from women with polycystic ovarian syndrome (Wood, Dumesic, Abbott, & Strauss, 2007). Our work found a higher relative abundance of MATER in oocytes matured with Noggin. This could be associated with a deficient cytoplasmic maturation; nevertheless, we did not find differences in blastocyst rates. On the contrary, in a recently published work, higher levels of MATER messengers in bovine oocytes have been associated with higher developmental rates after IVF (Urrego, Herrera-Puerta, Chavarria, & Camargo, 2015). These data demonstrate that is the cellular context rather than a particular value for a messenger level which finally defines the effects on bovine preimplantation embryos.

3.1.3 Embryo Development After Parthenogenic Activation or IVF

After parthenogenic activation, mBMP4 and mNoggin embryos had higher cleavage rates than mControl embryos though no effect on blastocyst production was found (Table 1). After IVF, mNoggin cleavage rate was lower than mBMP4 or mControl embryos. However, blastocyst rates were not affected indicating no effects of the treatments on advanced embryos. Similarly, the quantity of hatching embryos was not different among groups (Table 1).

No differences in embryonic cell numbers were found on day 2 (data not shown) in all groups, and blastocysts from the three groups had similar total cell numbers for PA and IVF (Table 1). These data suggest that the early kinetics of development of these embryos were not affected by BMP4 and Noggin during IVM.

Table 1 Effect of BMP4 and Noggin Added to *In Vitro* Maturation Medium on Development of Parthenogenic and *In Vitro* Fertilized Embryos

	Treatment	Total	Cleaved (%)	Blastocysts (%)	Hatching Blast. (%)	Blastocyst Cell Number ± SD
PA	Control	344	190 (55.2)^{b}	30 (8.7)		100 ± 33
	BMP4	274	180 (65.7)^{a}	22 (8.0)	1 (0.4)	88 ± 14
	Noggin	242	158 (65.3)^{a}	22 (9.0)		68 ± 8
IVF	Control	249	176 (70.7)$^{\alpha}$	35 (14.0)	5 (2.0)	90 ± 25
	BMP4	221	160 (72.4)$^{\alpha}$	28 (12.6)	4 (1.8)	120 ± 25
	Noggin	234	144 (61.5)$^{\beta}$	31 (13.2)	2 (0.8)	99 ± 8

Development of parthenogenic (PA) and *in vitro* fertilized (IVF) embryos produced from oocytes that were matured in the presence of 100 ng/ml of BMP4 or Noggin. Controls were matured in standard medium without supplementation. Different superscripts in PA or IVF indicate statistical differences (chi-square test, $p < 0.05$). For blastocyst cell numbers, one-way ANOVA was applied.

A higher proportion of Oct-4-expressing cells over total cells were observed in mBMP4-PA blastocysts compared with mControl and mNoggin PA blastocysts. No differences among groups were found for IVF embryos.

Supplementation of media with BMP4 or its inhibitor during IVM had different effects on early embryonic development depending on whether the embryos were parthenogenic or *in vitro* fertilized with no effect on subsequent development. Similar results were published by Fatehi et al. only for maturation with BMP4 prior to IVF (Fatehi et al., 2005). However, despite blastocysts' rates were similar, a higher proportion of Oct-4-expressing cells were observed in parthenogenic blastocysts obtained from oocytes matured with exogenous BMP4. In a recent work, García et al. found an increment in the relative messenger abundance of OCT-4 in eight-cell bovine embryos as well as in blastocysts after treatment of the embryos with BMP5 during the first 2 days of culture. Although they have not analyzed protein levels, this also demonstrates an effect of a BMP factor on stimulating later Oct-4 expression (García, Valdecantos, Barrera, Roldán-Olarte, & Miceli, 2014).

3.2 BMPs in the Oviduct

The oviduct is divided into three anatomical regions: the fimbria, the ampulla, and the isthmus. Each of them has particular characteristics and exerts different functions. Moreover, this organ undergoes physiological changes during the estrous cycle (García, Valdecantos, et al., 2014). Secretory cells lining the epithelium show numerous microvilli on their apical side and secrete products composing the oviductal fluid (Ulbrich, Zitta, Hiendleder, & Wolf, 2010). The composition of this fluid has been a matter of extensive research due to the importance for *in vitro* embryo culture. In this sense, Watson et al. demonstrated the expression of several growth factors by bovine and ovine epithelial oviductal cells in primary cultures. Within these growth factors, they found TGFβ mRNA expression (Watson, Hogan, Hahnel, Wiemer, & Schultz, 1992; Watson et al., 1994). Some years later, Chow found the presence of TGFβ type I and II receptors in mice oviductal cells, evidencing the importance of this signaling system in embryo–maternal cross talk (Chow, Lee, Chan, & Yeung, 2001). Different members of the BMP family are expressed in the bovine oviduct during the estrous cycle. RT-PCR analysis showed expression of BMP2, 3, 4, 7, 10, and 15 mRNA in oviduct epithelial cells from ampulla and isthmus. Interestingly, only BMP5 was exclusively detected in isthmus epithelial cells, showing a differential expression in the anatomic regions of the bovine oviduct. Regarding the estrous cycle, highest expression of BMP2, 4, and 7 factors was detected during the preovulatory stage. Considering that this phase is characterized by growth and differentiation of the oviduct mucosal cells to prepare for the presence of gametes and embryos, this suggests possible participation of these factors in the events around the time of ovulation. During the postovulatory and mid-luteal phase, gene expression of BMP5 in the isthmus gradually decreased, a similar regulation was found for BMP7. In contrast, gene expression of BMP2 and 4 decreased during the same phases in the ampulla but remained high in the isthmus. Nevertheless, when the oviductal fluids from the three phases were analyzed by Western blot, protein levels of BMP5 were found to stay relatively constant during the estrous cycle (García, Valdecantos, et al., 2014).

3.3 BMPs in Preimplantation Embryo Development

The effects of BMP4 and its inhibitor Noggin during *in vitro* culture of bovine embryos were assessed by our group. Although BMP4 has been well studied during organogenesis and extraembryonic differentiation, BMP4 function during *in vitro* embryo development to the blastocyst stage had not been previously studied in the bovine species.

3.3.1 Experimental Design

Embryos were produced by PA or IVF and then cultured in CR2 medium (cControl) or with the addition of BMP4 (cBMP4) or Noggin (cNoggin). To evaluate embryo developmental parameters, numbers of cleaved embryos and their cell numbers were recorded on day 2 and numbers of blastocysts and their cell numbers were recorded on day 9 postactivation/ fertilization. Oct-4 expression was studied in blastocysts by immunohistochemistry.

3.3.2 BMP4 or Noggin During Embryo Development *In Vitro*

In PA embryos, the cleavage rate and blastocyst formation were reduced by Noggin when compared to cBMP4 and cControl embryos (Table 2), showing the importance of the BMP system for early stages of development. No difference in blastocyst total cell number was found between groups (Table 2).

During culture of *in vitro* fertilized embryos, Noggin negatively affected cleavage, blastocyst, and hatching rates while BMP4 addition reduced blastocyst and hatching rates of IVF embryos (Table 2).

No differences in cell number of IVF embryos on day 2 of culture were seen, so neither acceleration nor retardation in development was observed. Similar to PA embryos, no differences in total cell number of IVF blastocysts were found among groups (Table 2).

Table 2 Effect of BMP4 and Noggin Added to Culture Medium on the Development of Parthenogenic and *In Vitro* Fertilized Embryos

	Treatment	Total	Cleaved (%)	Blastocysts (%)	Hatching Blast. (%)	Blastocyst Cell Number ± SD
PA	Control	354	237 (66.9)^{a}	48 (13.5)^{a}		68 ± 33
	BMP4	295	199 (67.4)^{a}	44 (14.9)^{a}	1 (0.3)	91 ± 43
	Noggin	269	154 (57.2)^{b}	19 (7.0)^{b}		71 ± 16
IVF	Control	218	138 (63.3)$^{\alpha}$	45 (20.6)$^{\alpha}$	10 (4.6)$^{\alpha}$	130 ± 47
	BMP4	217	146 (61.3)$^{\alpha\beta}$	22 (9.2)$^{\beta}$	3 (1.4)$^{\beta}$	117 ± 52
	Noggin	205	105 (51.2)$^{\beta}$	24 (11.7)$^{\beta}$	1 (0.5)$^{\beta}$	128 ± 21

Development of parthenogenic (PA) and *in vitro* fertilized (IVF) embryos cultured with 100 ng/ml of either BMP4 or Noggin. Controls were cultured without supplementation. Different superscripts in PA or IVF indicate statistical differences (chi-square test, $p < 0.05$). For blastocyst cell numbers, one-way ANOVA was applied.

When BMP4 and Noggin were included in the embryo culture medium, we found that Noggin decreased cleavage, blastocyst, and hatching rates of both PA and IVF embryos, showing the importance of BMP system for embryo development.

Proportions of Oct-4-expressing cells in IVF embryos were lower in cBMP4 and cNoggin groups than in cControls (Fig. 2). No differences were found in this proportion between treatments in PA embryos. The protein was localized in the ICM as well as in the trophoblast for all blastocysts.

In mice, Murohashi observed a lower ratio of MCI-derived cells over TE-derived cells in blastocysts previously treated with Noggin when compared with untreated embryos (Murohashi et al., 2010). Also in mice, Noggin supplementation of embryonic stem cells (ESCs) during 4 days resulted in an upregulated expression of Cdx2, an Oct-4 repressor

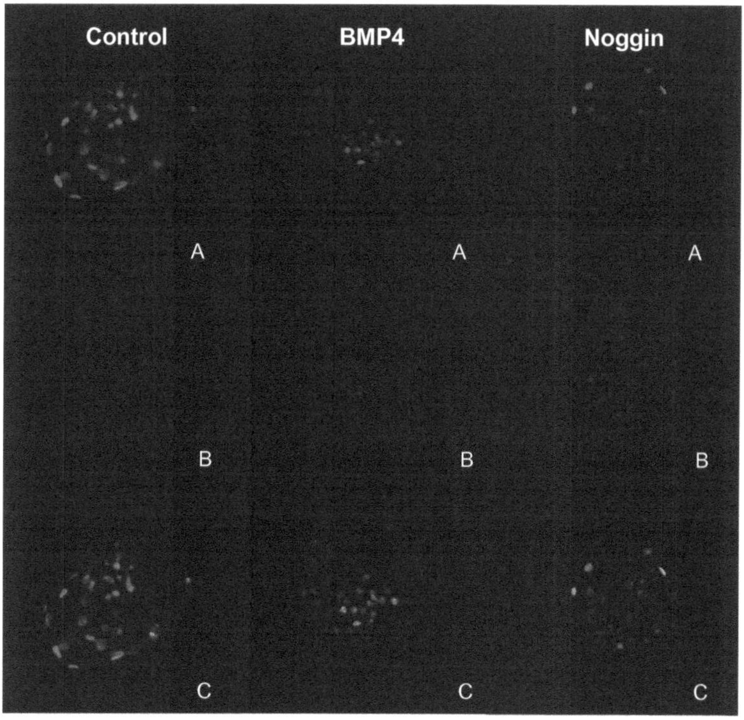

Figure 2 Oct-4 expression in bovine blastocysts. Oct-4 immunostaining of blastocysts cultured with 100 ng/ml of either BMP4 or Noggin; Controls were cultured without supplementation. Confocal microscopy, augmentation: 20 × and zoom: 2. (A) Oct-4-positive cells (green), (B) total nuclei (red) and (C) Oct-4-positive cells and total nuclei (merged). (See the color plate.)

(Peng, Hua, Cao, & Wang, 2010). In human ESCs, OCT-4 knockdown increases expression of BMP4 and its downstream target gene ID2, as well as reduces its antagonist CHORDIN expression (Babaie et al., 2007). These results evidence a negative feedback between Oct-4 and BMP4 in humans. It could be possible that our results were related to similar effects on the pluripotent capacity of bovine embryos. In this sense, an unbalanced BMP signaling during *in vitro* embryo development (i.e., with exogenous BMP4 or with Noggin) provokes changes in Oct-4 protein expression.

Although BMP4 addition to the culture medium did not affect cleavage rates for PA or IVF embryos, the rate of blastocyst production was reduced for IVF embryos, but not for PA embryos. These results suggest differences, in the BMP system, between PA and IVF embryos. This could be caused by the different imprinting pattern of both types of embryos (Ruddock et al., 2004). PA embryos only contain maternal genome which could, in turn, cause differences at many levels, e.g., expression of BMP ligands, receptors, inhibitors, and intracellular mediators.

The results obtained for IVF embryos are in concordance with Lee et al. who observed that members of the BMP signaling subfamily, such as BMP4, BMPR1A, and ID3, were highly expressed in degenerated embryos compared with blastocysts (Lee, Folger, et al., 2014). Although is not possible to ensure if it is the cause or a consequence of the degenerated condition, it is clear that BMP signaling system is playing a crucial role in bovine early embryo development. Another work supporting this statement is that of García et al. who have evaluated the effect of early supplementation of BMP5 on bovine embryo development and its influence in the expression of several genes. They found that supplementation of culture media with BMP5 up to day 2 increased blastocyst rate of the embryos. Also, increments in the relative abundance of certain transcripts have been observed in BMP5-treated embryos at different developmental stages, i.e., SOX2, DNA methyl transferase (DNMT) 1, and DNMT3B in two-cell stage embryos; ID2 and OCT4 in eight-cell stage embryos; and NANOG, OCT4, and DNMT1 in blastocysts. These results demonstrate that bovine embryos are affected by the presence of BMP5 in the culture media with effects on blastocyst rate and messenger expression of genes related to BMP pathway, pluripotency, and DNA methylation (García, Valdecantos, et al., 2014). This, together with previous observations that BMP5 is differentially expressed in the oviduct during the estrous cycle (see Section 3.2), suggests an important role of this factor and represents another example of the participation of this signaling system in embryo preimplantation development in the bovine species.

In mice only very recent works have studied the possible roles of BMPs during preimplantation development. Graham et al. studied the expression of BMPs in early mice embryos by analyzing the complete transcriptomes of 16-cell stage inside and outside cells of the embryos and have identified that components of the BMP signaling pathway are already differentially expressed at this early stage. Most notably, inside cells express the ligands BMP4 and BMP7 and outside cells express the type II receptor BMPR2. Differential expression of BMP ligands and receptors in different preimplantation cell lineages has, however, been observed before, with BMP4 and BMP7 found in the ICM and BMPR2 in the TE at the blastocyst stage. These authors show the same pattern of differential expression in the first distinct precursors of these lineages. They not only demonstrate gene expression but also show that nuclear pSmad1 is present already at the blastocyst stage, which indicates that the signaling is actually active (Graham et al., 2014). This group also developed an interesting strategy to determine the importance of active BMP system for preimplantation development. They inhibited the BMP signaling in half of the embryo from the two-cell stage; this resulted in a significant reduction in the number of TE and primitive endoderm (PE) cells in the injected clone, but did not affect the embryonic epiblast, suggesting that BMP signaling is important for the development of these two extraembryonic lineages. In further research, they knocked down expressions of Bmp4 and Bmp7 (ligands that are expressed exclusively by inside cells at the 16-cell stage and by the ICM), and they observed a dramatic reduction in the number of extraembryonic cells of TE and PE. This suggests that these inside-produced signaling molecules are the predominant ligands involved in extraembryonic development. So they concluded that BMP signaling system is important for extraembryonic tissue development (Graham et al., 2014). Later, Reyes de Mochel found that BMP signaling activity can be detected in all blastomeres as early as the four-cell stage and becomes spatially restricted by late blastocyst stage with ICM cells showing more intense pSmad1 signal than TE. Perturbation of BMP signaling in embryos by treatment with a pharmacologic inhibitor, by Noggin protein, or by overexpression of a dominant-negative BMP receptor, each delayed cell cleavage with no evidence of apoptosis. Based on these results, they propose that BMP signaling is required for normal cell cycle during development of the preimplantation mouse embryo (Reyes De Mochel et al., 2015).

3.4 BMPs in the Uterus

Mammalian embryos enter the uterus at the late morula stage. After hatching, primate and rodent embryos begin the implantation process while in ungulates this process occurs relatively later when embryos have reached the gastrula stage (Guillomot et al., 2004; Spencer et al., 2004). In those species different from primates and rodents, the polar trophoblast, also called "Rauber's layer," quickly disappears after blastocyst expansion and hatching. Subsequently, the trophoblast grows and differentiates from the cells which surround the blastocoelic cavity but not from the ones which cover the ICM. In ruminants, the trophoblast elongates exponentially and reaches more than 150 mm before implantation (Degrelle et al., 2005; Guillomot et al., 2004; Wintenberg-Torres & Flechòn, 1974). The same developmental pattern is observed in the porcine species (Blomberg, Hashizume, & Viebahn, 2008; Kirchhof et al., 2000). In ruminants, intense trophoblastic cell multiplication precedes cellular differentiation, and early elongation relies on cell growth and cell shape remodeling (Degrelle et al., 2005; Spencer & Bazer, 2004). On the other hand, Valdez Magaña et al. did not find evidence for such intense proliferation previous to shape remodeling in pigs (Valdez Magaña, Rodríguez, Zhang, Webb, & Alberio, 2014). As elongation of the spherical blastocyst continues, it develops first into an ovoid, then a tubular, and then into a filamentous conceptus (Wintenberg-Torres & Flechòn, 1974). Elongation of ruminant blastocyst is critical for developmentally regulated production of interferon tau (IFNt), a type I IFN that is the signal for maternal recognition of pregnancy in ruminants and acts in a paracrine manner on the endometrial epithelia to inhibit development of the luteolytic mechanism (Guillomot, Reinaud, La Bonnardiere, & Charpigny, 1998; Spencer & Bazer, 2004).

The structure of the endometrium has common features in all species. The endometrial mucosa is formed by a mono- or pseudostratified epithelium which is separated from the conjunctive stroma by a basal lamina. The stroma is highly vascularized and contains coiled and branched glands whose ducts open into the uterine lumen. The endometrial surface epithelium is composed of secretory cells with microvilli and ciliated cells, the latter being concentrated around the openings of the endometrial glands (Spencer et al., 2004). Progesterone induces differentiative events within all compartments of the endometrium, creating an environment receptive for blastocyst attachment and invasion. Endometrial epithelial glands undergo morphological and functional differentiation and commence active secretion of a complex nutritive and growth factor-rich media contributing to uterine fluid. This provides support to the preimplantation embryo, promoting growth and development before endometrial attachment (Jones, Stoikos, Findlay, & Salamonsen, 2006). Moreover, in species with relatively delayed implantation, the uterine fluid, also known as histotroph, plays a central role in maintaining embryo development for a longer period than in early implantation species (Wintenberg-Torres & Flechòn, 1974).

Although the stages of implantation are evolutionarily conserved, there are many differences in the process itself among species (Armant, 2005). As expressed before, differences could be related to the grade of invasion in maternal tissues which in turn differs with the type of placenta that will form a given species. In mammals with hemochorial placenta, the trophoblast invades the endometrium and is perfused with maternal blood (Arck, Hansen, Jericevic, Piccinni, & Szekeres-Bartho, 2007). In these species such as some primates, rodents, and some bats, stromal fibroblasts surrounding the developing spiral arterioles of the uterus begin to differentiate, or decidualize, eventually producing the decidua of pregnancy (Jones et al., 2006; Mansouri-Attia, Aubert, et al., 2009). On the contrary on those species with epitheliochorial placenta such as ungulates, trophectoderm cells are in contact to the epithelium of the endometrium, and so invasion is not deeper than fusion with epithelial cells (Arck et al., 2007; Bauersachs & Wolf, 2015). Interestingly, this evolutionary strategy of noninvasive placenta is a specialization acquired to solve the immunological problem of viviparity (Arck et al., 2007).

In species with epitheliochorial placenta, the uterus does not decidualize nevertheless is modified and differentiated by hormonal influences and by the presence of the embryo (Bazer, Burghardt, Johnson, Spencer, & Wu, 2008; Mansouri-Attia, Sandra, et al., 2009). In ruminants, endometrial glands are specifically localized in large endometrial areas, the intercaruncular zones, while small aglandular areas of stromal origin, the caruncles, are scattered over the endometrium surface. The caruncular areas form the maternal part of the placenta and fuse with the cotyledons in the conceptus forming the placentomes (Mansouri-Attia, Sandra, et al., 2009). In horses, the preimplantation phase is particularly long because placentation starts after day 42 of gestation. The equine conceptus is completely enveloped by a tough glycoprotein capsule between days 7 and 21 of gestation. The conceptus shows a constant, self-induced mobility throughout the uterine lumen which has been shown to be essential for maternal recognition of pregnancy. After day 16, the conceptus becomes immobilized at the base of one of the uterine horns by uterine contraction. Another unusual feature is the formation of the equine-specific endometrial cups beginning on day 35 which implies invasion of the chorionic girdles into the endometrial epithelium. The function of these structures is the secretion of equine chorionic gonadotropin (eCG) by binucleated trophoblastic cells (BNC) into the maternal circulation to further increase the production of progesterone in the ovaries (Bauersachs & Wolf, 2015).

In mice, TGFβ ligands and receptors are expressed in the epithelial and stromal cells of the uterus (Jones et al., 2006), indicating that this system participates in the embryo–maternal cross talk necessary for successful of pregnancy. In this regard, BMP2 is considered one of the master regulators of implantation also in humans (Clementi et al., 2013; Wetendorf & De Mayo, 2013). At the same time, other BMPs are expressed in murine uterus during the peri-implantation period (Clementi et al., 2013; Li et al., 2014). Nevertheless, species-specific differences in uterine BMP system are evidenced by a different expression pattern for Noggin. While it is expressed in stromal cells of murine uterus, it could not be detected in bovine endometrium around the time of implantation (Pennington & Ealy, 2012).

3.4.1 BMPs in Peri-Implantation Conceptus Development

Suzuki et al. have established trophoblastic cell lines from bovine blastocysts and observed that both the attachment and the segregation of the cells were improved when blastocysts had been previously cultured in the presence of BMP4 (Suzuki et al., 2011). This demonstrates that BMP4 affects the developmental capacity of the embryos.

Degrelle et al. found that BMP4 transcripts are detected in the extraembryonic tissues of bovine peri- and postimplantation conceptuses (from day 14 to day 25 p.i.). BMP4 expression is temporally regulated, and transcripts' levels increase as embryo development progresses. Furthermore, the author's hypothesis is that BMP4 level of expression in the studied tissues could serve as a discriminative marker for different developmental stages of conceptuses in the bovine species (Degrelle et al., 2011).

Later, and coincident with Degrelle et al., Pennington and Ealy describe the expression and potential actions of the BMP4 ligand and receptor system during the pre- and periattachment period of bovine conceptus development (day 17 p.i.). They have also studied this system in a bovine trophoblast cell line that produces IFNT but not BNC phenotype (CT1 cells). They have observed that transcripts for BMP2 and BMP4 were readily detectable in elongated conceptuses and CT1 cells. On the other hand, transcripts for

NOGGIN were detected in elongated conceptuses but were either absent or detected at very low levels in CT1 cells. Transcripts for various BMP2 and 4 receptors' subtypes (BMPR1A, BMPR1B, and BMPRII) also were detected in conceptuses and CT1 cells. When CT1 cells were treated with BMP2 or BMP4, the authors observed that BMP2 did not affect viable cell numbers while supplementation with BMP4 reduced CT1 cell number after 48 h of treatment. Cells treated with BMP4 did not display any overt microscopic evidence of apoptosis or necrosis (detached, swollen/shriveled, or punctuated cells). Nevertheless, studies to quantify apoptosis and proliferation rates were not completed in this work. Morphological and functional aspects were evaluated on treated cells. Microscopic examination of the cells in various studies failed to detect changes in the incidence of BNC formation. At the same time, exposure to BMP2 or 4 decreased the relative abundance of IFNT mRNA. Although the functional experiments were performed in a cell line cultured *in vitro*, the results strongly suggest that BMP system affects conceptus development at the time of peri-implantation (Pennington & Ealy, 2012).

In pigs, BMP4 protein is first detected by immunohistochemistry at the border between the epiblast and the TE of early ovoid embryos. In late ovoid embryos, the signal is more evident in the epiblast and extended beyond the limits of the embryonic disc in tubular and filamentous embryos. BMP2 is also detected from the ovoid stage and shows strong expression primarily in epiblast cells and some staining in the TE. Interestingly, phosphorylated Smad 1/5/8 signal is found in the TE of elongating porcine embryos, and also TE is able to respond to exogenous BMP4 *in vitro*. In summary, it has been shown that BMP4 produced by the embryo proper signal to the TE prior to elongation. Furthermore, the response to these signals seems to be different to that previously described for mice. This represents a complete new finding as it had been proposed that the epiblast, at the stage of embryo elongation, was too small to be able to produce a quantitative effect on trophoblast growth (Valdez Magaña et al., 2014).

An interesting study has been performed in equine conceptuses around the time of formation of endometrial cups (around day 30 of gestation). The authors found that chorionic girdle cells expressed BMP4 receptors *in vivo* and functionally differentiated in response to BMP4 *in vitro*. Moreover, they found that BMP4 was mainly expressed in the chorion and the yolk sac, *in vivo*. Finally, BMP4 treatment *in vitro* stimulates eCG secretion by cultured cells. Notably, this stimulation occurs in a very narrow period of time in which cells are sensitive to the factor. Their results suggest a role for

BMP4 in driving phenotypic change in the trophoblast, therefore leading to the attainment of a particular differentiation state that induces eCG expression instead of being, specifically, a direct inducer of eCG gene expression and/or secretion (Cabrera-Sharp et al., 2014).

4. CONCLUDING REMARKS

The reviewed information demonstrates a very important role for BMP signaling system at different stages of embryo preimplantation development from acquisition of gamete competence to regulation of trophoblast development and differentiation in mice as well as in ungulates.

Interestingly, this system participates in both sides of the embryo–maternal cross talk. As it has been shown, ligands and receptors are expressed by the epithelial cells of the oviduct as well as the embryo itself from very early stages. Exactly the same occurs in the uterus where members of this family are expressed by the uterus and by the blastocyst. And this seems to be conserved among species. Furthermore, in those species with relatively longer preimplantation periods, the system is implicated also in embryonic–extraembryonic communication and regulates trophoblast elongation and maternal recognition of pregnancy.

ACKNOWLEDGMENTS

The author wants to thank the University of Buenos Aires where this work has been developed. And especially thanks all the coauthors of the manuscript (La Rosa et al., 2011): Dr. Michelle Munk Pereira, Dr. Luiz Sergio Camargo, Dr. Dante Paz, Dr. Rafael Fernandez y Martín, and Dr. Daniel Salamone. Also thanks Ms. Vanesa García for unconditional sharing of information.

REFERENCES

Aparicio, I. M., Garcia-Herreros, M., O'Shea, L. C., Hensey, C., Lonergan, P., & Fair, T. (2011). Expression, regulation, and function of progesterone receptors in bovine cumulus oocyte complexes during in vitro maturation. *Biology of Reproduction*, *84*(5), 910–921. http://dx.doi.org/10.1095/biolreprod.110.087411.

Arck, P., Hansen, P. J., Jericevic, B. M., Piccinni, M. P., & Szekeres-Bartho, J. (2007). Progesterone during pregnancy: Endocrine-immune cross talk in mammalian species and the role of stress. *American Journal of Reproductive Immunology*, *58*, 268–279. http://dx.doi.org/10.1111/j.1600-0897.2007.00512.x.

Armant, D. R. (2005). Blastocysts don't go it alone. Extrinsic signals fine-tune the intrinsic developmental program of trophoblast cells. *Developmental Biology*, *280*(2), 260–280. http://dx.doi.org/10.1016/j.ydbio.2005.02.009.Blastocysts.

Babaie, Y., Herwig, R., Greber, B., Wasco, T. C., Brink, W., Groth, D., et al. (2007). Analysis of Oct4-dependent transcriptional networks regulating self-renewal and pluripotency in human embryonic stem cells. *Stem Cells (Dayton, Ohio)*, *25*(2), 500–510. http://dx.doi.org/10.1634/stemcells.2006-0426.

Bauersachs, S., & Wolf, E. (2015). Uterine responses to the preattachment embryo in domestic ungulates: Recognition of pregnancy and preparation for implantation. *Annual Review of Animal Biosciences*, *3*, 489–511. http://dx.doi.org/10.1146/annurev-animal-022114-110639.

Bazer, F. W., Burghardt, R. C., Johnson, G. A., Spencer, T. E., & Wu, G. (2008). Interferons and progesterone for establishment and maintenance of pregnancy: Interactions among novel cell signaling pathways. *Reproductive Biology*, *8*(3), 179–211. http://www.ncbi.nlm.nih.gov/pubmed/19092983.

Beddington, R. S. P., & Robertson, E. J. (1999). Axis development and early asymmetry in mammals. *Cell*, *96*(2), 195–209. http://dx.doi.org/10.1016/S0092-8674(00)80560-7.

Bedzhov, I., Graham, S. J. L., Leung, C. Y., & Zernicka-Goetz, M. (2014). Developmental plasticity, cell fate specification and morphogenesis in the early mouse embryo. *Philosophical Transactions of the Royal Society of London. Series B: Biological Sciences*, *369*(1657). http://dx.doi.org/10.1098/rstb.2013.0538.

Berg, D. K., Smith, C. S., Pearton, D. J., Wells, D. N., Broadhurst, R., Donnison, M., et al. (2011). Trophectoderm lineage determination in cattle. *Developmental Cell*, *20*(2), 244–255. http://dx.doi.org/10.1016/j.devcel.2011.01.003, Elsevier.

Blomberg, L., Hashizume, K., & Viebahn, C. (2008). Blastocyst elongation, trophoblastic differentiation, and embryonic pattern formation. *Reproduction (Cambridge, England)*, *135*(2), 181–195. http://dx.doi.org/10.1530/REP-07-0355.

Boroviak, T., & Nichols, J. (2014). The birth of embryonic pluripotency. *Philosophical Transactions of the Royal Society of London. Series B: Biological Sciences*, *369*(1657). http://dx.doi.org/10.1098/rstb.2013.0541.

Bristol-Gould, S., & Woodruff, T. K. (2006). Folliculogenesis in the domestic cat (Felis catus). *Theriogenology*, *66*(1), 5–13. http://dx.doi.org/10.1016/j.theriogenology.2006.03.019.

Brunet, L. J. (1998). Noggin, cartilage morphogenesis, and joint formation in the mammalian skeleton. *Science*, *280*(5368), 1455–1457. http://dx.doi.org/10.1126/science.280.5368.1455.

Cabrera-Sharp, V., Read, J. E., Richardson, S., Kowalski, A. A., Antczak, D. F., Cartwright, J. E., et al. (2014). SMAD1/5 signaling in the early equine placenta regulates trophoblast differentiation and chorionic gonadotropin secretion. *Endocrinology*, *155*(8), 3054–3064. http://dx.doi.org/10.1210/en.2013-2116.

Chen, D., Zhao, M., & Mundy, G. R. (2004). Bone morphogenetic proteins. *Growth Factors (Chur, Switzerland)*, *22*(4), 233–241. http://dx.doi.org/10.1080/08977190412331279890.

Chow, J. F. C., Lee, K.-f., Chan, S. T. H., & Yeung, W. S. B. (2001). Quantification of transforming growth factor β1 (TGFβ1) mRNA expression in mouse preimplantation embryos and determination of TGF B receptor (type I and type II) expression in mouse embryos and reproductive tract. *Molecular Human Reproduction*, 7(11), 1047–1056.

Clementi, C., Tripurani, S. K., Large, M. J., Edson, M. A., Creighton, C. J., Hawkins, S. M., et al. (2013). Activin-like kinase 2 functions in peri-implantation uterine signaling in mice and humans. *PLoS Genetics*, *9*(11), e1003863. http://dx.doi.org/10.1371/journal.pgen.1003863.

Comizzoli, P., Marquant-Le Guienne, B., Heyman, Y., & Renard, J. P. (2000). Onset of the first S-phase is determined by a paternal effect during the G1-phase in bovine zygotes. *Biology of Reproduction*, *62*(6), 1677–1684. http://www.ncbi.nlm.nih.gov/pubmed/10819771.

Degrelle, S. A., Campion, E., Cabau, C., Piumi, F., Reinaud, P., Richard, C., et al. (2005). Molecular evidence for a critical period in mural trophoblast development in bovine blastocysts. *Developmental Biology*, *288*(2), 448–460. http://dx.doi.org/10.1016/j.ydbio.2005.09.043.

Degrelle, S. A., Lê Cao, K.-A., Heyman, Y., Everts, R. E., Campion, E., Richard, C., et al. (2011). A small set of extra-embryonic genes defines a new landmark for bovine embryo staging. *Reproduction (Cambridge, England)*, *141*(1), 79–89. http://dx.doi.org/10.1530/REP-10-0174.

Elis, S., Dupont, J., Couty, I., Persani, L., Govoroun, M., Blesbois, E., et al. (2007). Expression and biological effects of bone morphogenetic protein-15 in the Hen ovary. *The Journal of Endocrinology*, *194*(3), 485–497. http://dx.doi.org/10.1677/JOE-07-0143.

Elvin, J. A., Yan, C., & Matzuk, M. M. (2000). Oocyte-expressed TGF-B superfamily members in female fertility. *Molecular and Cellular Endocrinology*, *159*(1–2), 1–5.

Erickson, G. F., & Shimasaki, S. (2003). The spatiotemporal expression pattern of the bone morphogenetic protein family in rat ovary cell types during the estrous cycle. *Reproductive Biology and Endocrinology*, *1*, 9.

Fabre, S., Pierre, A., Mulsant, P., Bodin, L., Di Pasquale, E., Persani, L., et al. (2006). Regulation of ovulation rate in mammals: Contribution of sheep genetic models. *Reproductive Biology and Endocrinology*, *4*, 20. http://dx.doi.org/10.1186/1477-7827-4-20.

Fatehi, A. N., van den Hurk, R., Colenbrander, B., Daemen, A. J. J. M., van Tol, H. T. A., Monteiro, R. M., et al. (2005). Expression of bone morphogenetic protein2 (BMP2), BMP4 and BMP receptors in the bovine ovary but absence of effects of BMP2 and BMP4 during IVM on bovine oocyte nuclear maturation and subsequent embryo development. *Theriogenology*, *63*(3), 872–889. http://dx.doi.org/10.1016/j.theriogenology.2004.05.013.

Florman, H. M., & Ducibella, T. (2006). Fertilization in mammals. In J. D. Neill (Ed.), *Knobil and Neill's physiology of reproduction* (3rd ed., pp. 55–111): Elsevier.

García, E. V., Rizo, G., Valdecantos, P. A., Miceli, D. C., & Barrera, D. (2014). Early supplementation of bmp5 to embryo culture medium and its effect on in vitro development and gene expression of bovine embryos. In *Second annual meeting of the Argentinian Embryo Transfer Society (SATE)*, Buenos Aires, Argentina.

García, E. V., Valdecantos, P. A., Barrera, D., Roldán-Olarte, M., & Miceli, D. C. (2014). Bone morphogenetic proteins in the bovine oviduct: Differential expression of BMP-5 in the isthmus during the estrous cycle. *Theriogenology*, *81*(8), 1032–1041. http://dx.doi.org/10.1016/j.theriogenology.2014.01.025, Elsevier.

Glister, C., Fred Kemp, C., & Knight, P. G. (2004). Bone morphogenetic protein (BMP) ligands and receptors in bovine ovarian follicle cells: Actions of BMP-4, -6 and -7 on granulosa cells and differential modulation of Smad-1 phosphorylation by follistatin. *Reproduction (Cambridge, England)*, *127*(2), 239–254. http://dx.doi.org/10.1530/rep.1.00090.

Graham, S. J. L., Wicher, K. B., Jedrusik, A., Guo, G., Herath, W., Robson, P., et al. (2014). BMP signalling regulates the pre-implantation development of extra-embryonic cell lineages in the mouse embryo. *Nature Communications*, *5*, 5667. http://dx.doi.org/10.1038/ncomms6667, Nature Publishing Group.

Groppe, J., Greenwald, J., Wiater, E., Rodriguez-leon, J., Economides, A. N., Kwiatkowski, W., et al. (2002). Structural basis of BMP signalling inhibition by the cystine knot protein Noggin. *Nature*, *420*, 636–642.

Guillomot, M., Reinaud, P., La Bonnardiere, C., & Charpigny, G. (1998). Characterization of conceptus-produced goat interferon and analysis of its temporal and cellular distribution during early pregnancy. *Journal of Reproduction and Fertility*, *112*, 149–156.

Guillomot, M., Turbe, A., Hue, I., & Renard, J.-P. (2004). Staging of ovine embryos and expression of the T-Box genes Brachyury and Eomesodermin around gastrulation. *Reproduction (Cambridge, England)*, *127*(4), 491–501. http://dx.doi.org/10.1530/rep.1.00057.

Hollnagel, A., Oehlmann, V., Heymer, J., Ru, U., Nordheim, A., & Tu, D. (1999). Id genes are direct targets of bone morphogenetic protein induction in embryonic stem cells. *The Journal of Biological Chemistry*, *274*(28), 19838–19845.

Hu, J., Chen, Y.-X., Wang, D., Qi, X., Li, T.-G., Hao, J., et al. (2004). Developmental expression and function of Bmp4 in spermatogenesis and in maintaining epididymal integrity. *Developmental Biology*, *276*(1), 158–171. http://dx.doi.org/10.1016/j.ydbio.2004.08.034.

Huminiecki, L., Goldovsky, L., Freilich, S., Moustakas, A., Ouzounis, C., & Heldin, C.-H. (2009). Emergence, development and diversification of the TGF-beta signalling pathway within the animal kingdom. *BMC Evolutionary Biology*, *9*, 28. http://dx.doi.org/10.1186/1471-2148-9-28.

Jones, R. L., Stoikos, C., Findlay, J. K., & Salamonsen, L. A. (2006). TGF-beta superfamily expression and actions in the endometrium and placenta. *Reproduction (Cambridge, England)*, *132*(2), 217–232. http://dx.doi.org/10.1530/rep.1.01076.

Juengel, J. L., Hudson, N. L., Berg, M., Hamel, K., Smith, P., Lawrence, S. B., et al. (2009). Effects of active immunization against growth differentiation factor 9 and/or bone morphogenetic protein 15 on ovarian function in cattle. *Reproduction (Cambridge, England)*, *138*(1), 107–114. http://dx.doi.org/10.1530/REP-09-0009.

Kirchhof, N., Carnwath, J. W., Lemme, E., Anastassiadis, K., Scho, H., & Niemann, H. (2000). Expression pattern of Oct-4 in preimplantation embryos of different species. *Biology of Reproduction*, *1705*, 1698–1705.

Knight, P. G., & Glister, C. (2006). TGF-beta superfamily members and ovarian follicle development. *Reproduction (Cambridge, England)*, *132*(2), 191–206. http://dx.doi.org/10.1530/rep.1.01074.

Knockaert, M., Sapkota, G., Alarcón, C., Massagué, J., & Brivanlou, A. H. (2006). Unique players in the BMP pathway: Small C-terminal domain phosphatases dephosphorylate Smad1 to attenuate BMP signaling. *Proceedings of the National Academy of Sciences of the United States of America*, *103*(32), 11940–11945. http://dx.doi.org/10.1073/pnas.0605133103.

Kowanetz, M., Valcourt, U., Bergstro, R., & Heldin, C.-h. (2004). Id2 and Id3 define the potency of cell proliferation and differentiation responses to transforming growth factor and bone morphogenetic protein. *Society*, *24*(10), 4241–4254. http://dx.doi.org/10.1128/MCB.24.10.4241.

Kregel, K. C. (2002). Heat shock proteins: Modifying factors in physiological stress responses and acquired thermotolerance. *Journal of Applied Physiology*, *92*(5), 2177–2186. http://dx.doi.org/10.1152/japplphysiol.01267.2001.

La Rosa, I., Camargo, L. S. A., Pereira, M. M., Fernandez-Martin, R., Paz, D. A., & Salamone, D. F. (2011). Effects of bone morphogenic protein 4 (BMP4) and its inhibitor, Noggin, on in vitro maturation and culture of bovine preimplantation embryos. *Reproductive Biology and Endocrinology*, *9*(1), 18. http://dx.doi.org/10.1186/1477-7827-9-18, BioMed Central Ltd.

Lee, J. H., Choi, S. S., Kim, H. W., Xiong, W. C., Min, C. K., & Lee, S. J. (2014). Neogenin as a receptor for early cell fate determination in preimplantation mouse embryos. *PLoS One*, *9*(7), e101989. http://dx.doi.org/10.1371/journal.pone.0101989.

Lee, K.-B., Folger, J. K., Rajput, S. K., & Smith, G. W. (2014). Temporal regulation of mRNAs for select bone morphogenetic proteins (BMP), BMP receptors and their associated SMAD proteins during bovine early embryonic development: Effects of exogenous BMP2 on embryo developmental progression. *Reproductive Biology and Endocrinology*, *12*(1), 67. http://dx.doi.org/10.1186/1477-7827-12-67.

Li, Y., Wei, Q.-W., Feng, J.-G., Xu, M.-L., Huang, R.-H., & Shi, F.-X. (2014). Expression of bone morphogenetic protein 2, 4, and related components of the BMP signaling pathway in the mouse uterus during the estrous cycle. *Journal of Zhejiang University Science B*, *15*(7), 601–610. http://dx.doi.org/10.1631/jzus.B1300288.

Madeja, Z. E., Sosnowski, J., Hryniewicz, K., Warzych, E., Pawlak, P., Rozwadowska, N., et al. (2013). Changes in sub-cellular localisation of trophoblast and inner cell mass specific transcription factors during bovine preimplantation development. *BMC Developmental Biology*, *13*(1), 32. http://dx.doi.org/10.1186/1471-213X-13-32.

Mansouri-Attia, N., Aubert, J., Reinaud, P., Giraud-Delville, C., Taghouti, G., Galio, L., et al. (2009). Gene expression profiles of bovine caruncular and intercaruncular endometrium at implantation. *Physiological Genomics*, *39*(1), 14–27. http://dx.doi.org/10.1152/physiolgenomics.90404.2008.

Mansouri-Attia, N., Sandra, O., Aubert, J., Degrelle, S., Everts, R. E., Giraud-Delville, C., et al. (2009). Endometrium as an early sensor of in vitro embryo manipulation technologies. *Proceedings of the National Academy of Sciences of the United States of America*, *106*(14), 5687–5692. http://dx.doi.org/10.1073/pnas.0812722106.

Massagué, J., Blain, S. W., & Lo, R. S. (2000). TGFbeta signaling in growth control, cancer, and heritable disorders. *Cell*, *103*(2), 295–309. http://www.ncbi.nlm.nih.gov/pubmed/11057902.

McNatty, K. P., Moore, L. G., Hudson, N. L., Quirke, L. D., Lawrence, S. B., Reader, K., et al. (2004). The oocyte and its role in regulating ovulation rate: A new paradigm in reproductive biology. *Reproduction (Cambridge, England)*, *128*(4), 379–386. http://dx.doi.org/10.1530/rep.1.00280.

Memili, E., & First, N. L. (1999). Control of gene expression at the onset of bovine embryonic development. *Biology of Reproduction*, *61*(5), 1198–1207.

Miyazono, K., Kusanagi, K., & Inoue, H. (2001). Divergence and convergence of TGF-beta/BMP signaling. *Journal of Cellular Physiology*, *187*(3), 265–276. http://dx.doi.org/10.1002/jcp.1080.

Montgomery, G. (2001). Genes controlling ovulation rate in sheep. *Reproduction*, *121*(6), 843–852. http://dx.doi.org/10.1530/reprod/121.6.843.

Mota, G. B., Batista, R. I., Serapião, R. V., Boité, M. C., Viana, J. H., Torres, C. A., et al. (2010). Developmental competence and expression of the MATER and ZAR1 genes in immature bovine oocytes selected by brilliant cresyl blue. *Zygote (Cambridge, England)*, *18*(3), 209–216. http://dx.doi.org/10.1017/S0967199409990219.

Mulsant, P., Pisselet, C., Riquet, J., Monniaux, D., Callebaut, I., Cribiu, E., et al. (2001). Mutation in bone morphogenetic protein receptor-IB is associated with increased ovulation rate in Booroola Merino ewes. *Proceedings of the National Academy of Sciences of the United States of America*, *98*(9), 5104–5109.

Murohashi, M., Nakamura, T., Tanaka, S., Ichise, T., Yoshida, N., Yamamoto, T., et al. (2010). An FGF4-FRS2alpha-Cdx2 axis in trophoblast stem cells induces Bmp4 to regulate proper growth of early mouse embryos. *Stem Cells (Dayton, Ohio)*, *28*(1), 113–121. http://dx.doi.org/10.1002/stem.247.

Nair, S. C., Rimerman, R. A., Toran, E. J., Chen, S., Prapapanich, V., Butts, R. N., et al. (1997). Molecular cloning of human FKBP51 and comparisons of immunophilin interactions with Hsp90 and progesterone receptor. *Molecular and Cellular Biology*, *17*(2), 594–603.

Paine-Saunders, S., Viviano, B. L., Economides, A. N., & Saunders, S. (2002). Heparan sulfate proteoglycans retain Noggin at the cell surface: A potential mechanism for shaping bone morphogenetic protein gradients. *The Journal of Biological Chemistry*, *277*(3), 2089–2096. http://dx.doi.org/10.1074/jbc.M109151200.

Peng, S., Hua, J., Cao, X., & Wang, H. (2010). Gelatin induces trophectoderm differentiation of mouse embryonic stem cells. *Cell Biology International*, *35*(6), 587–591.

Pennetier, S., Perreau, C., Uzbekova, S., Thélie, A., Delaleu, B., Mermillod, P., et al. (2006). MATER protein expression and intracellular localization throughout folliculogenesis and preimplantation embryo development in the bovine. *BMC Developmental Biology*, *6*, 26. http://dx.doi.org/10.1186/1471-213X-6-26.

Pennington, K. A., & Ealy, A. D. (2012). The expression and potential function of bone morphogenetic proteins 2 and 4 in bovine trophectoderm. *Reproductive Biology and Endocrinology*, *10*(1), 12. http://dx.doi.org/10.1186/1477-7827-10-12, BioMed Central Ltd.

Pesce, M., Gross, M. K., & Schöler, H. R. (1998). In line with our ancestors: Oct-4 and the mammalian germ. *BioEssays*, *20*, 722–732.

Reddi, A. H. (2005). BMPs: From bone morphogenetic proteins to body morphogenetic proteins. *Cytokine & Growth Factor Reviews*, *16*(3), 249–250. http://dx.doi.org/10.1016/j.cytogfr.2005.04.003.

Reyes De Mochel, N. S., Luong, M., Chiang, M., Javier, A. L., Luu, E., Fujimori, T., et al. (2015). BMP signaling is required for cell cleavage in preimplantation-mouse embryos. *Developmental Biology*, *397*(1), 45–55. http://dx.doi.org/10.1016/j.ydbio.2014.10.001.

Rossi, P., & Dolci, S. (2013). Paracrine mechanisms involved in the control of early stages of mammalian spermatogenesis. *Frontiers in Endocrinology*, *4*, 181. http://dx.doi.org/10.3389/fendo.2013.00181.

Ruddock, N. T., Wilson, K. J., Cooney, M. A., Korfiatis, N. A., Tayfur Tecirlioglu, R., & French, A. J. (2004). Analysis of imprinted messenger RNA expression during bovine preimplantation development. *Biology of Reproduction*, *70*(4), 1131–1135. http://dx.doi.org/10.1095/biolreprod.103.022236.

Sapkota, G., Alarcón, C., Spagnoli, F. M., Brivanlou, A. H., & Massagué, J. (2007). Balancing BMP signaling through integrated inputs into the Smad1 linker. *Molecular Cell*, *25*(3), 441–454. http://dx.doi.org/10.1016/j.molcel.2007.01.006.

Sieber, C., Kopf, J., Hiepen, C., & Knaus, P. (2009). Recent advances in BMP receptor signaling. *Cytokine & Growth Factor Reviews*, *20*(5–6), 343–355. http://dx.doi.org/10.1016/j.cytogfr.2009.10.007.

Silva, J. R. V., van den Hurk, R., van Tol, H. T. A., Roelen, B. A. J., & Figueiredo, J. R. (2005). Expression of growth differentiation factor 9 (GDF9), bone morphogenetic protein 15 (BMP15), and BMP receptors in the ovaries of goats. *Molecular Reproduction and Development*, *70*(1), 11–19. http://dx.doi.org/10.1002/mrd.20127.

Simmons, D. G., & Cross, J. C. (2005). Determinants of trophoblast lineage and cell subtype specification in the mouse placenta. *Developmental Biology*, *284*(1), 12–24. http://dx.doi.org/10.1016/j.ydbio.2005.05.010.

Spencer, T. E., & Bazer, F. W. (2004). Conceptus signals for establishment and maintenance of pregnancy. *Reproductive Biology and Endocrinology*, *2*, 49. http://dx.doi.org/10.1186/1477-7827-2-49.

Spencer, T. E., Johnson, G. A., Bazer, F. W., & Burghardt, R. C. (2004). Implantation mechanisms: Insights from the sheep. *Reproduction (Cambridge, England)*, *128*(6), 657–668. http://dx.doi.org/10.1530/rep.1.00398.

Spencer, T. E., Sandra, O., & Wolf, E. (2008). Genes involved in conceptus-endometrial interactions in ruminants: Insights from reductionism and thoughts on holistic approaches. *Reproduction*, *135*, 165–179. http://dx.doi.org/10.1530/REP-07-0327.

Suzuki, Y., Koshi, K., Imai, K., Takahashi, T., Kizaki, K., & Hashizume, K. (2011). Bone morphogenetic protein 4 accelerates the establishment of bovine trophoblastic cell lines. *Reproduction (Cambridge, England)*, *142*(5), 733–743. http://dx.doi.org/10.1530/REP-11-0275.

Tong, Z. B., Gold, L., Pfeifer, K. E., Dorward, H., Lee, E., Bondy, C. A., et al. (2000). Mater, a maternal effect gene required for early embryonic development in mice. *Nature Genetics*, *26*(3), 267–268. http://dx.doi.org/10.1038/81547.

Ulbrich, S. E., Zitta, K., Hiendleder, S., & Wolf, E. (2010). In vitro systems for intercepting early embryo-maternal cross-talk in the bovine oviduct. *Theriogenology*, *73*(6), 802–816. http://dx.doi.org/10.1016/j.theriogenology.2009.09.036, Elsevier.

Urrego, R., Herrera-Puerta, E., Chavarria, N. A., & Camargo, O. (2015). Theriogenology follicular progesterone concentrations and messenger RNA expression of MATER and OCT-4 in immature bovine oocytes as predictors of developmental competence. *Theriogenology*, 1–9. http://dx.doi.org/10.1016/j.theriogenology.2014.12.024, Elsevier.

Valdez Magaña, G., Rodríguez, A., Zhang, H., Webb, R., & Alberio, R. (2014). Paracrine effects of embryo-derived FGF4 and BMP4 during pig trophoblast elongation. *Developmental Biology, 387*(1), 15–27. http://dx.doi.org/10.1016/j.ydbio.2014.01.008, Elsevier.

Van Soom, A. V., Boerjan, M. L., Bols, P. E., Vanroose, G., Lein, A., & Coryn, M. (1997). Timing of compaction and inner cell allocation in bovine embryos produced in vivo after superovulation. *Biology of Reproduction, 1049*, 1041–1049.

Vitt, U. A. (2001). Evolution and classification of cystine knot-containing hormones and related extracellular signaling molecules. *Molecular Endocrinology, 15*(5), 681–694. http://dx.doi.org/10.1210/me.15.5.681.

Von Bubnoff, A., & Cho, K. W. (2001). Intracellular BMP signaling regulation in vertebrates: Pathway or network? *Developmental Biology, 239*(1), 1–14. http://dx.doi.org/10.1006/dbio.2001.0388.

Wagner, D. O., Sieber, C., Bhushan, R., Börgermann, J. H., Graf, D., & Knaus, P. (2010). BMPs: From bone to body morphogenetic proteins. *Science Signaling, 3*(107), mr1. http://dx.doi.org/10.1126/scisignal.3107mr1.

Watson, A. J., Hogan, A., Hahnel, A., Wiemer, K. E., & Schultz, G. A. (1992). Expression of growth factor ligand and receptor genes in the preimplantation bovine embryo. *Molecular Reproduction and Development, 31*(2), 87–95. http://dx.doi.org/10.1002/mrd.1080310202.

Watson, A. J., Watson, P. H., Arcellana-panlilio, M., Warnes, D., Walker, S. K., Schultz, G. A., et al. (1994). A growth factor phenotype map for ovine preimplantation development. *Biology of Reproduction, 50*(4), 725–733.

Wetendorf, M., & De Mayo, F. J. (2013). The progesterone receptor regulates implantation, decidualization, and glandular development via a complex paracrine signaling network. *Molecular and Cellular Endocrinology, 357*(713), 108–118. http://dx.doi.org/10.1016/j.mce.2011.10.028.

Wijgerde, M., Karp, S., McMahon, J., & McMahon, A. P. (2005). Noggin antagonism of BMP4 signaling controls development of the axial skeleton in the mouse. *Developmental Biology, 286*(1), 149–157. http://dx.doi.org/10.1016/j.ydbio.2005.07.016.

Wintenberg-Torres, S., & Flechòn, J. E. (1974). Ultrastructural evolution of the trophoblast cells of the pre-implantation sheep blastocyst from day 8 to day 18. *Journal of Anatomy, 118*(1), 143–153.

Wood, J. R., Dumesic, D. A., Abbott, D. H., & Strauss, J. F. (2007). Molecular abnormalities in oocytes from women with polycystic ovary syndrome revealed by microarray analysis. *The Journal of Clinical Endocrinology and Metabolism, 92*(2), 705–713. http://dx.doi.org/10.1210/jc.2006-2123.

Yamanaka, Y., Ralston, A., Stephenson, R. O., & Rossant, J. (2006). Cell and molecular regulation of the mouse blastocyst. *Developmental Dynamics: An Official Publication of the American Association of Anatomists, 235*(9), 2301–2314. http://dx.doi.org/10.1002/dvdy.20844.

Yao, Y., Watson, A. D., Ji, S., & Boström, K. I. (2009). Heat shock protein 70 enhances vascular bone morphogenetic protein-4 signaling by binding matrix Gla protein. *Circulation Research, 105*(6), 575–584. http://dx.doi.org/10.1161/CIRCRESAHA.109.202333.

CHAPTER EIGHT

Strategies of Manipulating BMP Signaling in Microgravity to Prevent Bone Loss

Jamila H. Siamwala*, Saranya Rajendran†, Suvro Chatterjee†,‡,1
*Department of Orthopaedic Surgery, University of California, San Diego, California, USA
†Vascular Biology Lab, AU-KBC Research Centre, Anna University, MIT Campus, Chennai, India
‡Department of Biotechnology, Anna University, Chennai, India
[1]Corresponding author: e-mail address: soovro@yahoo.ca

Contents

Abstract

Bone structure and function is shaped by gravity. Prolonged exposure to microgravity leads to 1–2% bone loss per month in crew members compared to 1% bone loss per year in postmenopausal women. Exercise countermeasures developed to date are ineffective in combating bone loss in microgravity. The search is on for alternate therapies to prevent bone loss in space.

Vitamins and Hormones, Volume 99
ISSN 0083-6729
http://dx.doi.org/10.1016/bs.vh.2015.05.004

Microgravity is an ideal stimulus to understand bone interactions at different levels of organizations. Spaceflight experiments are limited by high costs and lack of opportunity. Ground-based microgravity analogs have proven to simulate biological responses in space. Mice experiments have given important signaling clues in microgravity-associated bone loss, but are restricted by numbers and human application. Cell-based systems provide initial clues to signaling changes; however, the information is simplistic and limited to the cell type. There is a need to integrate information at different levels and provide a complete picture which will help develop a unique strategy to prevent bone weakening.

Limited exposure to simulated microgravity using random positioning machine induces proliferation and differentiation of bipotential murine oval liver stem cells. Bone morphogenetic proteins (BMPs) are the prototypal osteogenic signaling molecule with multitude of bone protective functions. In this chapter, we discuss the basic BMP structure, its significance in bone repair, and stem cell differentiation in microgravity. Based on the current information, we propose a model for BMP signaling in space. Development of new technologies may help osteoporosis patients, bedridden people, spinal injuries, or paralytic patients.

ABBREVIATIONS

ACTR activin receptor
BMD bone mass density
BMOL bipotential murine oval liver stem cells
BMP bone morphogenetic protein
CTGF connective tissue growth factor
LMHF low magnitude and high frequency
MSCs mesenchymal stem cells
TGF-β transforming growth factor beta

1. INTRODUCTION

Life has evolved in the presence of uniform Earth's gravity. Earth's gravity literally shapes the structure and function of all organisms. Human skeleton experiences large mechanical loads when performing tasks such as walking, lifting against gravity. The skeleton is sensitive to changes in gravity with altering mechanical loads. The bone structure is thus adapted to the mechanical function and varies according to the distribution of loads. The two main bone classifications based on the porosity and microstructure are cortical bone or the compact bone and the trabecular bone or the cancellous or spongy bone. The cortical bone is found in the shaft of the long bones and forms the outer shell around the trabecular bone at the end of the joints and the vertebrae. Although there are different proportions of cortical bone, trabecular bone, and marrow, the basic characteristics of bone circulation are the same. The vasculature comprises the principal nutrient artery in the center, small nutrient arteries in the periphery, and a capillary net in the bone substance and profuse sinusoids in the contained marrow (McCarthy, 2006). The bone circulation is important in fracture repair and osteoporosis, yet is poorly understood due to the nature of the bone itself and the paucity of techniques available to measure bone circulation. The bone and the bone circulation (both microvascular and large arterial flow) in gravity depends on two main factors: (1) mechanical loading and (2) distribution of tissue fluid pressures. In unit gravity, there is a gradient of tissue fluid pressures due to the hydrostatic pressures. In reduced gravity, redistribution of tissue pressures has a significant impact on the cardiovascular system and is manifested as decreased blood flow to the legs and increased intracranial pressures, facial puffiness. The changes in the bone density have come from measurements on space station (Lang et al., 2004), clinostat models to simulate microgravity, hindlimb suspension in mice (Morey-Holton & Globus, 2002), and prolonged bed rest study (Leblanc, Schneider, Evans, Engelbretson, & Krebs, 1990). These studies show that gravity causes bone remodeling and lack of gravity can lead to inhibition of bone formation and increased bone resorption which is evidenced as accelerated bone loss and increase in risk for fractures.

Bone health depends on proper functioning of a number of systemic or circulating hormones including growth and sex hormones. The list includes parathyroid hormone, calcitriol (active vitamin D), calcitonin (Sexton, Findlay, & Martin, 1999), estrogen, testosterone (Falahati-Nini et al., 2000), growth hormone/insulin-like growth factor (Wang, Zhou, Cheng, Kopchick, & Bondy, 2004), thyroid hormone (Abe et al., 2003), and cortisol (Canalis, 2005). Growth factors such as bone morphogenetic proteins (BMP1–BMP8a) are known to be involved in bone and cartilage development. It was shown that parathyroid hormone enhanced bone morphogenetic protein (BMP) signaling to induce differentiation of mesenchymal stromal/stem cells to osteoblast lineage (Yu et al., 2012). Calcitriol treatment in rats enhanced markers of chondrocyte differentiation, such as vitamin D receptor and BMP7 (Sanchez & He, 2007). Similarly, estrogens are reported to activate BMP2 gene transcription in mouse mesenchymal stem cells (MSCs; Zhou et al., 2003). Hence, it is important to focus on the BMP signaling and its probable remedial role in bone fracture, bone loss on earth and space.

BMPs are responsible for a variety of functions during development. Specifically, these proteins have critical roles in skeletal development, bone formation, and bone repair (Beederman et al., 2013). The BMP subfamily comprises more than 20 proteins. BMPs have highly conserved amino acid sequence, which is considered to be as old as 600 million years. For this reason, human recombinant BMPs are highly effective in lower life forms, including fruit flies. BMPs are a group of growth factors also known as cytokines and as metabologens (Gilbert, 2006). BMPs are responsible for a group of pivotal morphogenetic signals, orchestrating tissue architecture throughout the body (Jain, Pundir, & Sharma, 2013). Studies of astronauts who spent many months on space station Mir demonstrated that astronauts can lose 1–2% of bone mass each month (Keyak, Koyama, LeBlanc, Lu, & Lang, 2009). Although weakening of the bones due to the progressive loss of bone mass is a serious health complication in extended spaceflight, understanding of BMP's implications in bone pathologies is poorly understood. In this chapter, we attempt to consolidate recent research in the BMP signaling in space conditions, implications, scopes in the development of a suitable countermeasure, Earth-based applications, and future directions.

1.1 BMP Structure and Its Significance in Bone Induction

BMPs are multifunctional growth factors belonging to transforming growth factor beta (TGF-β) superfamily. It is a dimeric molecule linked by disulfide bonds. This dimerization is a prerequisite for bone induction. Carboxy-terminal ends of BMPs contain seven cysteine residues which are highly conserved among all the reported members of the TGF-β superfamily (Wozney et al., 1988). BMPs are synthesized intracellularly in a precursor form with a hydrophobic stretch of about 50–100 amino acids. Prior to secretion, BMPs consist of a signal peptide, prodomain, and mature peptide. Upon the cleavage of signal peptide, glycosylation and dimerization of precursor protein take place. The prodomain is cleaved, following the secretion of mature bioactive dimeric BMP as either heterodimers (two different chains) or homodimers (two identical chains), by the cell. Due to their ubiquitous expression and importance as regulators throughout the body, deficiency in BMP production or functionality usually leads to marked defects or severe pathologies (Wang et al., 2014).

1.2 BMP Signaling in Bone Formation and Repair

The BMP signaling is the key pathway in bone formation and is involved in multiple stages of the developmental process, including osteoblast differentiation (Chen, Deng, & Li, 2012), mesoderm patterning (Nikaido, Tada, Saji, & Ueno, 1997), bone formation (Chen et al., 2012; Nie, Luukko, & Kettunen, 2006), and craniofacial (Ahn, Mishina, Hanks, Behringer, & Crenshaw, 2001; Kanzler, Foreman, Labosky, & Mallo, 2000; Luo et al., 2010), and limb development (Bandyopadhyay et al., 2006). Several knockout models have provided insight into the critical mechanisms responsible for skeletal development (Boyle, Simonet, & Lacey, 2003; Dudas, Sridurongrit, Nagy, Okazaki, & Kaartinen, 2004; McPherron, Lawler, & Lee, 1999; Shu et al., 2011). Disruptions of TGF-β/BMP signaling have been implicated in multiple bone diseases, including tumor metastasis, brachydactyly type A2, and osteoarthritis (Miyazono, Maeda, & Imamura, 2005; Papachroni, Karatzas, Papavassiliou, Basdra, & Papavassiliou, 2009; Siegel & Massague, 2003). Proper bone formation requires the differentiation of osteoblasts from MSC precursors, a process mediated in part by BMP signaling. BMP2 and BMP4 deletion or deletion of BMP2 alone results in a severe chondrodysplastic phenotype, while deletion of BMP4 alone has minor effects on cartilage development. Double knockout and BMP2 knockout results in disorganization of chondrocytes within the growth plate, decreases cell proliferation, poor differentiation, and increases apoptosis (Beederman et al., 2013).

2. BONE REMODELING IN THE ABSENCE OF GRAVITY

Bone undergoes constant remodeling and this process is maintained by complex interactions between the two different cell types: (1) osteoblast and (2) osteoclast. The osteoblast arises from the local osteoprogenitor cells and the osteoclast occurs from a fusion of mononuclear precursors. The osteoblast is responsible for depositing bone matrix and regulation of bone mineralization. Osteoclasts resorb bone thereby maintaining a balance between the bone formation and bone resorption. The bone remodeling involves osteoclast activation, bone resorption, and subsequent bone formation. The regulatory cytokines released by the osteoclast during bone resorption, receptor activation of nuclear factor-κB ligand, and osteoprotegerin produced by the osteoblasts regulate the processes of osteoblastogenesis and osteoclastogenesis (Boyle et al., 2003). The uncoupling of bone remodeling with increased bone resorption markers and decreased bone formation markers may be responsible for bone loss observed in space. Although there are early indications of increases in bone resorption markers such as deoxypyridinoline and the C-terminal peptide of type I collagen after 180 days of spaceflight (Caillot-Augusseau et al., 1998), research has focused on the reduction in the bone formation during spaceflight. During 6-month spaceflight, serum levels of bone formation markers such as alkaline phosphatase, osteocalcin, and type I procollagen propeptide were reduced by 28%, 27%, and 38%, respectively (Caillot-Augusseau et al., 1998). One-month spaceflight also resulted in decrease in bone formation markers along with bone loss. There is a 2.27% reduction in trabecular bone mass in the tibia of a cosmonaut (Collet et al., 1997). A dual-energy X-ray absorptiometry (DEXA) scan of 16 individuals who were in space for 16–28 weeks had an overall 3.5% reduction in bone mass. Site-specific bone loss measurements show that 97% bone loss occurs in the weight-bearing bones such as hip and pelvis (Leblanc et al., 1990). Spaceflight did not alter the bone mass in the nonweight-bearing regions such as the radius. Bone loss may also involve compromised angiogenesis or vascular remodeling since bone formation or osteogenesis requires blood vessels to supply nutrients to the developing bone. Angiogenesis precedes osteogenesis and this empirical observation has led to the suggestion that angiogenesis plays an active role in the process of osteogenesis and not a passive role. The amount of mineralization and alkaline phosphatase activity is higher in fetal calvarial cell and endothelial cell coculture than cells cultured separately (Villanueva & Nimni, 1990). Further, the microvessel cells have a mitogenic effect on osteoblast-enriched calvarial cells (Jones, Clark, & Brighton, 1995). Endothelium damage is an important factor in the failure of vascularized bone grafts (McCarthy, Andhoga, Batten, & Mathie, 1997). Nitric oxide, a potent angiogenesis factor, is released by the osteoblasts in response to fluid shear stress and also mechanical stimulation of bone (Briggs, Moran, & Wood, 1998). Bone perfusion, interstitial fluid pressures, and direct vascular-bone cell coupling mechanisms thus modulate bone remodeling (Brandi & Collin-Osdoby, 2006; Colleran et al., 2000; Farhat & Cauley, 2008; Hillsley & Frangos, 1994).

3. BONE PERFUSION IN THE ABSENCE OF GRAVITY

In the conditions of absence of mechanical loads (e.g., spaceflights, prolonged bed rest, spinal cord injury), the bone is the most impacted. Spaceflight is the ultimate model to study the effects of gravity on the human skeleton (Vico et al., 2000). Spaceflights pose a myriad problems to astronauts in the form of fluid shifts (Hargens & Richardson, 2009) and postflight orthostatic hypotension (Buckey et al., 1996); altered cardiac susceptibility to ventricular arrhythmias (Fritsch-Yelle et al., 1998); reduced cardiac muscle mass and cardiac function (Levine, Zuckerman, & Pawelczyk, 1997); neurovestibular problems (Oman, Lichtenberg, Money, & McCoy, 1986); infections (Taylor, 1993); sleep problems (Gundel, Polyakov, & Zulley, 1997), loss of muscle mass, strength, and endurance (Baldwin, 1996; Edgerton et al., 1995; Fitts, Riley, & Widrick, 2000); and bone loss. Astronauts or cosmonauts experience 1–2% bone loss per month compared to 1% bone loss postmenopausal women per year (Grigoriev et al., 1998; Vico et al., 2000). The osteopenic effects appeared in cosmonauts after the Vostok 2 and 3 mission (Goode & Rambaut, 1985). Later, significant loss of calcaneus mass was observed in astronauts after the Gemini, Apollo, and Skylab flights (Goode & Rambaut, 1985). Bone mass density (BMD) measurements during the Mir spaceflight of 4.5–14 months indicate an average loss of $5.6 \pm 0.8\%$ from the lumbar spine, $11 \pm 1.4\%$ from the pelvis, and $8.1 \pm 1.2\%$ from the proximal femur (Grigoriev et al., 1998). BMD measurements of the distal radius and tibia in 15 cosmonauts of the Russian Mir space station show -1.7% in the cancellous tibia at the end of 1-month flight period and -1.9% in tibial cortex in 1 cosmonaut (Vico et al., 2000). Further, there are indications of inhibition of bone growth in mice during flight; however, bone growth returns to normal, 27 days after returning from spaceflight (Yagodovsky, Triftanidi, & Gorokhova, 1976). At this rate, there is an increase in the risk of fractures or injury to soft connective tissue after extended spaceflight if long-term Mars mission for 3.5 years is undertaken (White & Averner, 2001). The bone loss during spaceflight is comparable to spinal cord injury, prolonged bed rest, and muscle paralysis on Earth. This is evident from similar increases in bone resorption, decreased bone mass, and an increased risk of fracture (Ausk et al., 2012; Battaglino, Lazzari, Garshick, & Morse, 2012) comparable to spinal cord injury, prolonged bed rest, and muscle paralysis on Earth. Bone loss in space is chiefly attributed to mechanical unloading, increased osteoclast activity, and altered skeletal perfusion.

4. MECHANISMS OF BONE LOSS IN MICROGRAVITY

Studies have shown a profound reduction in bone and marrow blood flow during hindlimb unloading (HU) experiments in mice (Colleran et al., 2000; Stabley, Prisby, Behnke, & Delp, 2013) and a blunting of the reloading hyperemia in bone and marrow on mechanical loading. This is due to the increase in vascular resistance due to remodeling of vasculature and an enhanced vasoconstrictor responsiveness of bone-resistant arteries and diminished vasodilator responsiveness of the resistance vasculature. Two-week hindlimb unloaded mice showed an increased vascular resistance of bone-resistant arteries, decreased endothelium-dependant vasodilator responses, and decreased skeletal perfusion (Prisby, Behnke, Allen, & Delp, 2015). There is a growing body of evidence that distribution of the bone density changes may be related to fluid shifts observed in microgravity (McCarthy, 2005). Spaceflight causes a headward fluid shift leading to high intracranial pressure, facial puffiness, and headaches. micro-CT and mechanical testing studies' analysis of spaceflight mice calvaria showed an increase in calvarial bone volume (Zhang, Cory, Bhattacharya, Sah, & Hargens, 2013). Gene expression studies of bone markers in the calvaria showed that BMP plays a major role in modulating osteoblast differentiation and calvarial remodeling (unpublished data). The problem of bone loss in space has been explained previously with different propositions, hypotheses, and schemes. However, three prominent propositions emerging from all the elaborations so far are summarized below.

4.1 Bone Loss Due to Mechanical Unloading

The mechanisms underlying skeletal adaptations to spaceflight have come from the rats sent to orbit on board Soviet Cosmos Biosatellites and American Spacelab 3. These rats have reduced bone mass density (BMD) in the tibial metaphysis (Yagodovsky et al., 1976) and reduction of the periosteal bone formation of the tibial and humeral shafts (Wronski & Morey, 1983). Bone loss and disturbed bone geometry in space have been attributed to mechanical unloading (Sievanen, 2010). Mechanical loading of the bone with exercise is proposed to improve bone formation mass and geometry. Since cancellous bone compartments are more prone to disuse bone loss, exercise regimes have targeted the cancellous region to study exercise-associated improvements in bone health. The mechanical loading of bone occurs via the tensile forces produced by the contracting muscles during exercise. The external signal is tranduced by the osteocytes in the bone. Mice HU studies showed that with exercise, the sclerostin protein expression in the osteoclasts decreases, Wnt signaling is upregulated which in turn activates the osteoblast activity and new bone formation (Robling et al., 2008). The bone mass thereby increases, geometry is restored, and strength is recovered.

4.2 Increased Osteoclast Activity

Osteoclasts flown aboard Foton M3 satellite show an increase in number of discrete resorption pits compared to ground controls indicating an increased bone resorption activity. Further, osteoblasts exposed to 5 days of microgravity had changes in cell surface adhesion, actin cytoskeleton, and stress fibers. The nucleus appears condensed or fragmented and cells are elongated (Nabavi, Khandani, Camirand, & Harrison, 2011).

One of the other suggested mechanisms of bone loss in space is the action of locally acting cytokines which results in an increased osteoclast activity and bone resorption. Studies in mice hindlimb suspension have shown arterial vasculature remodeling in skeletal muscle (Delp, Colleran, Wilkerson, McCurdy, & Muller-Delp, 2000) and cerebral tissue (Wilkerson, Muller-Delp, Colleran, & Delp, 1999). If these changes also occur in the hindlimb bones, the mechanical environment of the osteoprogenitor cells, osteoblasts, and osteoclasts will also change. This, in turn, will change the regulation of these cells tipping the balance between bone formation and bone resorption or osteoclast activity.

4.3 Alterations in Skeletal Perfusion as a Mechanism of Bone Loss

The cephalic fluid shift due to weightlessness may affect the bone blood flow and alter the hydrostatic, intramuscular, and interstitial pressures (Colleran et al., 2000). On Earth, in an upright posture, hydrostatic pressure influenced by gravity ensures blood circulation to the lower part of the body. In the absence of gravity, the hydrostatic pressure is reduced and there is headward fluid shift. As a result, the blood volume in the legs decreases, the astronauts develop "bird legs"(Hargens & Watenpaugh, 1996). Eventually, reduction in the blood flow and volume leads to increased osteoclast activity and bone loss. Decreased blood flow is attributed to the myogenic mechanism described by Bayliss in 1902. In weightlessness, the transmural pressure across the blood vessel wall increases, leading to vasoconstriction and decreased blood flow. Conversely, as the transmural pressure decreases, there is vasodilation and blood flow increases. These are referred to as the myogenic responses and this mechanism primarily alters blood circulation in the lower part of the body resulting in bone loss. Needless to mention that the myogenic effects cannot be completely delineated from the sympathetic nervous system effects in the observed space effects on the calf. Having established the bone problems in the absence of gravity and the mechanism of microgravity-induced bone loss, it is important to understand how the BMP signaling is altered in microgravity. There are very few flight-based experiments and limited ground-based studies that can explain BMP changes in microgravity and its use as therapeutic agent.

5. BMP SIGNALING IN SPACE

The recent work of Aleshcheva et al. (2015) suggested that short-term microgravity affects the gene expression of distinct proteins (Aleshcheva et al., 2015). The upregulation of BMP2, TGF-β1, and SOX9 in chondrocytes may play a key role in preventing cytoskeletal alterations (Patel et al., 2009). A study which uses HU, as a simulation of microgravity, demonstrated that HU for 28 days resulted in the decrease of vinculin-containing focal adhesion in MSCs. During HU, the interval intramuscular injection of BMP2 or FGF2 alone could increase the osteogenic potential of MSCs and the expression of osteoblast gene marker mRNA. A study by Patel et al. (2009) also found that low magnitude and high frequency (LMHF) loading alone can induce mineralization of 2T3 cells in a magnitude-dependent manner, and up to 20 ng/ml of BMP4 or BMP2 supplementation further increased this response in an additive manner. More importantly, the study confirmed that treatment with Noggin, the BMP antagonist, completely blocked mineralization induced by the LMHF stimulus (Patel et al., 2009). Hence, this study indicates a possible loss of BMP signaling in microgravity conditions (Qian et al., 2014). The work of Cao et al. (2003) during 21-day tail-suspension immunohistochemical results of rat tibia showed that both BMP and TGF-β declined in tissues from all regions of the tibia (Cao et al., 2003). Similarly, another study also showed that BMP2 can stimulate the differentiation of ROS17/2.8 cells in 1 G condition and this process is reduced under simulated weightlessness (Wang, Zhang, & Wu, 2004). However, osteoblasts showed high levels of alkaline phosphatase expression and activity, and elevated expression of osteopontin, osteocalcin, and bone morphogenetic protein 4 (BMP4) when the cells were placed in rotating wall vessel (RWV) (Rucci, Migliaccio, Zani, Taranta, & Teti, 2002). These results indicate a BMP paradox in microgravity conditions.

We perceive that the expression and activity of BMPs depend on cellular and molecular cross talks in bone niche and the microenvironment. This highly interactive three-dimensional bone milieu would define the expression pattern of BMPs in different bone cells. This could be the reason why

BMP expression is low in mature bone cells and tissues in microgravity conditions. However, in naïve stem cells, and limited microgravity, BMP can promote stem cell proliferation and differentiation. BMP signaling varies with cell and organ type and degree of exposure to microgravity. In our study, we exposed liver stem cell or bipotential murine oval liver (BMOL) stem cells to 2 h of microgravity in a 3D clinostat and followed the BMP4 expression pattern for 48 h of BMOL cell differentiation.

6. BMP4 DIFFERENTIATES STEM CELLS IN SIMULATED MICROGRAVITY CONDITIONS

6.1 Simulation of Microgravity Conditions

Simulated microgravity is induced using indigenously fabricated 3D clinostat machine as described earlier by our group (Siamwala et al., 2010a, 2010b).

The formula for microgravity (g') is

$$g' = \frac{\omega^2 R}{g}$$

where $g = 9.8\ \mathrm{m/s^2}$; ω, angular velocity in rad/s; R = radius from the center of rotation, the random positioning machine (RPM) rotates with a constant angular velocity ($\omega = \theta/\pi$, where angular velocity is equal to angular displacement in radians/time taken).

$$T = \frac{\theta}{\tau} = 2\,\mathrm{rads/s}$$

The sample holders are mounted on the center of the platform located on the inner frame and can experience 10^{-3} gravity. Control samples are kept in customized vertical shaker to provide movements. However, the movements are not randomized as in case of RPM and this set was referred as gravity-treated cells. To avoid any nonspecific effect due to treatment conditions, both vertical shaker and RPM machine were kept in a glass chamber maintaining temperature at 37 °C and relative humidity (Fig. 1A).

6.2 Microgravity Induces Differentiation of BMOL Cells

BMOL cells were plated in 12-well plates and maintained for 10 days unless it reaches 90% confluence. Next, the cells were treated under microgravity for 2 h followed by incubation with normal and differentiation media for the next 48 h. Samples were prepared from the treated cells for studying HNF4α

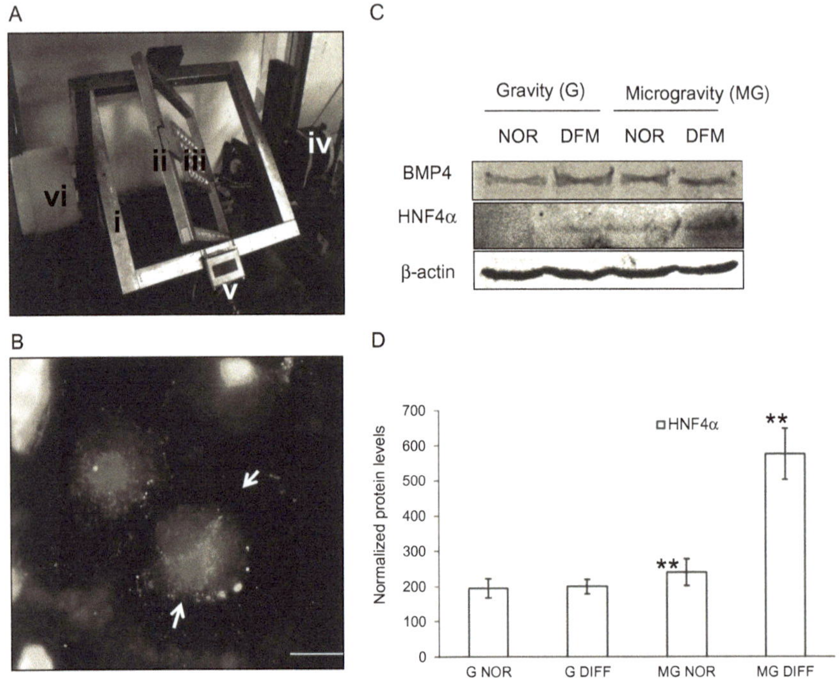

Figure 1 (A) Clinostat for inducing microgravity. Outer frame (i) rotates unidirectionally at 20 rpm and an inner frame (ii) with a sample holder (iii) which rotates bidirectionally at 10 rpm. Motor (iv) rotates the outer frame and the inner frame (v), and its rotational direction and velocity are controlled by the controller (vi) connected to a computer with a clinostat control program. (B) HNF4α expression in BMOL cells. Differentiated BMOL cells subjected to microgravity are stained with HNF4α-fluorescein isothiocyanate (FITC) primary antibody (green) and counterstained with 4′,6-diamidino-2-phenylindole (DAPI), a nuclear stain (blue). White arrows indicate binding of antibodies to HNFα in differentiated BMOL cells. (C) Western blot analysis of BMP4, HNF4α, and β-actin performed on BMOL cells. The samples are probed with BMP4, HNF4α, and β-actin and the band intensities are estimated using densitometry quantification function of ImageJ. (D) The differences in protein expression are expressed as mean ± SE, $^{**}p < 0.001$ by one-way ANOVA. (See the color plate.)

expression, a marker of hepatocytes. Samples prepared after 0 and 48 h of microgravity treatment did not show HNF4α expression indicating absence of differentiated cells (Fig. 1C). After 48 h of microgravity treatment, the BMOL cells showed 1.3-fold higher level of HNF4α expression with normal media, while microgravity-treated cells incubated with differentiation media showed 2.9-fold higher HNF4α expression compared to gravity controls (Fig. 1D). Similarly, there was a 2- and 1.3-fold increase in BMP4 expression level in microgravity-treated cells cultured with and without differentiation media, respectively (Fig. 1C). Knocking out BMP4 expression using an BMP4 inhibitor dorsomorphin, or chordin-conditioned media from chordin-transfected cells attenuated microgravity-mediated differentiation of stem cells (Majumder et al., 2011). These results strongly emphasize that microgravity increases BMP4 expression and is a key player in stem cell proliferation and differentiation.

7. AN ANTICIPATED SCHEME FOR BMP SIGNALING IN SPACE

Based on the existing knowledge base of BMPs implicating in bone physiology and pathology, we propose a scheme which elaborates possible scenarios of BMP perturbations in microgravity conditions and a possible therapeutic approach for bone repair. The expression pattern of BMPs and modulation of BMP signaling in microgravity conditions are not conclusive rather contradictory on many occasions. This scheme explains the possibilities of BMPs being multitarget and adds to the layers of complications. We hereby put forward a scheme that states three possible mechanisms of bone signaling in space: (1) microgravity suppresses natural inhibitors of BMPs, (2) microgravity dampens natural inhibitors of BMP receptors, and (3) microgravity reduces the expression and secretion of BMPs (Fig. 2).

8. STRATEGIES TO MANIPULATE BMP SIGNALING IN SPACE

Three main strategies proposed to increase bone strength and bone formation in space are (1) targeting BMP ligands, (2) targeting BMP inhibitors, and (3) targeting BMP receptors. The possible mechanisms of bone repair by targeting BMP inhibitors such as sclerostin, chordin, connective tissue growth factor (CTGF), follistatin, and gremlin are summarized below.

8.1 Natural Inhibitors Targeting BMPs

8.1.1 Sclerostin

Sclerostin is a monomeric glycoprotein containing a cysteine knot-like domain with homology to the Cerberus/DAN family of BMP antagonists. It antagonizes Wnt/β-catenin signaling (Li et al., 2005) in osteoblasts by binding to low-density lipoprotein (LDL) receptor-related proteins 5 and 6 (LRP6 and LRP6), thereby inhibiting osteoblast differentiation, activity,

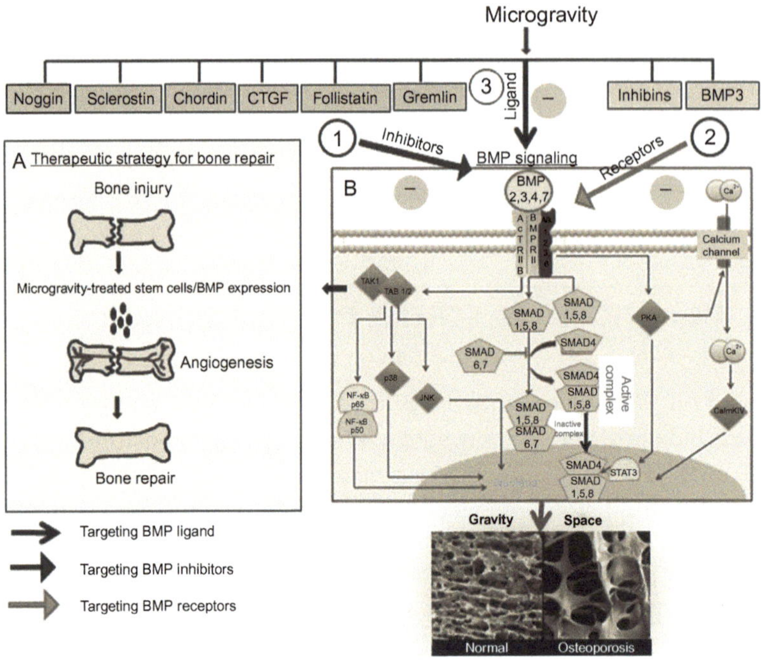

Figure 2 The schematic diagram represents (A) possible therapeutic approach to facilitate bone repair in space or Earth through BMP signaling and angiogenesis (left). (B) Probable BMP signaling events leading to bone weakening in space. The main strategies highlighted in the box (right) are (1) suppression of natural inhibitors of BMPs, (2) suppression of natural inhibitors of BMP receptors, and (3) reduction of BMP expression and secretion. *The BMP pathway is reproduced with permission from Affymetrix, Inc., Copyright (2012), USA and NASA, USA.* (See the color plate.)

and survival (Baron & Rawadi, 2007). Targeted deletion of the sclerostin gene in mice results in increased bone formation and bone strength (Li et al., 2008).

8.1.2 Chordin

Chordin is a 120 kDa protein containing four small (about 70 amino acids each) cysteine-rich domains (CRDs). It dorsalizes the developing embryo by binding BMPs (Pappano et al., 1998). The biological activity of chordin resides in the CRs, especially in CR1 and CR3, which have dorsalizing activity and bind BMPs with dissociation constants in the nanomolar range (Larrain et al., 2000). Because of chordin's BMP-binding activity, its regulation by growth factors, its location, and its increased levels, they probably play a role in the remodeling and repair of osteoarthritic cartilage (Tardif et al., 2006).

8.1.3 Connective Tissue Growth Factor

CTGF, also known as CCN2, is a matricellular protein. CTGF potently antagonizes the signaling activity of BMP4 in osteogenesis assays and in embryonic patterning (Abreu, Ketpura, Reversade, & De Robertis, 2002). Upregulation of CTGF inhibits BMP7 signal transduction in the diabetic kidney (Nguyen et al., 2008). However, Safadi et al. clearly demonstrated the role of CTGF in osteoblast differentiation and proliferation *in vitro* and bone formation *in vivo* (Safadi et al., 2003).

8.1.4 Follistatin

Follistatin is an autocrine glycoprotein encoded by the FST gene. Its primary function is binding and bioneutralization of members of the TGF-β superfamily. Follistatin inhibits BMP signaling by binding to BMPs, forming a trimeric complex (Iemura et al., 1998). It binds BMP2, BMP4, and BMP7, with a higher affinity for BMP7. Also, follistatin-like 1 (Fstl1) is a secreted protein of the BMP inhibitor class. It is shown that Fstl1 is a crucial regulator in BMP signaling during mouse development and in mouse lung development (Geng et al., 2011; Sylva et al., 2011). Inoue et al. detected the follistatin mRNA in the callus of repairing bone after fracture thus emphasizing the importance of follistatin in bone system *in vivo* (Inoue et al., 1994).

8.1.5 Gremlin

Gremlin (Grem), also known as Drm, is a highly conserved 20.7 kDa, 184 amino acid glycoprotein part of the DAN family and is a cysteine knot-secreted protein. Grem 1 inhibits BMP2, -4, and -7. Inhibition of BMPs by grem 1 in mice allows the expression of fibroblast growth factors (FGFs) 4 and 8 and Sonic hedgehog (SHH) which are necessary for proper limb development (Gazzerro et al., 2005). Deletion of grem 1 in mice causes increased bone formation and increased trabecular bone volume, whereas overexpression causes inhibition of bone formation and osteopenia (Gazzerro et al., 2005, 2007). In addition to the inhibition of BMP, grem 1 inhibits activation of Wnt/β-catenin signaling activity. On the other hand, recent reports suggest that gremlin 1 defines a population of osteochondroreticular stem cells in the bone marrow. These stem cells have self-renewal capacity and generate osteoblasts, chondrocytes, and reticular marrow stromal cells and help in bone development, bone remodeling, and fracture repair, thus opening a new strategy for bone repair (Worthley et al., 2015).

8.2 Natural Inhibitors Targeting BMP Receptors

Another strategy of improving bone strength and bone formation is targeting BMP receptors using inhibitors such as Noggins and Inhibins.

8.2.1 *Noggin*

Noggin binds several BMPs with very high affinities, with a marked preference for BMP2 and BMP4 over BMP7. By binding tightly to BMPs, Noggin prevents BMPs from binding their receptors. Inhibition of endogenous BMP signaling by Noggin is essential for elaboration of the vertebrate body plan (McMahon et al., 1998). Peng et al. (2005) showed that the stem cells engineered to express Noggin led to regeneration of bone that is similar to normal bone. In addition, Yoshimura showed colocalization of Noggin and BMP4 during fracture repair process, emphasizing the importance of the balance between these factors and receptors during fracture healing (Yoshimura et al., 2001).

8.2.2 *Inhibins*

Inhibins are endogenous antagonists of activin signaling and in concert with its coreceptor, betaglycan, it competes with activin for binding to type II activin receptors (ACTRs) and, thus, prevents activin signaling. Because BMPs also utilize type II ACTRs, they act as antagonists of BMP responses by competing with BMPs for binding to type II receptors (Wiater & Vale, 2003). The importance of BMP3 as a mediator of BMP signaling in the adult skeleton was well known. Targeted disruption of BMP3 in mice results in increased trabecular bone formation, whereas transgenic overexpression of BMP3 in skeletal cells leads to spontaneous fracture, consistent with BMP3 having a negative role in bone mass regulation (Kokabu et al., 2012).

9. CONCLUSIONS AND FUTURE DIRECTIONS

Most of the countermeasures such as exercise regimes, dietary calcium, vitamin D supplements have proved to be ineffective to prevent bone loss (Smith et al., 1999). The only countermeasure used successfully so far is exercise. Initially on the Gemini capsule, bungee cords strapped to hands and feet were used as countermeasure, although its efficacy was not documented. Next the Soyuz 9 flight bungee experiments showed the need to develop effective countermeasures to combat muscle deconditioning. Later, the bungee cords gave way to passive treadmill exercises, bicycle ergometer, and gravity simulation suit (Gazenko et al., 1976). The skylab astronauts used unusual form of tethered locomotion and mini gym exercise device which benefited the arms and trunk. Vibration isolation and stabilization system, a cycle ergometer with vibration isolation and the interim resistive exercise device, is currently being used with no data of these interventions preventing bone loss. Lang et al. (2004) showed that these devices were not effective countermeasures during long-duration spaceflight for bone loss (Lang et al., 2004) . Prolonged bed rest is used as an analog of spaceflight. Currently, countermeasures such as lower body negative pressure which can restore the hydrostatic gradient and pull the fluid back to the feet are being developed. This is the only available countermeasure, so far, apart from the exercises. Although, there are no inflight experiments conducted to test the efficacy of this countermeasure. The space industry is now looking at pharmaceutical options that are used to treat or delay postmenopausal osteoporosis. Research is also underway to use a specific exercise protocol along with a therapeutic agent. Stem cell therapy to combat bone loss is another unexplored option on Earth or in space.

There is a growing interest in the genetic and molecular mechanisms of bone and muscle interaction in relation to aging-associated bone weakening. Decline in bone strength leads to reduced mobility and morbidity and an increase in propensity of falls and fractures. Low rate of bone recovery is heightened by increase in risk of mortality in otherwise healthy aged individuals. This has led to acceleration of development of interventions that can be used to improve bone strength in both space and on Earth. The ideal candidate identified so far is the osteogenic protein, BMP4 which interacts with growth factors such as vascular endothelial growth factor (VEGF) to promote bone formation and bone healing by increasing cell recruitment, increasing cell viability, and enhancing angiogenesis and perfusion. This leads to cartilage formation, decrease in bone resorption which has been shown to accelerate cartilage healing, and improved bone mineralization. However, it is yet early to label BMPs as the ultimate bone and fracture healers due to the different levels of BMP interaction with other proteins. BMP4 by itself produced a lesser effect on bone healing compared to BMP4 in combination with VEGF, which has double the effect on the amount of bone formed The early-phase BMPs elicit different responses compared to late-phase BMP proteins. The lack of weight-bearing forces on the bone makes microgravity an ideal stimulus to understand bone cell responses and bone mineralization. Microgravity exposure leads to evident bone loss and aberrant bone signaling in the absence of gravity. The specific responses of BMP to microgravity or space conditions is an unexplored territory. Studying BMP on spaceflight is limited by high costs and lack of opportunities. Thus, ground-based model of microgravity, such as parabolic flight, clinostat, RPM, horizontal rotating vessel, and diamagnetic levitation, is developed to study the mechanisms of different physiological conditions. Animal models such as HU limit comparisons to human responses, although most of the mechanistic insights regarding bone loss in space have come from HU models. The closest system is probably human cell culture model and they may provide the basic mechanism at one level. Unfortunately, BMPs interact at different levels of cell and tissue organization making it difficult to obtain the overall BMP picture in one system, such as cells. Therefore, there is a need to develop *in vivo* and 3D multilayer cell culture models to study plan for studying BMP's behavior in space and the possibility of using the BMPs as preventive or to accelerate bone healing in space and Earth. BMP–stem cell therapy is another challenging but promising countermeasure and a deeper understanding of BMP–stem cell responses to changes in environmental stress, BMP–stem cell interactions with the environment micromilieu, BMP receptor interaction, receptor recycling, and overall systemic effect of BMP is the need of the hour. Development of new technologies that can proliferate and differentiate stem cells without making any biochemical change or disturbing the internal or external cell milieu is the need of the hour. Simulating microgravity using RPM may provide an unperturbed environment for the growth, proliferation, and maintenance of the therapeutic stem cells. These therapeutic stem cells may then be applied to the affected region by different homing technologies. A 3D matrix or supporting osteoblast cells may be provided to mimic physiology as closely as possible. These strategies would benefit not only space travelers and dwellers but also the aging population, particularly postmenopausal women and osteoporosis patients on the Earth.

ACKNOWLEDGMENTS

This work was partially supported by grants from University Grant Commission Faculty Recharge Program (UGC-FRP) and Department of Biotechnology (Grant # BT/PR3731/MED/32/209/2011), Government of India to S.C.

REFERENCES

Abe, E., Marians, R. C., Yu, W., Wu, X. B., Ando, T., Li, Y., et al. (2003). TSH is a negative regulator of skeletal remodeling. *Cell*, *115*(2), 151–162.

Abreu, J. G., Ketpura, N. I., Reversade, B., & De Robertis, E. M. (2002). Connective-tissue growth factor (CTGF) modulates cell signalling by BMP and TGF-beta. *Nature Cell Biology*, *4*(8), 599–604. http://dx.doi.org/10.1038/ncb826.

Ahn, K., Mishina, Y., Hanks, M. C., Behringer, R. R., & Crenshaw, E. B., 3rd. (2001). BMPR-IA signaling is required for the formation of the apical ectodermal ridge and dorsal-ventral patterning of the limb. *Development, 128*(22), 4449–4461.

Aleshcheva, G., Wehland, M., Sahana, J., Bauer, J., Corydon, T. J., Hemmersbach, R., et al. (2015). Moderate alterations of the cytoskeleton in human chondrocytes after short-term microgravity produced by parabolic flight maneuvers could be prevented by up-regulation of BMP-2 and SOX-9. *The FASEB Journal, 29*(6), 2303–2314. http://dx.doi.org/10.1096/fj.14-268151.

Ausk, B. J., Huber, P., Poliachik, S. L., Bain, S. D., Srinivasan, S., & Gross, T. S. (2012). Cortical bone resorption following muscle paralysis is spatially heterogeneous. *Bone, 50*(1), 14–22. http://dx.doi.org/10.1016/j.bone.2011.08.028.

Baldwin, K. M. (1996). Effect of spaceflight on the functional, biochemical, and metabolic properties of skeletal muscle. *Medicine and Science in Sports and Exercise, 28*(8), 983–987.

Bandyopadhyay, A., Tsuji, K., Cox, K., Harfe, B. D., Rosen, V., & Tabin, C. J. (2006). Genetic analysis of the roles of BMP2, BMP4, and BMP7 in limb patterning and skeletogenesis. *PLoS Genetics, 2*(12), e216. http://dx.doi.org/10.1371/journal.pgen.0020216.

Baron, R., & Rawadi, G. (2007). Targeting the Wnt/beta-catenin pathway to regulate bone formation in the adult skeleton. *Endocrinology, 148*(6), 2635–2643. http://dx.doi.org/10.1210/en.2007-0270.

Battaglino, R. A., Lazzari, A. A., Garshick, E., & Morse, L. R. (2012). Spinal cord injury-induced osteoporosis: Pathogenesis and emerging therapies. *Current Osteoporosis Reports, 10*(4), 278–285. http://dx.doi.org/10.1007/s11914-012-0117-0.

Beederman, M., Lamplot, J., Nan, G., Wang, J., Liu, X., Yin, L., et al. (2013). BMP signaling in mesenchymal stem cell differentiation and bone formation. *Journal of Biomedical Science and Engineering, 6*, 32–52. http://dx.doi.org/10.4236/jbise.2013.68A1004.

Boyle, W. J., Simonet, W. S., & Lacey, D. L. (2003). Osteoclast differentiation and activation. *Nature, 423*(6937), 337–342. http://dx.doi.org/10.1038/nature01658.

Brandi, M. L., & Collin-Osdoby, P. (2006). Vascular biology and the skeleton. *Journal of Bone and Mineral Research, 21*(2), 183–192. http://dx.doi.org/10.1359/JBMR.050917.

Briggs, P. J., Moran, C. G., & Wood, M. B. (1998). Actions of endothelin-1, 2, and 3 in the microvasculature of bone. *Journal of Orthopaedic Research, 16*(3), 340–347. http://dx.doi.org/10.1002/jor.1100160310.

Buckey, J. C., Jr., Lane, L. D., Levine, B. D., Watenpaugh, D. E., Wright, S. J., Moore, W. E., et al. (1996). Orthostatic intolerance after spaceflight. *Journal of Applied Physiology (Bethesda, Md.: 1985), 81*(1), 7–18.

Caillot-Augusseau, A., Lafage-Proust, M. H., Soler, C., Pernod, J., Dubois, F., & Alexandre, C. (1998). Bone formation and resorption biological markers in cosmonauts during and after a 180-day space flight (Euromir 95). *Clinical Chemistry, 44*(3), 578–585.

Canalis, E. (2005). Mechanisms of glucocorticoid action in bone. *Current Osteoporosis Reports, 3*(3), 98–102.

Cao, X. S., Yang, L. J., Wu, X. Y., Wu, Y. H., Zhang, L. N., & Zhang, L. F. (2003). Changes of bone morphogenesis proteins and transforming growth factor-beta in hind-limb bones of 21 d tail-suspended rats. *Space Medicine & Medical Engineering (Beijing), 16*(4), 269–271.

Chen, G., Deng, C., & Li, Y. P. (2012). TGF-beta and BMP signaling in osteoblast differentiation and bone formation. *International Journal of Biological Sciences, 8*(2), 272–288. http://dx.doi.org/10.7150/ijbs.2929.

Colleran, P. N., Wilkerson, M. K., Bloomfield, S. A., Suva, L. J., Turner, R. T., & Delp, M. D. (2000). Alterations in skeletal perfusion with simulated microgravity: A possible mechanism for bone remodeling. *Journal of Applied Physiology (Bethesda, Md: 1985), 89*(3), 1046–1054.

Collet, P., Uebelhart, D., Vico, L., Moro, L., Hartmann, D., Roth, M., et al. (1997). Effects of 1- and 6-month spaceflight on bone mass and biochemistry in two humans. *Bone, 20*(6), 547–551.

Delp, M. D., Colleran, P. N., Wilkerson, M. K., McCurdy, M. R., & Muller-Delp, J. (2000). Structural and functional remodeling of skeletal muscle microvasculature is induced by simulated microgravity. *American Journal of Physiology. Heart and Circulatory Physiology, 278*(6), H1866–H1873.

Dudas, M., Sridurongrit, S., Nagy, A., Okazaki, K., & Kaartinen, V. (2004). Craniofacial defects in mice lacking BMP type I receptor Alk2 in neural crest cells. *Mechanisms of Development, 121*(2), 173–182. http://dx.doi.org/10.1016/j.mod.2003.12.003.

Edgerton, V. R., Zhou, M. Y., Ohira, Y., Klitgaard, H., Jiang, B., Bell, G., et al. (1995). Human fiber size and enzymatic properties after 5 and 11 days of spaceflight. *Journal of Applied Physiology (Bethesda, Md.: 1985), 78*(5), 1733–1739.

Falahati-Nini, A., Riggs, B. L., Atkinson, E. J., O'Fallon, W. M., Eastell, R., & Khosla, S. (2000). Relative contributions of testosterone and estrogen in regulating bone resorption and formation in normal elderly men. *The Journal of Clinical Investigation, 106*(12), 1553–1560. http://dx.doi.org/10.1172/JCI10942.

Farhat, G. N., & Cauley, J. A. (2008). The link between osteoporosis and cardiovascular disease. *Clinical Cases in Mineral and Bone Metabolism, 5*(1), 19–34.

Fitts, R. H., Riley, D. R., & Widrick, J. J. (2000). Physiology of a microgravity environment invited review: Microgravity and skeletal muscle. *Journal of Applied Physiology (Bethesda, Md.: 1985), 89*(2), 823–839.

Fritsch-Yelle, J. M., Leuenberger, U. A., D'Aunno, D. S., Rossum, A. C., Brown, T. E., Wood, M. L., et al. (1998). An episode of ventricular tachycardia during long-duration spaceflight. *American Journal of Cardiology, 81*(11), 1391–1392.

Gazenko, O. G., Gurovsky, N. N., Genin, A. M., Bryanov, I. I., Eryomin, A. V., & Egorov, A. D. (1976). Results of medical investigations carried out on board the Salyut orbital stations. *Life Sciences and Space Research, 14*, 145–152.

Gazzerro, E., Pereira, R. C., Jorgetti, V., Olson, S., Economides, A. N., & Canalis, E. (2005). Skeletal overexpression of gremlin impairs bone formation and causes osteopenia. *Endocrinology, 146*(2), 655–665. http://dx.doi.org/10.1210/en.2004-0766.

Gazzerro, E., Smerdel-Ramoya, A., Zanotti, S., Stadmeyer, L., Durant, D., Economides, A. N., et al. (2007). Conditional deletion of gremlin causes a transient increase in bone formation and bone mass. *The Journal of Biological Chemistry, 282*(43), 31549–31557. http://dx.doi.org/10.1074/jbc.M701317200.

Geng, Y., Dong, Y., Yu, M., Zhang, L., Yan, X., Sun, J., et al. (2011). Follistatin-like 1 (Fstl1) is a bone morphogenetic protein (BMP) 4 signaling antagonist in controlling mouse lung development. *Proceedings of the National Academy of Sciences of the United States of America, 108*(17), 7058–7063. http://dx.doi.org/10.1073/pnas.1007293108.

Gilbert, S. F. (2006). *Developmental Biology* (8 th ed.). Sunderland, Massachusetts: Sinauer.

Goode, A. W., & Rambaut, P. C. (1985). The skeleton in space. *Nature, 317*(6034), 204–205.

Grigoriev, A. I., Oganov, V. S., Bakulin, A. V., Poliakov, V. V., Voronin, L. I., Morgun, V. V., et al. (1998). Clinical and physiological evaluation of bone changes among astronauts after long-term space flights. *Aviakosmicheskaia i Ekologicheskaia Meditsina, 32*, 21–25.

Gundel, A., Polyakov, V. V., & Zulley, J. (1997). The alteration of human sleep and circadian rhythms during spaceflight. *Journal of Sleep Research, 6*(1), 1–8.

Hargens, A. R., & Richardson, S. (2009). Cardiovascular adaptations, fluid shifts, and countermeasures related to space flight. *Respiratory Physiology & Neurobiology, 169*(Suppl. 1), S30–S33. http://dx.doi.org/10.1016/j.resp.2009.07.005.

Hargens, A. R., & Watenpaugh, D. E. (1996). Cardiovascular adaptation to spaceflight. *Medicine and Science in Sports and Exercise*, *28*(8), 977–982.
Hillsley, M. V., & Frangos, J. A. (1994). Bone tissue engineering: The role of interstitial fluid flow. *Biotechnology and Bioengineering*, *43*(7), 573–581. http://dx.doi.org/10.1002/bit.260430706.
Iemura, S., Yamamoto, T. S., Takagi, C., Uchiyama, H., Natsume, T., Shimasaki, S., et al. (1998). Direct binding of follistatin to a complex of bone-morphogenetic protein and its receptor inhibits ventral and epidermal cell fates in early Xenopus embryo. *Proceedings of the National Academy of Sciences of the United States of America*, *95*(16), 9337–9342.
Inoue, S., Nomura, S., Hosoi, T., Ouchi, Y., Orimo, H., & Muramatsu, M. (1994). Localization of follistatin, an activin-binding protein, in bone tissues. *Calcified Tissue International*, *55*(5), 395–397.
Jain, A. P., Pundir, S., & Sharma, A. (2013). Bone morphogenetic proteins: The anomalous molecules. *Journal of Indian Society of Periodontology*, *17*(5), 583–586. http://dx.doi.org/10.4103/0972-124X.119275.
Jones, A. R., Clark, C. C., & Brighton, C. T. (1995). Microvessel endothelial cells and pericytes increase proliferation and repress osteoblast phenotypic markers in rat calvarial bone cell cultures. *Journal of Orthopaedic Research*, *13*(4), 553–561. http://dx.doi.org/10.1002/jor.1100130410.
Kanzler, B., Foreman, R. K., Labosky, P. A., & Mallo, M. (2000). BMP signaling is essential for development of skeletogenic and neurogenic cranial neural crest. *Development*, *127*(5), 1095–1104.
Keyak, J. H., Koyama, A. K., LeBlanc, A., Lu, Y., & Lang, T. F. (2009). Reduction in proximal femoral strength due to long-duration spaceflight. *Bone*, *44*(3), 449–453. http://dx.doi.org/10.1016/j.bone.2008.11.014.
Kokabu, S., Gamer, L., Cox, K., Lowery, J., Tsuji, K., Raz, R., et al. (2012). BMP3 suppresses osteoblast differentiation of bone marrow stromal cells via interaction with Acvr2b. *Molecular Endocrinology*, *26*(1), 87–94. http://dx.doi.org/10.1210/me.2011-1168.
Lang, T., LeBlanc, A., Evans, H., Lu, Y., Genant, H., & Yu, A. (2004). Cortical and trabecular bone mineral loss from the spine and hip in long-duration spaceflight. *Journal of Bone and Mineral Research*, *19*(6), 1006–1012. http://dx.doi.org/10.1359/JBMR.040307.
Larrain, J., Bachiller, D., Lu, B., Agius, E., Piccolo, S., & De Robertis, E. M. (2000). BMP-binding modules in chordin: A model for signalling regulation in the extracellular space. *Development*, *127*(4), 821–830.
Leblanc, A. D., Schneider, V. S., Evans, H. J., Engelbretson, D. A., & Krebs, J. M. (1990). Bone mineral loss and recovery after 17 weeks of bed rest. *Journal of Bone and Mineral Research*, *5*(8), 843–850. http://dx.doi.org/10.1002/jbmr.5650050807.
Levine, B. D., Zuckerman, J. H., & Pawelczyk, J. A. (1997). Cardiac atrophy after bed-rest deconditioning: A nonneural mechanism for orthostatic intolerance. *Circulation*, *96*(2), 517–525.
Li, X., Ominsky, M. S., Niu, Q. T., Sun, N., Daugherty, B., D'Agostin, D., et al. (2008). Targeted deletion of the sclerostin gene in mice results in increased bone formation and bone strength. *Journal of Bone and Mineral Research*, *23*(6), 860–869. http://dx.doi.org/10.1359/jbmr.080216.
Li, X., Zhang, Y., Kang, H., Liu, W., Liu, P., Zhang, J., et al. (2005). Sclerostin binds to LRP5/6 and antagonizes canonical Wnt signaling. *The Journal of Biological Chemistry*, *280*(20), 19883–19887. http://dx.doi.org/10.1074/jbc.M413274200.
Luo, J., Tang, M., Huang, J., He, B. C., Gao, J. L., Chen, L., et al. (2010). TGFbeta/BMP type I receptors ALK1 and ALK2 are essential for BMP9-induced osteogenic signaling in

mesenchymal stem cells. *The Journal of Biological Chemistry*, *285*(38), 29588–29598. http://dx.doi.org/10.1074/jbc.M110.130518.

Majumder, S., Siamwala, J. H., Srinivasan, S., Sinha, S., Sridhara, S. R. C., Soundararajan, G., et al. (2011). Simulated microgravity promoted differentiation of bipotential murine oval liver stem cells by modulating BMP4/Notch1 signaling. *Journal of Cellular Biochemistry*, *112*, 1898–1908.

McCarthy, I. D. (2005). Fluid shifts due to microgravity and their effects on bone: A review of current knowledge. *Annals of Biomedical Engineering*, *33*(1), 95–103.

McCarthy, I. (2006). The physiology of bone blood flow: A review. *Journal of Bone and Joint Surgery (American Volume)*, *88*(Suppl. 3), 4–9. http://dx.doi.org/10.2106/jbjs.f.00890.

McCarthy, I. D., Andhoga, M., Batten, J. J., & Mathie, R. T. (1997). Endothelium-dependent vasodilatation produced by the L-arginine/nitric oxide pathway in normal and ischemic bone. *Acta Orthopaedica Scandinavica*, *68*(4), 361–368.

McMahon, J. A., Takada, S., Zimmerman, L. B., Fan, C. M., Harland, R. M., & McMahon, A. P. (1998). Noggin-mediated antagonism of BMP signaling is required for growth and patterning of the neural tube and somite. *Genes & Development*, *12*(10), 1438–1452.

McPherron, A. C., Lawler, A. M., & Lee, S. J. (1999). Regulation of anterior/posterior patterning of the axial skeleton by growth/differentiation factor 11. *Nature Genetics*, *22*(3), 260–264. http://dx.doi.org/10.1038/10320.

Miyazono, K., Maeda, S., & Imamura, T. (2005). BMP receptor signaling: transcriptional targets, regulation of signals, and signaling cross-talk. *Cytokine & Growth Factor Reviews*, *16*, 251–263.

Morey-Holton, E. R., & Globus, R. K. (2002). Hindlimb unloading rodent model: Technical aspects. *Journal of Applied Physiology (Bethesda, Md.: 1985)*, *92*(4), 1367–1377. http://dx.doi.org/10.1152/japplphysiol.00969.2001.

Nabavi, N., Khandani, A., Camirand, A., & Harrison, R. E. (2011). Effects of microgravity on osteoclast bone resorption and osteoblast cytoskeletal organization and adhesion. *Bone*, *49*(5), 965–974. http://dx.doi.org/10.1016/j.bone.2011.07.036.

Nguyen, T. Q., Roestenberg, P., van Nieuwenhoven, F. A., Bovenschen, N., Li, Z., Xu, L., et al. (2008). CTGF inhibits BMP-7 signaling in diabetic nephropathy. *Journal of the American Society of Nephrology*, *19*(11), 2098–2107. http://dx.doi.org/10.1681/ASN.2007111261.

Nie, X., Luukko, K., & Kettunen, P. (2006). BMP signalling in craniofacial development. *The International Journal of Developmental Biology*, *50*(6), 511–521. http://dx.doi.org/10.1387/ijdb.052101xn.

Nikaido, M., Tada, M., Saji, T., & Ueno, N. (1997). Conservation of BMP signaling in zebrafish mesoderm patterning. *Mechanisms of Development*, *61*(1–2), 75–88.

Oman, C. M., Lichtenberg, B. K., Money, K. E., & McCoy, R. K. (1986). M.I.T./Canadian vestibular experiments on the Spacelab-1 mission: 4. Space motion sickness: Symptoms, stimuli, and predictability. *Experimental Brain Research*, *64*(2), 316–334.

Papachroni, K. K., Karatzas, D. N., Papavassiliou, K. A., Basdra, E. K., & Papavassiliou, A. G. (2009). Mechanotransduction in osteoblast regulation and bone disease. *Trends in Molecular Medicine*, *15*, 208–216.

Pappano, W. N., Scott, I. C., Clark, T. G., Eddy, R. L., Shows, T. B., & Greenspan, D. S. (1998). Coding sequence and expression patterns of mouse chordin and mapping of the cognate mouse chrd and human CHRD genes. *Genomics*, *52*(2), 236–239. http://dx.doi.org/10.1006/geno.1998.5474.

Patel, M. J., Chang, K. H., Sykes, M. C., Talish, R., Rubin, C., & Jo, H. (2009). Low magnitude and high frequency mechanical loading prevents decreased bone formation responses of 2T3 preosteoblasts. *Journal of Cellular Biochemistry*, *106*(2), 306–316. http://dx.doi.org/10.1002/jcb.22007.

Peng, H., Usas, A., Hannallah, D., Olshanski, A., Cooper, G. M., & Huard, J. (2005). Noggin improves bone healing elicited by muscle stem cells expressing inducible BMP4. *Molecular Therapy, 12*(2), 239–246.

Prisby, R. D., Behnke, B. J., Allen, M. R., & Delp, M. D. (2015). Effects of skeletal unloading on the vasomotor properties of the rat femur principal nutrient artery. *Journal of Applied Physiology (Bethesda, Md.: 1985), 118*, 980–988. http://dx.doi.org/10.1152/japplphysiol.00576.2014.

Qian, X., Zhang, C., Chen, G., Tang, Z., Liu, Q., Chen, J., et al. (2014). Effects of BMP-2 and FGF2 on the osteogenesis of bone marrow-derived mesenchymal stem cells in hindlimb-unloaded rats. *Cell Biochemistry and Biophysics, 70*(2), 1127–1136. http://dx.doi.org/10.1007/s12013-014-0032-3.

Robling, A. G., Niziolek, P. J., Baldridge, L. A., Condon, K. W., Allen, M. R., Alam, I., et al. (2008). Mechanical stimulation of bone in vivo reduces osteocyte expression of Sost/sclerostin. *Journal of Biological Chemistry, 283*(9), 5866–5875. http://dx.doi.org/10.1074/jbc.M705092200.

Rucci, N., Migliaccio, S., Zani, B. M., Taranta, A., & Teti, A. (2002). Characterization of the osteoblast-like cell phenotype under microgravity conditions in the NASA-approved rotating wall vessel bioreactor (RWV). *Journal of Cellular Biochemistry, 85*(1), 167–179.

Safadi, F. F., Xu, J., Smock, S. L., Kanaan, R. A., Selim, A. H., Odgren, P. R., et al. (2003). Expression of connective tissue growth factor in bone: Its role in osteoblast proliferation and differentiation in vitro and bone formation in vivo. *Journal of Cellular Physiology, 196*(1), 51–62.

Sanchez, C. P., & He, Y. Z. (2007). Bone growth during daily or intermittent calcitriol treatment during renal failure with advanced secondary hyperparathyroidism. *Kidney International, 72*(5), 582–591. http://dx.doi.org/10.1038/sj.ki.5002375.

Sexton, P. M., Findlay, D. M., & Martin, T. J. (1999). Calcitonin. *Current Medicinal Chemistry, 6*(11), 1067–1093.

Shu, B., Zhang, M., Xie, R., Wang, M., Jin, H., Hou, W., et al. (2011). BMP2, but not BMP4, is crucial for chondrocyte proliferation and maturation during endochondral bone development. *Journal of Cell Science, 124*(Pt. 20), 3428–3440. http://dx.doi.org/10.1242/jcs.083659.

Siamwala, J. H., Reddy, S. H., Majumder, S., Kolluru, G. K., Muley, A., Sinha, S., et al. (2010a). Simulated microgravity perturbs actin polymerization to promote nitric oxide associated migration in human immortalized Eahy926 cells. *Protoplasma, 242*(1-4), 3–12.

Siamwala, J. H., Majumder, S., Tamilarasan, K. P., Muley, A., Reddy, S. H., Kolluru, G. K., et al. (2010b). Simulated microgravity promotes nitric oxide-supported angiogenesis via the iNOS-cGMP-PKG pathway in macrovascular endothelial cells. *FEBS Letters, 4, 584*(15), 3415–3423.

Siegel, P. M., & Massague, J. (2003). Cytostatic and apoptotic actions of TGF-beta in homeostasis and cancer. *Nature Reviews Cancer, 3*, 807–821.

Sievanen, H. (2010). Immobilization and bone structure in humans. *Archives of Biochemistry and Biophysics, 503*(1), 146–152. http://dx.doi.org/10.1016/j.abb.2010.07.008.

Smith, S. M., Wastney, M. E., Morukov, B. V., Larina, I. M., Nyquist, L. E., Abrams, S. A., et al. (1999). Calcium metabolism before, during, and after a 3-mo spaceflight: Kinetic and biochemical changes. *American Journal of Physiology, 277*(1 Pt. 2), R1–R10.

Stabley, J. N., Prisby, R. D., Behnke, B. J., & Delp, M. D. (2013). Chronic skeletal unloading of the rat femur: Mechanisms and functional consequences of vascular remodeling. *Bone, 57*(2), 355–360. http://dx.doi.org/10.1016/j.bone.2013.09.003.

Sylva, M., Li, V. S., Buffing, A. A., van Es, J. H., van den Born, M., van der Velden, S., et al. (2011). The BMP antagonist follistatin-like 1 is required for skeletal and lung organogenesis. *PloS One, 6*(8), e22616. http://dx.doi.org/10.1371/journal.pone.0022616.

Tardif, G., Pelletier, J. P., Hum, D., Boileau, C., Duval, N., & Pelletier, J. M. (2006). Differential regulation of the bone morphogenic protein antagonist chordin in human normal and osteoarthritic chondrocytes. *Annals of the Rheumatic Diseases*, *65*(2), 261–264.

Taylor, G. R. (1993). Overview of spaceflight immunology studies. *Journal of Leukocyte Biology*, *54*(3), 179–188.

Vico, L., Collet, P., Guignandon, A., Lafage-Proust, M. H., Thomas, T., Rehaillia, M., et al. (2000). Effects of long-term microgravity exposure on cancellous and cortical weight-bearing bones of cosmonauts. *Lancet*, *355*(9215), 1607–1611.

Villanueva, J. E., & Nimni, M. E. (1990). Promotion of calvarial cell osteogenesis by endothelial cells. *Journal of Bone and Mineral Research*, *5*(7), 733–739. http://dx.doi.org/10.1002/jbmr.5650050710.

Wang, R. N., Green, J., Wang, Z., Deng, Y., Qiao, M., Peabody, M., et al. (2014). Bone morphogenetic protein (BMP) signaling in development and human diseases. *Genes & Diseases*, *1*(1), 87–105. http://dx.doi.org/10.1016/j.gendis.2014.07.005.

Wang, B., Zhang, S., & Wu, X. Y. (2004). Effects of BMP-2 on the gene expression of rat osteosarcoma cells under simulated weightlessness. *Space Medicine & Medical Engineering (Beijing)*, *17*(3), 176–179.

Wang, J., Zhou, J., Cheng, C. M., Kopchick, J. J., & Bondy, C. A. (2004). Evidence supporting dual, IGF-I-independent and IGF-I-dependent, roles for GH in promoting longitudinal bone growth. *The Journal of Endocrinology*, *180*(2), 247–255.

White, R. J., & Averner, M. (2001). Humans in space. *Nature*, *409*(6823), 1115–1118. http://dx.doi.org/10.1038/35059243.

Wiater, E., & Vale, W. (2003). Inhibin is an antagonist of bone morphogenetic protein signaling. *The Journal of Biological Chemistry*, *278*(10), 7934–7941. http://dx.doi.org/10.1074/jbc.M209710200.

Wilkerson, M. K., Muller-Delp, J., Colleran, P. N., & Delp, M. D. (1999). Effects of hindlimb unloading on rat cerebral, splenic, and mesenteric resistance artery morphology. *Journal of Applied Physiology (Bethesda, Md.: 1985)*, *87*(6), 2115–2121.

Worthley, D. L., Churchill, M., Compton, J. T., Tailor, Y., Rao, M., Si, Y., et al. (2015). Gremlin 1 identifies a skeletal stem cell with bone, cartilage, and reticular stromal potential. *Cell*, *160*(1–2), 269–284. http://dx.doi.org/10.1016/j.cell.2014.11.042.

Wozney, J. M., Rosen, V., Celeste, A. J., Mitsock, L. M., Whitters, M. J., Kriz, R. W., et al. (1988). Novel regulators of bone formation: Molecular clones and activities. *Science*, *242*(4885), 1528–1534.

Wronski, T. J., & Morey, E. R. (1983). Effect of spaceflight on periosteal bone formation in rats. *American Journal of Physiology*, *244*(3), R305–R309.

Yagodovsky, V. S., Triftanidi, L. A., & Gorokhova, G. P. (1976). Space flight effects on skeletal bones of rats (light and electron microscopic examination). *Aviation, Space, and Environmental Medicine*, *47*(7), 734–738.

Yoshimura, Y., Nomura, S., Kawasaki, S., Tsutsumimoto, T., Shimizu, T., & Takaoka, K. (2001). Colocalization of noggin and bone morphogenetic protein-4 during fracture healing. *Journal of Bone and Mineral Research*, *16*(5), 876–884.

Yu, B., Zhao, X., Yang, C., Crane, J., Xian, L., Lu, W., et al. (2012). Parathyroid hormone induces differentiation of mesenchymal stromal/stem cells by enhancing bone morphogenetic protein signaling. *Journal of Bone and Mineral Research*, *27*(9), 2001–2014. http://dx.doi.org/10.1002/jbmr.1663.

Zhang, B., Cory, E., Bhattacharya, R., Sah, R., & Hargens, A. R. (2013). Fifteen days of microgravity causes growth in calvaria of mice. *Bone*, *56*(2), 290–295. http://dx.doi.org/10.1016/j.bone.2013.06.009.

Zhou, S., Turgeman, G., Harris, S. E., Leitman, D. C., Komm, B. S., Bodine, P. V., et al. (2003). Estrogens activate bone morphogenetic protein-2 gene transcription in mouse mesenchymal stem cells. *Molecular Endocrinology*, *17*(1), 56–66. http://dx.doi.org/10.1210/me.2002-0210.

CHAPTER NINE

Bmpr1a Signaling in Cartilage Development and Endochondral Bone Formation

Junjun Jing*, Robert J. Hinton†, Jian Q. Feng†,[1]
*State Key Laboratory of Oral Diseases, West China Hospital of Stomatology, Sichuan University, Chengdu, China
†Department of Biomedical Sciences, Texas A&M Baylor College of Dentistry, Dallas, Texas, USA
[1]Corresponding author: e-mail address: jfeng@bcd.tamhsc.edu

Contents

Abstract

The type IA bone morphogenetic protein receptor (*Bmpr1a*), encoded by 11 exons and spanning about 40 kb on chromosome 14 in mice and chromosome 10 in human (Derynck & Feng, 1997; Mishina, Hanks, Miura, Tallquist, & Behringer, 2002), is an essential receptor for BMP signaling. This chapter focuses on the current understanding of the role of *Bmpr1a* in cartilage development and endochondral ossification, including formation of the mesenchymal condensation, chondrocyte differentiation and maturation, and endochondral bone development.

1. INTRODUCTION

BMPs were originally identified as proteins that induce bone and cartilage formation. It is now clear that they play much broader roles in different developmental processes. Thus, some investigators have even suggested that

Vitamins and Hormones, Volume 99
ISSN 0083-6729
http://dx.doi.org/10.1016/bs.vh.2015.06.001

they be named body morphogenetic proteins (Wagner et al., 2010). BMPs comprise an extensive group of phylogenetically conserved growth factors of which over 20 members have been identified to date, and they constitute the largest subgroup of the transforming growth factor beta (TGF-β) superfamily (Anderson, Schneider, Morisato, Jin, & Ferguson, 1992; Ferguson & Anderson, 1992; Lavery, Swain, Falb, & Alaoui-Ismaili, 2008; Zwijsen, Verschueren, & Huylebroeck, 2003). BMPs are first translated as large preproproteins, containing a signal peptide, prodomain, and mature domain. After removal of the signal peptide, the proproteins undergo dimerization, after which the specific proteolytic enzymes cleave the dimerized proprotein to generate the biologically active dimeric mature protein (Chen, Deng, & Li, 2012).

BMPs can be divided into four distinct subfamilies according to their sequences and functions: BMPs 2 and 4; BMPs 5, 6, 7, 8a, and 8b; BMPs 9 and 10; and BMPs 12, 13, and 14. BMP2, BMP4, and BMP7 display osteogenic and dentinogenic potential. BMPs 8 and 9 show distinct actions in chondrogenesis and BMP10 plays a part in heart, and oocyte development, and in follicular formation (Carreira et al., 2014). Although they share some fundamental similarities with other members of the TGF-β superfamily, members of the BMP family contain two conserved cysteines in their structure, and the complexity of levels at which their signaling function is both determined and modulated far exceeds that in the other members of the superfamily. As the number of BMPs increased to accommodate the increasing complexity of the organisms whose formation they directed, functional constraints increased in parallel. The action of BMPs is time dependent, triggering a sequential cascade of events, which leads to chondrogenesis, osteogenesis, angiogenesis, and controlled synthesis of extracellular matrix by modulation of several genes. BMPs promote almost every aspect of chondrogenesis, from commitment to terminal differentiation (Yoon & Lyons, 2004).

Like other TGF-β superfamily members, BMPs bind to two major types of membrane-bound serine/threonine kinase receptors, the type I and type II receptors (Wan and Cao, 2005). BMP signaling first transmits signals across the plasma membrane through the formation of heteromeric complexes of specific type I and type II serine/threonine kinase receptors. The type I receptor is phosphorylated following the activation of the specific type II receptor. Activated type I receptors initiate intracellular signaling through phosphorylation of specific Smad proteins, R-Smads. Activated R-Smads form a complex with co-Smad and Smad4, and then translocate into the nucleus to direct the transcriptional response. Five known BMP type I receptors (BRI) function with the multiplicity of ligands described above: ALK1 (Acvrl1), ALK2 (ActR1A), ALK3 (BMPR1A), ALK4 (ActR1b), and ALK6 (BMPR1B). In addition, there are three type II receptors: BRII, ActRIIa, and ActRIIb (Bragdon et al., 2011; Nohe, Keating, Knaus, & Petersen, 2004). BMPR1A is more widely expressed than BMPR1B in various tissues (Ikeda et al., 1996).

2. CARTILAGE DEVELOPMENT AND ENDOCHONDRAL BONE FORMATION

The skeleton in mammals is formed via two distinct processes during embryogenesis: intramembranous ossification and endochondral ossification (Karsenty, 2003; Komori, 2011; Kronenberg, 2003; Long & Ornitz, 2013). Many of the craniofacial bones are formed directly from mesenchymal condensations via intramembranous bone formation. In contrast, endochondral ossification, the principal process responsible for forming much of the mammalian skeleton, generates bone via a cartilage intermediate. During endochondral ossification, multipotent undifferentiated mesenchymal cells differentiate into proliferative chondrocytes, prehypertrophic chondrocytes, and hypertrophic chondrocytes; subsequently, the cartilage tissues formed by these chondrocytes are replaced by bone (De Crombrugghe et al., 2000; Nishimura et al., 2008). The transition from cartilage to bone is tightly coupled with chondrocyte, osteoblast, and vascular differentiation. Endochondral ossification is apparently more complex than intramembranous ossification.

Endochondral bone development begins with the condensation of mesenchymal cells of either neural crest origin in the craniofacial region or mesodermal origin elsewhere in the body. Formation of the mesenchymal condensation is a prerequisite for subsequent chondrogenic differentiation. In the limb bud, the condensation forms mainly through active congregation of cells without changes in cell proliferation. These condensations can be visualized by the dense packing of cells, the high affinity to the lectin peanut agglutinin; and the transient upregulation of versican, tenascin, syndecan, N-CAM, and N-cadherin (N-cad) (Hall & Miyake, 2000). Following mesenchymal condensation, cells in the core of the condensations differentiate into chondrocytes that secrete a cartilage matrix rich in type II and type X collagen and specific proteoglycans such as aggrecan, whereas the expression of type I collagen is suppressed. Cells at the periphery of the condensation form the perichondrium which continues to express type I collagen and demarcates the developing skeletal element from the surrounding mesenchyme (Caplan & Pechak, 1987). After formation of the cartilage primordia, all chondrocytes undergo rapid proliferation that drives the linear growth of the skeletal elements. At a certain stage specific for each element, chondrocytes in the center undergo progressive maturation. These cells eventually exit the cell cycle and undergo hypertrophy associated with the secretion of type X collagen. The terminal hypertrophic chondrocytes express additional molecular markers such as matrix metalloproteinase 13 (MMP13); most if not all, are generally believed to undergo apoptosis. These changes are accompanied by vascular invasion of the hypertrophic cartilage and by differentiation of the inner perichondrium cells into osteoblasts, which secrete bone matrix to form the bone collar. Vascular invasion of the hypertrophic cartilage leads to further degradation of the mineralized matrix by MMP9 secreted by a resorptive cell type (the so-called chondroclast) (Stickens et al., 2004), whose formation may be stimulated in turn by RANKL (receptor activator of NF-kB ligand) produced by hypertrophic chondrocytes (Xiong et al., 2011). Vascular invasion also brings with it osteoprogenitors that differentiate into osteoblasts, which establish the primary ossification center to generate the trabecular bone (Maes et al., 2010).

The orderly maturation of chondrocytes in the embryonic growth plate produces zones of proliferation, hypertrophy, and bone formation, linearly progressing from the epiphysis to the shaft (diaphysis) of the skeletal element. The proliferative zone chondrocytes can be further divided into regions of round versus columnar chondrocytes based on their morphology. Continued proliferation of the less mature chondrocytes at the epiphysis, followed by their hypertrophy and their eventual replacement by trabecular bone near the diaphysis, results in a distal displacement of the growth plate and longitudinal growth of each skeletal element (Long et al., 2004).The orderly progression of the growth plate is critical for proper development of the endochondral skeleton and is subject to regulation by key extracellular signals and nuclear factors.

3. *Bmpr1a* SIGNALING IN CHONDROGENIC COMMITMENT

BMPR1A null mice die at embryonic day 8.0 without mesoderm specification, demonstrating its vital role in gastrulation (Mishina et al.,

2004). *In vitro* systems have been used to investigate the role of BMP pathways in the earliest stages of chondrogenesis: commitment and condensation. BMPs induce differentiation of pluripotent mesenchymal cell lines, such as C3H10T1/2, into chondrocytes when grown at high density (Ju et al., 2000; Kramer et al., 2000). High-density culture mimics the condensation event that precedes chondrogenesis *in vivo*. One mechanism by which BMPs induce chondrogenesis in this system is through upregulation of N-cad function.

The requirement for BMP pathways in the formation of precartilaginous condensations *in vivo* has been demonstrated by studies in the chick limb. The use of the secreted BMP inhibitor noggin permits antagonism of endogenously produced BMPs while avoiding potential artifacts arising from overexpression of dominant-negative (DN) receptors. Overexpression of noggin blocks condensation, leading to a total absence of cartilage. On the other hand, overexpression of constitutively active (CA) BMP receptors results in expansion of cartilage at the expense of muscle and soft tissues (Majumdar, Wang, & Morris, 2001). These and related studies establish that BMP signaling is required for, and acts as part of, an instructive signal to promote commitment to the chondrogenic lineage. The newborn mice of the BMPR1A mutant with Prx1-Cre exhibited under-mineralized parietal and interparietal bones, absence of multiple phalanges, dysmorphic shortening of all remaining limb elements, as well as a partially split sternum. *Bmpr1a* signaling is required for precartilaginous mesenchymal condensation during the limb bud formation (Lim et al., 2015; Logan et al., 2002).

The DN forms of BMP receptors were introduced into immature and mature chondrocytes isolated from lower and upper portions of chick embryo sternum to examine the role of BMP receptors in chondrocytes during endochondral ossification. Expression of a DN-type II BMP receptor (termed DN-BMPR-II) in immature lower sternal (LS) chondrocytes led to a loss of differentiated functions; compared with control cells, the LS chondrocytes expressing DN-BMPRII proliferated more rapidly, acquired a fibroblastic morphology, showed little expression of type II collagen and aggrecan genes, and upregulated type I collagen gene expression. DN-BMPRIA had little effect, while DN-BMPRIB had similar but weaker effects compared with those of DN-BMPRII. We conclude that BMP signaling, particularly that mediated by the type II BMP receptor, is required for maintenance of the differentiated phenotype, control of cell proliferation, and expression of hypertrophic phenotype (Enomoto-Iwamoto et al., 1998).

Conversely, the overexpression of a DN-type I BMP receptor or treatment with Noggin reduces cartilage formation and type II collagen production (Pizette & Niswander, 2000). The clearest *in vivo* evidence demonstrating that continuous BMP signaling is required in chondrogenesis comes from studies in chick limbs. Precartilaginous cells in condensations do not differentiate into chondrocytes in the absence of BMP signaling.

Type I BMP receptors have both distinct and overlapping expression patterns in the chondrogenic lineage. In addition to overlapping expression of *Bmpr1a* and *1b* in prechondrogenic condensations, *Bmpr1a* is expressed in prehypertrophic and hypertrophic chondrocytes, while *Bmp1b* is found throughout the growth plate and Alk-2 is expressed primarily in resting and proliferating chondrocytes (Shukunami, Akiyama, Nakamura, & Hiraki, 2000). That BMP receptors have overlapping functions in mammals is shown by the more severe phenotype of mice lacking both *Bmp7* and *Bmp1a* (Zou, Wieser, Massague, & Niswander, 1997). In these mice, a number of skeletal elements are severely reduced or absent, demonstrating that BMP7 can activate BMPR1A and/or ALK-2 *in vivo*, and that these receptors have synergistic functions with BMPR1B.

Bmpr1a and *Bmpr1b* are functionally redundant during early growth plate chondrogenesis. Mice deficient in type I receptors *Bmpr1a* or *Bmpr1b* in cartilage are able to form intact cartilaginous elements, while double mutants develop a severe generalized chondrodysplasia. The majority of skeletal elements that form through endochondral ossification are absent, and the ones that form are rudimentary. The few cartilage condensations that form in double mutants are delayed in the prechondrocytic state and never form an organized growth plate. The reduced size of mutant condensations results from increased apoptosis and decreased proliferation. Moreover, the expression of cartilage-specific extracellular matrix proteins is severely reduced in mutant elements. The defect in chondrocyte differentiation can be attributed to lack of Sox9, L-Sox5, and Sox6 expression in precartilaginous condensations in double mutants (Yoon et al., 2005).

Previous studies suggest that BMPs are required to maintain the chondrocyte phenotype and this implies that BMP signaling regulates the distinct patterns of gene expression characteristic of the different populations of chondrocytes within the growth plate. Identification of the signaling pathways through which BMPs act to control distinct aspects of chondrocyte proliferation and differentiation is an area of intense interest. Overexpression of BMPR1A or 1B in chick limb buds results in identical expansions of cartilaginous elements and chondrocyte proliferation (Yi, Daluiski, Pederson, Rosen, & Lyons, 2000).

4. BMPR1A IN ARTICULAR CHONDROCYTE DIFFERENTIATION AND LIMB JOINT DEVELOPMENT

The synovial joints including the elbow, the knee, or interphalangeal joints in the limbs are intricate and diverse organs. They are composed of reciprocally shaped surfaces covered by articular cartilage, stabilized mechanically by intrajoint and peri-joint ligaments, and insulated from the body environment by the synovial lining and a thick surrounding synovial capsule (Archer, Caterson, Benjamin, & Ralphs, 1999). Articular cartilage itself is rather complex and is organized in histologically and phenotypically distinct zones (Warman et al., 2011). The superficial zone contains elongated and flat-shaped cells oriented parallel to the articular surface, held together by a largely collagenous matrix, and producing lubricin, hyaluronate, and other antiadhesive macromolecules essential for frictionless joint movement (Hunziker, Kapfinger, & Geiss, 2007). Articular chondrocytes in the middle zone are round in shape, usually organized in small vertical rows, and produce and maintain important extracellular components—particularly collagen II and aggrecan—that confer on the tissue its key biomechanical feature: resilience. The chondrocytes in the bottom zone tend to be larger in size, are also active in matrix production and maintenance, and face the critical tissue boundary—referred to as the tidemark between articular cartilage and underlying subchondral bone (Decker, Koyama, & Pacifici, 2014; Jay, Tantravahi, Britt, Barrach, & Cha, 2001).

Nishihara and coworkers (Fujii et al., 1999) used a recombinant adenovirus system to examine whether type I receptors for the members of the BMP family modulate differentiation of articular chondrocytes induced by retinoic acid. Constitutively active forms of type I receptors were infected in bovine articular chondrocytes. The intensity of Alcian blue staining revealed that of the six constitutively active type I receptors, ALK-1, ALK-2, ALK-3, and ALK-6 maintained the chondrocytic phenotype. BMPs bind to ALK-3 and ALK-6. Moreover, BMP-6, BMP-7, and probably other BMPs bind to ALK-2, which is structurally similar to ALK-1 (Ebisawa et al., 1999; Macías-Silva, Hoodless, Tang, Buchwald, & Wrana, 1998; Ten Dijke et al., 1994). In contrast to these receptors, ALK-4 and ALK-5, type I receptors for activin and TGF-β, respectively, did not help maintain chondrocytic phenotype. It has been reported that in the chondrocyte cell line ATDC5 constitutively active forms of ALK-3 and ALK-6 (but not those of ALK-1 or ALK-2) stimulated chondrogenesis (Nishihara, Fujii, Sampath, Miyazono, & Reddi, 2003). This difference in findings is probably due to differences in cell types tested, i.e., primary chondrocytes or the cell line used in the study.

Rountree and coworkers used the Cre/loxP system and joint-specific Gdf5-Cre (Rountree et al., 2004) to remove the *Bmpr1a* gene in mouse limb joint. Tests with reporter mice show that this system is capable of modifying genes in all of the structures of the mature synovial joint, including the ligaments of the joint capsule, the synovial membrane, and the articular cartilage. Gdf5-Cre recombination bypasses the early embryonic lethality of null mutations in *Bmpr1a* and shows that this receptor is required for early joint formation at some locations and for initiation of programmed cell death in webbing between digits. Moreover, *Bmpr1a* is also required for postnatal maintenance of articular cartilage throughout most of the skeleton. In Gdf5-Cre/*Bmpr1a* mutant mice, articular cartilage initially forms normally, but subsequently loses expression of several key cartilage markers after birth. It ultimately fibrillates and degenerates, resulting in severe osteoarthritis and loss of mobility. These data suggest that *Bmpr1a* signaling is required for normal maintenance of postnatal articular cartilage and that modulation of the *Bmpr1a* signaling pathway may play an important role in joint disease.

5. *Bmpr1a* SIGNALING IN MECKEL'S CARTILAGE AND TEMPOROMANDIBULAR JOINT DEVELOPMENT

Meckel's cartilage, which is an intermediate structure in the development of the embryonic mandible in mammals, disappears by undergoing different cell fates along the distal-proximal axis, with the majority (middle portion) undergoing degeneration and chondroclastic resorption (Ito et al., 2002). To investigate the function of BMP signaling in Meckel's cartilage development, Wang et al. used Noggin mutant mice as a model for gain-of-BMP signaling function, with a focus on the middle portion (Wang, Zheng, Chen, & Chen, 2013). They showed that *Bmp2* and *Bmp7* are expressed in early developing Meckel's cartilage, but their expression disappears thereafter. In contrast, Noggin is expressed constantly in Meckel's cartilage throughout the entire gestation period. In the absence of Noggin, Meckel's cartilage is significantly thickened due to a dramatically elevated cell proliferation rate associated with enhanced phosphorylated *Smad*1/5/8 expression. Importantly, chondrocyte-specific expression of a constitutively active form of *Bmpr1a* leads to enlargement of Meckel's cartilage, phenocopying the consequence of Noggin deficiency. This indicated that elevated BMP signaling prevents degeneration and leads to endochondral ossification of Meckel's cartilage and suggests that withdrawal of *Bmpr1a* signaling may be required for normal Meckel's cartilage development and subsequent degeneration.

The temporomandibular joint (TMJ) is an important contributor to the growth of the mandible in length and height, resulting from endochondral ossification at the deep surface of the mandibular condylar cartilage (MCC). Because its morphogenesis occurs late in prenatal development, the MCC has been designated as a secondary cartilage in contradistinction to primary cartilages of the limbs and cranial base (Beresford, 1981), and it develops independently of Meckel's cartilage. Because of its persistence and role in the growth of the mandible, the MCC is one of the most important secondary cartilages. A secondary cartilage, such as the MCC, differs from primary cartilages most clearly in its superficial layers, comprising a perichondrium in which the cells that are relatively undifferentiated (prechondroblastic) secrete a matrix rich in type I collagen rather than the type II collagen matrix secreted by chondrocytes (Mizoguchi, Nakamura, Takahashi, Kagayama, & Mitani, 1990). It is these relatively undifferentiated cells of the prechondroblastic zone of the perichondrium, not the chondrocytes in deeper layers, that proliferate and mature to effect growth at the MCC. Unlike the proliferative chondrocytes of primary cartilaginous joints, the prechondrocytes in the condylar cartilage exhibit a dual potential, forming either cartilage or bone, depending on the mechanical forces impinging on the tissue (Lydiatt & Davis, 1985).

In light of this unusual manner of development and the distinctive characteristics of the dividing cells in the MCC, it would not be surprising if the molecular determinants of development and growth of the TMJ were somewhat different from those in primary cartilaginous joints. However, studies of the specific genes regulating TMJ morphogenesis and growth have only begun to appear in the literature within the last decade, although many of these developmental and structural differences have been known for years (Hinton, 2014).

Gu et al. (2014) investigated the function of BMPR1A-mediated signaling in TMJ development in mice by transgenic loss-of- and gain-of-function approaches. They found that BMPR1A is expressed in the cranial neural crest (CNC)-derived developing condyle and glenoid fossa, major components of TMJ, as well as the interzone mesenchymal cells. Wnt1-Cre-mediated tissue-specific inactivation of *Bmpr1a* in CNC lineage led to defective TMJ development, including failure of articular disk separation from a hypoplastic condyle, persistence of interzone cells, and failed formation of a functional fibrocartilage layer on the articular surface of the glenoid fossa and condyle, which could be at least partially attributed to the downregulation of Indian hedgehog (IHH) in the developing condyle and inhibition of apoptosis in the interzone. On the other hand, augmented

BMPR1A signaling by *Wnt1-Cre* driven expression of a CA form of *Bmprla* (ca*Bmprla*) inhibited osteogenesis of the glenoid fossa and converted the condylar primordium from secondary cartilage to primary cartilage associated with ectopic activation of Smad-dependent pathway but inhibition of JNK pathway, leading to TMJ agenesis. This clearly indicates that the *Bmpr1a*-mediated signaling plays an essential role in TMJ development.

To examine the effect of *Bmpr1a* on the postnatal growth of the TMJ, Jing and coworkers ablated *Bmpr1a* in the condylar cartilage of neonatal mice and assessed the consequences for mandibular condyle growth and organization at intervals over the ensuing 4 weeks. *Bmpr1a* deficiency caused significant chondrodysplasia and almost eliminated the chondrocytic phenotype in the TMJ. Many important markers such as Sox9, collagen II, proteoglycan were dramatically reduced with little sign of cell proliferation in the conditional knockout cartilage tissue (Fig. 1). In addition, subchondral

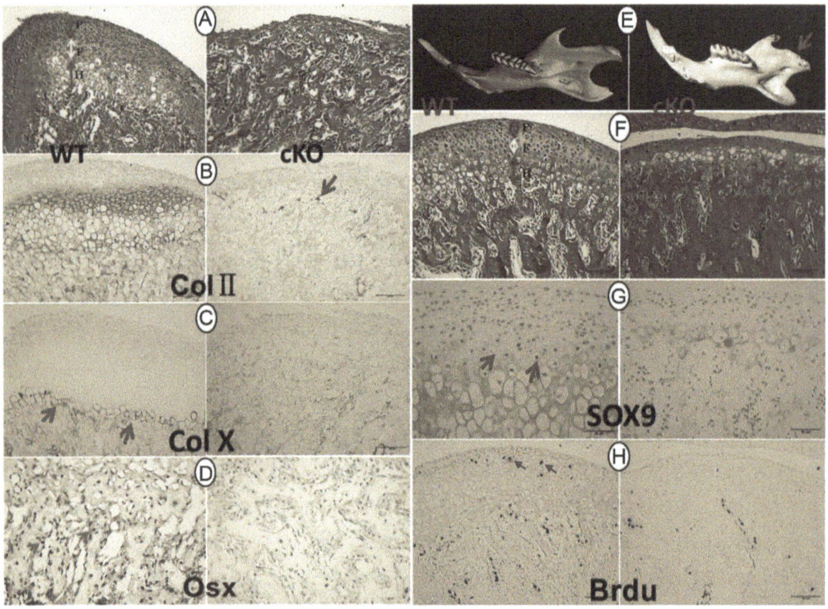

Figure 1 Phenotype of mandibular condyle in *Bmpr1a*$^{fl/fl}$; *Aggrecan-CreERT2* mice. (A) No indication of proteoglycans in the knockout MCC (2 weeks of age). (B) Col II immunoreactivity. (C) col X immunoreactivity. (D) Osterix immunoreactivity (small red arrows). (E) Micro-CT images demonstrating the shorter (red arrow) condyle in *Bmpr1a*-deficient mice (4 weeks of age). (F) Safranin O staining. (G) SOX9 immunoreactivity. (H) BrdU immunoreactivity (red arrows). P, polymorphic layer; F, flattened layer; H, hypertrophic layer. *Reproduced with permissions from Jing J, Hinton RJ, Mishina Y, Liu Y, Zhou X, Feng JQ. Critical role of Bmpr1a in mandibular condyle growth, Connective Tissue Research, 2014 Aug;55 Suppl 1:73-8.* (See the color plate.)

bone formation was also disturbed and was accompanied by reduced Osterix expression (Jing et al., 2014). These findings strongly suggest that *Bmpr1a* is critical for the postnatal growth of the mandibular condyle via its effect on proliferation of prechondroblasts and chondrocyte differentiation.

6. *Bmpr1a* SIGNALING IN GROWTH PLATE CHONDROGENIC DIFFERENTIATION AND GROWTH

While BMP signaling is clearly essential for the initiation of chondrogenesis in growth plate (Section 3), the effect of *Bmpr1a* signaling on the subsequent growth and differentiation of the growth plate has also been studied using gain- and loss-of-function approaches. Jing et al. took advantage of an inducible Aggrecan CreERT2 knock-in allele mouse line (Agc1CreERT2) in which the Cre recombinase is expressed in all the cartilage layers from the resting zone to the hypertrophic zone when tamoxifen is injected (Jing et al., 2013). *Bmpr1a* cKO mice were generated through crossing this line to the homozygous *Bmpr1a flox/flox* line (Mishina, Hanks, Miura, Tallquist, & Behringer, 2002) with a one-time tamoxifen injection at birth and growth plates were harvested separately at different time points. There was essentially no long bone growth after *Bmpr1a* is removed postnatally from the progenitor cells in the growth plate. There was a sharp reduction in bone length with a small metaphyseal bone mass in the null femur at 1 month of age (Fig. 2A). Numerous calcified trabeculae with no signs of osteocytes in the control femur metaphyses were detected using SEM, and they are connected to the chondrocyte columns (Fig. 2A, lower panels). In contrast, in the null metaphysis, there were no chondrocyte columns but numerous mineral spheres (red arrowheads) that directly formed a few large clusters of bone. These histological images disclosed a lack of mature chondrocytes and proteoglycan production. Safranin O staining revealed little production of proteoglycans and a lack of endochondral bone formation at 1 month null mice (Fig. 2B).

The failure of long bone growth indicated that there were no new progenitor cells added after deletion of the *Bmpr1a* gene by one-time injection of tamoxifen in newborns. Surprisingly, the null growth plate stopped normal chondrogenesis and was gradually replaced by bone and fibrous tissues as shown by FITC stain. In addition to this mixture, there were also numerous blood vessels (arrows) in the null growth plate (Fig. 2C). It is known that SOX9 (Bi et al., 2001; Henry, Liang, Akdemir, & de Crombrugghe, 2012) and IHH (Hilton, Tu, Cook, Hu, & Long, 2005; Long et al.,

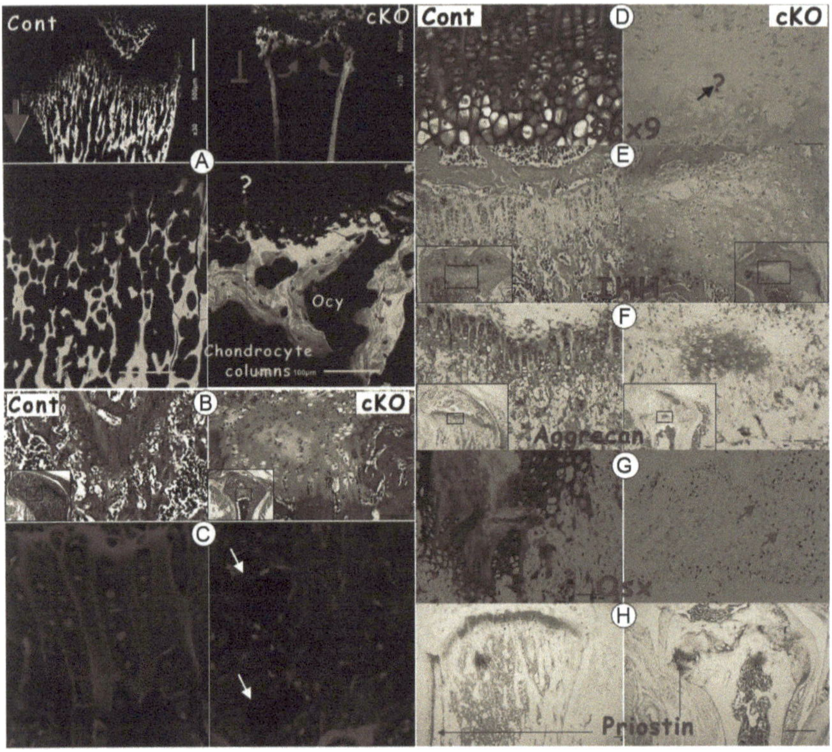

Figure 2 Long bone phenotype in *Bmpr1a*$^{fl/fl}$; *Aggrecan-CreERT2* mice. Representative backscattered SEM (A) images showed a lack of chondrocyte columns and trabeculae in the cKO femur. Instead, there were numerous spherical mineral bodies adjacent to a few large clusters of mature bones, in which the osteocytes (Ocy) were well formed; the Safranin O stained images revealed lack of proteoglycan in the mutant growth plate (B) and the FITC confocal images revealed blood vessels and fibrous-like tissues in the cKO growth plate (C); IHC-stained images in the cKO growth plate indicated either a lack of signals of SOX9 (D) or IHH (E) separately, or faint signals of Aggrecan (F), or ectopic expressions of OSX (G), and Periostin (H). Cont, control; cKO, conditional knockout. *Reproduced with permissions from Jing J, Ren Y, Zong Z, Chuanju Liu C, Nobuhiro Kamiya N, Mishina Y, Liu Y, Zhou X, and Feng JQ. BMP Receptor 1A determines the cell fate of the postnatal growth plate Int. J. Biol. Sci. 2013; 9(9):895-906.* (See the color plate.)

2004) are critical for cartilage formation, so the failure of chondrogenesis in the *Bmpr1a* null growth plate could be the consequence of reductions in these two factors (Fig. 2D–E). There was only patchy expression of aggrecan in the null growth plate, which likely reflects the remains of this extracellular matrix protein from the existing cartilage prior to *Bmpr1a* removal (Fig. 2F). Perhaps, the most surprising finding is the ectopic expression of bone and fibrous markers in the null growth plate, including OSX and Periostin (highly expressed in periosteum) (Fig. 2G–H). These findings demonstrate a remarkable plasticity of the progenitor cells in the growth plate.

The effect of CA *Bmpr1a* (*caBmpr1a*) has been studied by overexpression in chondrocytes using two systems: *caBmpr1a* was directly driven by a rat type II collagen promoter in a conventional transgenic system and indirectly driven in a UASGal4 binary system. *CaBmpr1a* expression caused shortening of the columnar layer of proliferating chondrocytes and upregulation of maturation markers, suggesting acceleration of differentiation of proliferating chondrocytes toward hypertrophic chondrocytes. In addition to the acceleration of chondrocyte differentiation, conventional transgenic mice showed widening of cartilage elements and morphological alteration of perichondrial cells, possibly due to stimulation of differentiation of prechondrogenic cells. Moreover, bigenic expression of *caBmpr1a* rescued the differentiation defect of prechondrogenic cells in *Bmpr1b*-null phalanges. This finding indicates that BMP signaling is necessary for phalangeal prechondrogenic cells to differentiate into chondrocytes and that signaling of *Bmpr1b* in this context is replaceable by that of a CA *Bmpr1a*. Thus, BMP signaling stimulates chondrogenic differentiation in prechondrogenic cells while promoting maturation of growth plate chondrocytes toward hypertrophy (Kobayashi, Lyons, McMahon, & Kronenberg, 2005).

7. INTERACTIONS OF *Bmpr1a* SIGNALING WITH OTHER SIGNALING PATHWAYS

The interaction of BMPR1A with other regulators that are functionally important in growth plate chondrogenesis has also been studied. Fibroblast growth factors (FGFs) and their receptors (FGFRs) play significant roles in vertebrate organogenesis and morphogenesis. Yoon and coworkers provided a genetic *in vivo* demonstration that the progression of chondrocytes through the growth plate is controlled by antagonistic BMP and FGF signaling pathways, as they showed that BMP signaling is required to promote IHH expression, and to inhibit activation of STAT and ERK1/2 MAPK, key effectors of FGF signaling. BMP pathways inhibit FGF signaling, at least in part, by inhibiting the expression of FGF receptor 1 (FGFR1) (Yoon et al., 2006). Another FGF receptor, FGF receptor 3 (FGFR3) is a negative regulator of chondrogenesis and constitutive activity of FGFR3 results in achondroplasia, one of the most common dwarfisms in humans.

Chondrocyte-specific deletion of *Bmpr1a* could rescue the bone overgrowth phenotype observed in *Fgfr3*-deficient mice by reducing chondrocyte differentiation. Using an *in vitro* chondrogenic differentiation assay system, FGFR3 has been shown to inhibit BMPR1A-mediated chondrogenic differentiation through the Smurf1-mediated ubiquitination pathway (Qi et al., 2014).

Bmpr1a has also been reported to coordinate with Activin receptor type IA (ACVR1/ALK2), which is another type I BMP receptor. *Acvr1*CKO mice are viable but exhibit defects in the development of cranial and axial structures. Mutants exhibit a shortened cranial base, and cervical vertebrae are hypoplastic. *Acvr1*CKO adult mice develop progressive kyphosis. These morphological defects were associated with decreased levels of Smad1/5 and p38 activation and with reduced rates of chondrocyte proliferation in vertebral cartilage. *Acvr1/Bmpr1a* and *Acvr1/Bmpr1b* mutant mice exhibited generalized perinatal lethal chondrodysplasia that was much more severe than in any of the corresponding mutant strains. ACVR1 is required for chondrocyte proliferation and differentiation, particularly in craniofacial and axial elements, but exerts coordinated functions with both BMPR1A and BMPR1B throughout the developing endochondral skeleton (Rigueur et al., 2015).

BMPR1A was reported to have some interaction with other genes which are critical in the limb development. TGF-β activated kinase 1 (TAK1), a member of the MAPKKK family, controls diverse functions ranging from innate and adaptive immune system activation to vascular development and apoptosis. Tak1 mutation in cartilage displayed severe chondrodysplasia with runting, impaired formation of secondary centers of ossification, and joint abnormalities including elbow dislocation and tarsal fusion. This phenotype resembled that of BMPR1A and *Gdf5*-deficient mice. The authors also demonstrated that BMPR1A binds to XIAP which in turn mediates the interaction between BMPR1A and TAB1. The interaction of BMPR1A with XIAP occurs normally in TAK1-deficient chondrocytes and TAK1 functions downstream of the BMPR–XIAP complex (Shim et al., 2009).

8. CONCLUSIONS

This review has shown that BMPR1A is clearly instrumental for the initiation of chondrogenesis, for regulating differentiation along the chondrogenic lineage, and for endochondral bone formation. However, several recent studies in growth plate have challenged the dogma that the osteogenic cells derived from bone marrow are responsible for depositing the new bone at the cartilage-bone interface during endochondral ossification (Yang, Tsang, Tang, Chan, & Cheah, 2014; Yang, Zhu, et al., 2014; Zhou et al., 2014). Rather, these studies have shown using cell lineage tracing that many, possibly a majority, of osteogenic cells arise via the direct transformation of hypertrophic chondrocytes into osteoblasts or osteocytes. This new concept raises the possibility that regulators of chondrogenesis and chondrogenic differentiation—such as BMPR1A—may regulate endochondral bone formation indirectly via their effects on chondrocytes. Although a direct test of BMPR1A's role has not been conducted, the BMPR1A conditional knockout mice exhibit not only a major reduction in chondrogenesis but also an almost complete cessation of endochondral bone formation deep to the growth plate (Jing et al., 2013) and MCC (Jing et al., 2014).

REFERENCES

Anderson, K. V., Schneider, D. S., Morisato, D., Jin, Y., & Ferguson, E. L. (1992). Extracellular morphogens in Drosophila embryonic dorsal-ventral patterning. *Cold Spring Harbor Symposia on Quantitative Biology*, *57*, 409–417.

Archer, C. W., Caterson, B., Benjamin, M., & Ralphs, J. R. (1999). *The biology of the synovial joint.* London: Harwood Academics.

Beresford, W. A. (1981). *Chondroid bone, secondary cartilage, and metaplasia* (p. 454) Baltimore, MD: Urban and Schwartzenberg.

Bi, W., Huang, W., Whitworth, D. J., Deng, J. M., Zhang, Z., Behringer, R. R., et al. (2001). Haploinsufficiency of Sox9 results in defective cartilage primordia and premature skeletal mineralization. *Proceedings of the National Academy of Sciences of the United States of America*, *98*, 6698–6703.

Bragdon, B., Moseychuk, O., Saldanha, S., King, D., Julian, J., & Nohe, A. (2011). Bone morphogenetic proteins: A critical review. *Cellular Signalling*, *23*(4), 609–620.

Caplan, A. I., & Pechak, D. G. (1987). The cellular and molecular embryology of bone formation. In W. A. Peck (Ed.), *Bone and mineral research: Vol. 5* (pp. 117–183). New York: Elsevier.

Carreira, A. C., Lojudice, F. H., Halcsik, E., Navarro, R. D., Sogayar, M. C., & Granjeiro, J. M. (2014). Bone morphogenetic proteins: Facts, challenges, and future perspectives. *Journal of Dental Research*, *93*(4), 335–345.

Chen, G., Deng, C., & Li, Y. P. (2012). TGF-β and BMP signaling in osteoblast differentiation and bone formation. *International Journal of Biological Sciences*, *8*(2), 272–288.

De Crombrugghe, B., Lefebvre, V., Behringer, R. R., Bi, W., Murakami, S., & Huang, W. (2000). Transcriptional mechanisms of chondrocyte differentiation. *Matrix Biology*, *19*, 389–394.

Decker, R. S., Koyama, E., & Pacifici, M. (2014). Genesis and morphogenesis of limb synovial joints and articular cartilage. *Matrix Biology*, *39*, 5–10.

Derynck, R., & Feng, X. H. (1997). TGF-beta receptor signaling. *Biochimica et Biophysica Acta*, *1333*(2), F105–F150.

Ebisawa, T., Tada, K., Kitajima, I., Tojo, K., Sampath, T. K., Kawabata, M., et al. (1999). Characterization of bone morphogenetic protein-6 signaling pathways in osteoblast differentiation. *Journal of Cell Science*, *112*(Pt 20), 3519–3527.

Enomoto-Iwamoto, M., Iwamoto, M., Mukudai, Y., Kawakami, Y., Nohno, T., Higuchi, Y., et al. (1998). Bone morphogenetic protein signaling is required for maintenance of differentiated phenotype, control of proliferation, and hypertrophy in chondrocytes. *Journal of Cell Biology*, *140*(2), 409–418.

Ferguson, E. L., & Anderson, K. V. (1992). Localized enhancement and repression of the activity of the TGF-beta family member, decapentaplegic, is necessary for dorsal-ventral pattern formation in the Drosophila embryo. *Development*, *114*(3), 583–597.

Fujii, M., Takeda, K., Imamura, T., Aoki, H., Sampath, T. K., Enomoto, S., et al. (1999). Roles of bone morphogenetic protein type I receptors and Smad proteins in osteoblast and chondroblast differentiation. *Molecular Biology of the Cell*, *10*(11), 3801–3813.

Gu, S., Wu, W., Liu, C., Yang, L., Sun, C., Ye, W., et al. (2014). BMPRIA mediated signaling is essential for temporomandibular joint development in mice. *PloS One*, *9*(8), e101000.

Hall, B. K., & Miyake, T. (2000). All for one and one for all: Condensations and the initiation of skeletal development. *Bioessays*, *22*, 138–147.

Henry, S. P., Liang, S., Akdemir, K. C., & de Crombrugghe, B. (2012). The postnatal role of Sox9 in cartilage. *Journal of Bone and Mineral Research*, *27*, 2511–2525.

Hilton, M. J., Tu, X., Cook, J., Hu, H., & Long, F. (2005). Ihh controls cartilage development by antagonizing Gli3, but requires additional effectors to regulate osteoblast and vascular development. *Development*, *132*, 4339–4351.

Hinton, R. J. (2014). Genes that regulate morphogenesis and growth of the temporomandibular joint: A review. *Developmental Dynamics*, *243*(7), 864–874.

Hunziker, E. B., Kapfinger, E., & Geiss, M. D. (2007). The structural architecture of adult mammalian articular cartilage evolves by a synchronized process of tissue resorption and neoformation during postnatal development. *Osteoarthritis Cart*, *15*, 403–413.

Ikeda, T., Takahashi, H., Suzuki, A., Ueno, N., Yokose, S., Yamaguchi, A., et al. (1996). Cloning of rat type I receptor cDNA for bone morphogenetic protein-2 and bone morphogenetic protein-4, and the localization compared with that of the ligands. *Developmental Dynamics*, *206*(3), 318–329.

Ito, Y., Bringas, P., Jr., Mogharei, A., Zhao, J., Deng, C., & Chai, Y. (2002). Receptor-regulated and inhibitory Smads are critical in regulating transforming growth factor beta-mediated Meckel's cartilage development. *Developmental Dynamics*, *224*(1), 69–78.

Jay, G. D., Tantravahi, U., Britt, D. E., Barrach, H. J., & Cha, C. J. (2001). Homology of lubricin and superficial zone protein (SZP): Products of megakaryocyte stimulating factor (MSF) gene expression by human synovial fibroblasts and articular chondrocytes localized to chromosome 1q25. *Journal of Orthopaedic Research*, *19*, 677–687.

Jing, J., Hinton, R. J., Mishina, Y., Liu, Y., Zhou, X., & Feng, J. Q. (2014). Critical role of Bmpr1a in mandibular condyle growth. *Connective Tissue Research*, *55*(Suppl. 1), 73–78.

Jing, J., Ren, Y., Zong, Z., Liu, C., Kamiya, N., Mishina, Y., et al. (2013). BMP receptor 1A determines the cell fate of the postnatal growth plate. *International Journal of Biological Sciences*, *9*(9), 895–906.

Ju, W., Hoffmann, A., Verschueren, K., Tylzanowski, P., Kaps, C., Gross, G., et al. (2000). The bone morphogenetic protein 2 signaling mediator Smad1 participates predominantly in osteogenic and not in chondrogenic differentiation in mesenchymal progenitors C3H10T1/2. *Journal of Bone and Mineral Research*, *15*(10), 1889–1899.

Karsenty, G. (2003). The complexities of skeletal biology. *Nature*, *423*, 316–318.

Kobayashi, T., Lyons, K. M., McMahon, A. P., & Kronenberg, H. M. (2005). BMP signaling stimulates cellular differentiation at multiple steps during cartilage development.

Proceedings of the National Academy of Sciences of the United States of America, *102*(50), 18023–18027.

Komori, T. (2011). Signaling networks in RUNX2-dependent bone development. *Journal of Cellular Biochemistry*, *112*, 750–755.

Kramer, J., Hegert, C., Guan, K., Wobus, A. M., Müller, P. K., & Rohwedel, J. (2000). Embryonic stem cell derived chondrogenic differentiation *in vitro*: Activation by BMP-2 and BMP-4. *Mechanisms of Development*, *92*(2), 193–205.

Kronenberg, H. M. (2003). Developmental regulation of the growth plate. *Nature*, *423*, 332–336.

Lavery, K., Swain, P., Falb, D., & Alaoui-Ismaili, M. H. (2008). BMP-2/4 and BMP-6/7 differentially utilize cell surface receptors to induce osteoblastic differentiation of human bone marrow-derived mesenchymal stem cells. *Journal of Biological Chemistry*, *283*(30), 20948–20958.

Lim, J., Tu, X., Choi, K., Akiyama, H., Mishina, Y., & Long, F. (2015). BMP-Smad4 signaling is required for precartilaginous mesenchymal condensation independent of Sox9 in the mouse. *Developmental Biology*, *400*(1), 132–138.

Logan, M., Martin, J. F., Nagy, A., Lobe, C., Olson, E. N., & Tabin, C. J. (2002). Expression of Cre Recombinase in the developing mouse limb bud driven by a Prxl enhancer. *Genesis*, *2*, 77–80.

Long, F., Chung, U. I., Ohba, S., McMahon, J., Kronenberg, H. M., & McMahon, A. P. (2004). Ihh signaling is directly required for the osteoblast lineage in the endochondral skeleton. *Development*, *131*, 1309–1318.

Long, F., & Ornitz, D. M. (2013). Development of the endochondral skeleton. *Cold Spring Harbor Perspectives in Biology*, *5*(1), a008334.

Lydiatt, D. D., & Davis, L. F. (1985). The effects of immobilization on the rabbit temporomandibular joint. *Journal of Oral and Maxillofacial Surgery*, *43*, 188–193.

Macías-Silva, M., Hoodless, P. A., Tang, S. J., Buchwald, M., & Wrana, J. L. (1998). Specific activation of Smad1 signaling pathways by the BMP7 type I receptor, ALK2. *Journal of Biological Chemistry*, *273*(40), 25628–25636.

Maes, C., Kobayashi, T., Selig, M. K., Torrekens, S., Roth, S. I., Mackem, S., et al. (2010). Osteoblast precursors, but not mature osteoblasts, move into developing and fractured bones along with invading blood vessels. *Developmental Cell*, *19*, 329–344.

Majumdar, M. K., Wang, E., & Morris, E. A. (2001). BMP-2 and BMP-9 promotes chondrogenic differentiation of human multipotential mesenchymal cells and overcomes the inhibitory effect of IL-1. *Journal of Cellular Physiology*, *189*(3), 275–284.

Mishina, Y., Hanks, M. C., Miura, S., Tallquist, M. D., & Behringer, R. R. (2002). Generation of Bmpr/Alk3 conditional knockout mice. *Genesis*, *32*, 69–72.

Mishina, Y., Starbuck, M. W., Gentile, M. A., Fukuda, T., Kasparcova, V., Seedor, J. G., et al. (2004). Bone morphogenetic protein type IA receptor signaling regulates postnatal osteoblast function and bone remodeling. *Journal of Biological Chemistry*, *279*(26), 27560–27566.

Mizoguchi, I., Nakamura, M., Takahashi, I., Kagayama, M., & Mitani, H. (1990). An immunohistochemical study of localization of type I and type II collagens in mandibular condylar cartilage compared with tibial growth plate. *Histochemistry*, *93*, 593–599.

Nishihara, A., Fujii, M., Sampath, T. K., Miyazono, K., & Reddi, A. H. (2003). Bone morphogenetic protein signaling in articular chondrocyte differentiation. *Biochemical and Biophysical Research Communications*, *301*(2), 617–622.

Nishimura, R., Hata, K., Ikeda, F., Ichida, F., Shimoyama, A., Matsubara, T., et al. (2008). Signal transduction and transcriptional regulation during mesenchymal cell differentiation. *Journal of Bone and Mineral Metabolism*, *26*, 203–212.

Nohe, A., Keating, E., Knaus, P., & Petersen, N. O. (2004). Signal transduction of bone morphogenetic protein receptors. *Cellular Signalling*, *16*(3), 291–299.

Pizette, S., & Niswander, L. (2000). BMPs are required at two steps of limb chondrogenesis: Formation of prechondrogenic condensations and their differentiation into chondrocytes. *Developmental Biology, 219*(2), 237–249.

Qi, H., Jin, M., Duan, Y., Du, X., Zhang, Y., Ren, F., et al. (2014). FGFR3 induces degradation of BMP type I receptor to regulate skeletal development. *Biochimica et Biophysica Acta, 1843*(7), 1237–1247.

Rigueur, D., Brugger, S., Anbarchian, T., Kim, J. K., Lee, Y., & Lyons, K. M. (2015). The type I BMP receptor ACVR1/ALK2 is required for chondrogenesis during development. *Journal of Bone and Mineral Research, 30*(4), 733–741.

Rountree, R. B., Schoor, M., Chen, H., Marks, M. E., Harley, V., Mishina, Y., et al. (2004). BMP receptor signaling is required for postnatal maintenance of articular cartilage. *PLoS Biology, 2*(11), e355.

Shim, J. H., Greenblatt, M. B., Xie, M., Schneider, M. D., Zou, W., Zhai, B., et al. (2009). TAK1 is an essential regulator of BMP signalling in cartilage. *EMBO Journal, 28*(14), 2028–2041.

Shukunami, C., Akiyama, H., Nakamura, T., & Hiraki, Y. (2000). Requirement of autocrine signaling by bone morphogenetic protein-4 for chondrogenic differentiation of ATDC5 cells. *FEBS Letters, 469*(1), 83–87.

Stickens, D., Behonick, D. J., Ortega, N., Heyer, B., Hartenstein, B., Yu, Y., et al. (2004). Altered endochondral bone development in matrix metalloproteinase 13-deficient mice. *Development, 131*, 5883–5895.

Ten Dijke, P., Yamashita, H., Ichijo, H., Franzén, P., Laiho, M., Miyazono, K., et al. (1994). Characterization of type I receptors for transforming growth factor-beta and activin. *Science, 264*(5155), 101–104.

Wagner, D. O., Sieber, C., Bhushan, R., Börgermann, J. H., Graf, D., & Knaus, P. (2010). BMPs: From bone to body morphogenetic proteins. *Science Signaling, 3*(107), mr1.

Wan, M., & Cao, X. (2005). BMP signaling in skeletal development. *Biochemical and Biophysical Research Communications, 328*(3), 651–657.

Wang, Y., Zheng, Y., Chen, D., & Chen, Y. (2013). Enhanced BMP signaling prevents degeneration and leads to endochondral ossification of Meckel's cartilage in mice. *Developmental Biology, 381*(2), 301–311.

Warman, M. L., Cormier-Daire, V., Hall, C., Krakow, D., Lachman, R., LeMerrer, M., et al. (2011). Nosology and classification of genetic skeletal disorders: 2010 revision. *American Journal of Medical Genetics. Part A, 155A*, 943–968.

Xiong, J., Onal, M., Jilka, R. L., Weinstein, R. S., Manolagas, S. C., & O'Brien, C. A. (2011). Matrix-embedded cells control osteoclast formation. *Nature Medicine, 17*, 1235–1241.

Yang, L., Tsang, K. Y., Tang, H. C., Chan, D., & Cheah, K. S. E. (2014). Hypertrophic chondrocytes can become osteoblasts and osteocytes in endochondral bone formation. *Proceedings of the National Academy of Sciences, 111*(33), 12097–12102.

Yang, G., Zhu, L., Hou, N., Lan, Y., Wu, X.-M., Zhou, B., et al. (2014). Osteogenic fate of hypertrophic chondrocytes. *Cell Research, 24*, 1266–1269.

Yi, S. E., Daluiski, A., Pederson, R., Rosen, V., & Lyons, K. M. (2000). The type I BMP receptor BMPRIB is required for chondrogenesis in the mouse limb. *Development, 127*(3), 621–630.

Yoon, B. S., & Lyons, K. M. (2004). Multiple functions of BMPs in chondrogenesis. *Journal of Cellular Biochemistry, 93*(1), 93–103.

Yoon, B. S., Ovchinnikov, D. A., Yoshii, I., Mishina, Y., Behringer, R. R., & Lyons, K. M. (2005). Bmpr1a and Bmpr1b have overlapping functions and are essential for chondrogenesis *in vivo*. *Proceedings of the National Academy of Sciences of the United States of America, 102*(14), 5062–5067.

Yoon, B. S., Pogue, R., Ovchinnikov, D. A., Yoshii, I., Mishina, Y., Behringer, R. R., et al. (2006). BMPs regulate multiple aspects of growth-plate chondrogenesis through opposing actions on FGF pathways. *Development, 133*(23), 4667–4678.

Zhou, X., von der Mark, K., Henry, S., Norton, W., Adams, H., & de Crombrugghe, B. (2014). Chondrocytes transdifferentiate into osteoblasts in endochondral bone during development, postnatal growth and fracture. *PLoS Genetics, 10*(12), e1004820.

Zou, H., Wieser, R., Massague, J., & Niswander, L. (1997). Distinct roles of type I bone morphogenetic protein receptors in the formation and differentiation of cartilage. *Genes & Development, 11*(17), 2191–2203.

Zwijsen, A., Verschueren, K., & Huylebroeck, D. (2003). New intracellular components of bone morphogenetic protein/Smad signaling cascades. *FEBS Letters, 546*(1), 133–139.

CHAPTER TEN

Bone Morphogenetic Proteins: Promising Molecules for Bone Healing, Bioengineering, and Regenerative Medicine

Ana Claudia Oliveira Carreira*, Willian Fernando Zambuzzi[†], Mariana Correa Rossi[†], Renato Astorino Filho*, Mari Cleide Sogayar*[,‡], José Mauro Granjeiro[§,¶,1]

*NUCEL-NETCEM (Cell and Molecular Therapy Center), Internal Medicine Department, School of Medicine, University of São Paulo, São Paulo, Brazil
[†]Department of Chemistry and Biochemistry, Biosciences Institute, UNESP, Universidade Estadual Paulista, Botucatu, Brazil
[‡]Chemistry Institute, Biochemistry Department, São Paulo, Brazil
[§]Bioengineering Division, National Institute of Metrology, Quality, and Technology, Duque de Caxias, Brazil
[¶]Department of Dental Materials, Dental School, Fluminense Federal University, Niteroi, Brazil
[1]Corresponding author: e-mail addresses: jmgranjeiro@inmetro.gov.br; jmgranjeiro@gmail.com

Contents

Abstract

Bone morphogenetic proteins (BMPs), glycoproteins secreted by some cells, are members of the TGF-β superfamily that have been implicated in a wide variety of roles. Currently, about 20 different BMPs have been identified and grouped into subfamilies, according to similarities with respect to their amino acid sequences. It has been shown that BMPs are secreted growth factors involved in mesenchymal stem cell differentiation, also being reported to control the differentiation of cancer stem cells. BMPs initiate

Vitamins and Hormones, Volume 99
ISSN 0083-6729
http://dx.doi.org/10.1016/bs.vh.2015.06.002

signaling from the cell surface by binding to two different receptors (R: Type I and II). The heterodimeric formation of type I R and II R may occur before or after BMP binding, inducing signal transduction pathways through SMADs. BMPs may also signal through SMAD-independent pathways via mitogen-activated protein kinases (ERK, p38MAPKs, JNK). BMPs may act in an autocrine or paracrine manner, being regulated by specific antagonists, namely: noggin and chordin. Genetic engineering allows the production of large amounts of BMPs for clinical use, and clinical trials have shown the benefits of FDA-approved recombinant human BMPs 2 and 7. Several materials from synthetic to natural sources have been tested as BMP carriers, ranging from hydroxyapatite, and organic polymers to collagen. Bioactive membranes doped with BMPs are promising options, acting to accelerate and enhance osteointegration. The development of smart materials, mainly based on biopolymers and bone-like calcium phosphates, appears to provide an attractive alternative for delivering BMPs in an adequately controlled fashion. BMPs have revealed a promising future for the fields of Bioengineering and Regenerative Medicine. In this chapter, we review and discuss the data on BMP structure, mechanisms of action, and possible clinical applications.

1. INTRODUCTION

Bone morphogenetic proteins (BMPs) are members of the transforming growth factor β (TGF-β) superfamily which have been implicated a wide variety of roles, including cell proliferation, apoptosis, differentiation, and morphogenesis. The first evidences for the morphogenetic role of BMPs were described in 1965, when the group of Prof. Urist discovered that acellular, devitalized, and demineralized bone matrix (DBM) could induce cartilage and bone formation upon implantion under the skin or into the muscle of adult rodents. They attributed this osteoinductivity role to BMPs and later on other studies indicated that BMP could be solubilized and extracted from the living bone matrix. Since then, several studies have been made relating to BMP functions, as we can find those reporting BMP involvement with cancer progression. A number of osteogenic proteins, or BMPs, have been discovered and, based on sequence homology, most of the GDFs (growth/differentiation factors) have been added to the BMP family (Carreira, Alves, Zambuzzi, Sogayar, & Granjeiro, 2014; Carreira, Lojudice, et al., 2014; Mi et al., 2015).

Mechanistically, BMPs are glycoproteins secreted by the cell, acting as ligands for receptors present on the plasma membrane of different cell types, thereby exerting paracrine and autocrine effects. The transduction of this signal occurs through a cascade of specific molecules, initiating with the serine/threonine kinase receptors and propagating through the signal SMADS proteins to the nucleus. Clinical studies and meta-analysis demonstrate the

BMPs efficiency in the induction of bone lesions repair, such as pseudoarthrosis (Giannoudis et al., 2009), arthrodesis (Dawson, Bae, Burkus, Stambough, & Glassman, 2009), and alveolar bone (Carreira, Alves, et al., 2014; Jung et al., 2009).

This chapter aims to assemble and discuss the structural, molecular, and fundamental cellular actions of BMPs, with possible clinical applications and prospects for their use in the near future.

2. BMPs: STRUCTURAL ASPECTS

Currently, about 20 different BMPs have been identified and grouped into subfamilies according to their similarity with respect to their amino acid sequences. BMP-1 does not belong to the TGF-β superfamily, being a metalloprotease that cleaves the C-terminus of procollagen I, II, and III and being capable of inducing cartilage formation *in vivo* (Kessler, Takahara, Biniaminov, Brusel, & Greenspan, 1996). BMPs have structural similarities to TGF-β in the order of 30–50% of its primary structure (Mi et al., 2015; Wozney et al., 1988). It is also known that TGF-β is found in the bone matrix in larger amounts than BMPs, being classified as cytokines, rather than morphogenetic proteins. Thus, while BMP-2 can induce or increase the expression of osteocalcin and alkaline phosphatase activity (classical markers of osteoblast differentiation), TGF-β1 significantly inhibits expression and activity of these markers (Mi et al., 2015; Spinella-Jaegle et al., 2001). In nature, more specifically in mammals, there is a need for fine control over both the activity and the expression of these genes, accurately directing the molecular events that orchestrate bone formation.

The 30–38 kDa homodimers BMPs/GDFs are synthesized as large precursor proteins of approximately 400–525 amino acids (aa), with the cleavage of the variable length prosegment occurring prior to secretion. It is known that the secretion of the 100–140 aa C-terminal mature segment forms a dimer, sometimes nondisulfide-linked (GDF-3, -9, and BMP-15). Upon dimerization, they are proteolytically cleaved at an Arg-X-X-Arg consensus site to yield carboxy-terminal mature dimers (Fig. 1).

In general, BMPs have one more conserved cysteine, involving intermolecular disulfide bonding. The BMPs/GDFs have been grouped into subsets based on amino acid residues sequence homology, as follows: (1) BMP-2 and BMP-4; (2) BMP-3 and BMP-3b; (3) BMP-5, BMP-6, BMP-7, and BMP-8; (4) BMP-9 and BMP-10; (5) BMP-12, BMP-13,

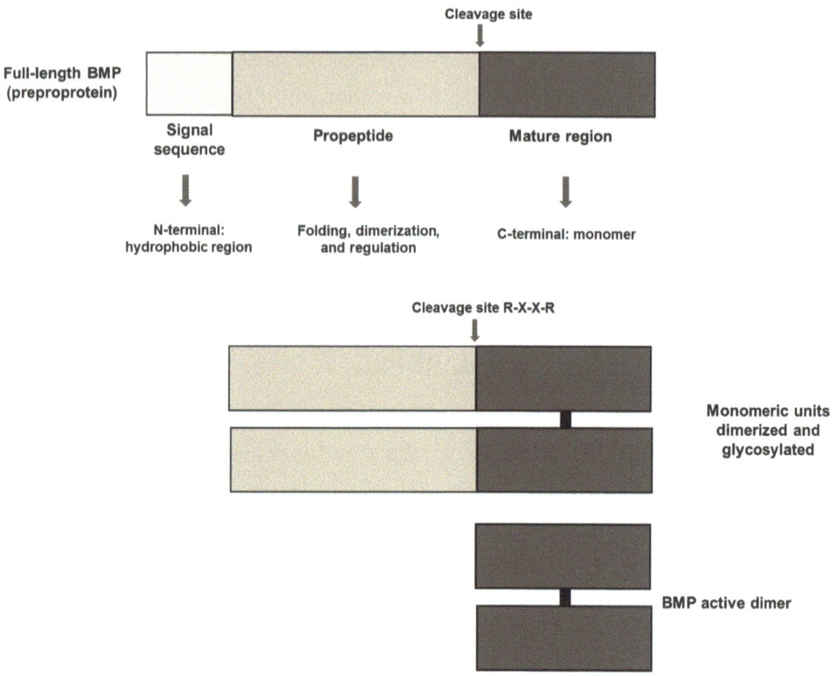

Figure 1 Structural organization of BMPs. Full-length BMP consists of a signal sequence, propeptide, and a mature region. The mature protein is secreted as a homodimer linked by disulfide bonds. The C-terminal mature protein is proteolytically cleaved by proteases from the prodomain at an R-X-X-R sequence before dimerization (Carreira, Alves, et al., 2014).

and BMP-14; and (6) BMP-11 and GDF-8 (Table 1). Some BMP/GDF family of cytokines had their structure solved as shown in Table 2.

The three-dimensional structure of BMP-7 protein or osteogenic protein-1 (OP-1) is available (Griffith, Keck, Sampath, Rueger, & Carlson, 1996; Mi et al., 2015; Spinella-Jaegle et al., 2001). Thus, although the structural similarity with TGF-β2 is less than 38%, it is known that OP-1 is similar to BMP-2 and BMP-5 (around 60% and 85%, respectively). This monomer has three disulfide bridges (S38–S104, S67–S136, S71–S138), forming the core, while four secondary structures of antiparallel beta-sheets form two finger-like projections (Fig. 2).

3. INTRACELLULAR PATHWAYS CONTROL BMP SIGNALING

BMPs play fundamental roles in development and tissue homeostasis; on the other hand, aberrant BMP signaling is associated with several human

Table 1 Characteristics of Different BMP Family Members

BMP	Names	Subcellular Location	Functions	Year of Description/PMID
BMP-2	BMP-2A, BDA2	Secreted	Induces bone and cartilage formation	1988 (3201241)
BMP-3A	Osteogenin, BMP-3	Secreted	Negatively regulates bone density; antagonizes the ability of certain osteogenic BMPs to induce osteoprogenitor differentiation and ossification	1988 (3201241)
BMP-3B	GDF-10	Secreted	Regulates cell growth and differentiation in both embryonic and adult tissues	1995 (8679252)
BMP-4	BMP-2B, BMP-2B, MCOPS6, OFC11, ZYME	Secreted: extracellular matrix	Induces cartilage and bone formation; acts in mesoderm induction, tooth development, limb formation, and fracture repair; acts in concert with PTHLH/PTHRP to stimulate ductal outgrowth during embryonic mammary development and to inhibit hair follicle induction (by similarity)	1988 (3201241)
BMP-5	–	Secreted	Induces cartilage and bone formation	1990 (2263636)
BMP-6	VGR, VGR-1	Secreted (by similarity)	Induces cartilage and bone formation; proposed role in early development	1990 (2263636)
BMP-7	OP-1	Secreted	Induces cartilage and bone formation; may be the osteoinductive factor responsible for the phenomenon of epithelial osteogenesis; plays a role in calcium regulation and bone homeostasis	1990 (2263636)

Continued

Table 1 Characteristics of Different BMP Family Members—cont'd

BMP	Names	Subcellular Location	Functions	Year of Description/PMID
BMP-8A	OP-2	Secreted	Induces cartilage and bone formation; may be the osteoinductive factor responsible for the phenomenon of epithelial osteogenesis; plays a role in calcium regulation and bone homeostasis (by similarity)	2002 (1460021)
BMP-8B	BMP-8	Secreted	Induces cartilage and bone formation; may be the osteoinductive factor responsible for the phenomenon of epithelial osteogenesis; plays a role in calcium regulation and bone homeostasis (by similarity)	1992 (12477932)
BMP-9	GDF-2	Secreted	Potent circulating inhibitor of angiogenesis; could be involved in bone formation; signals through the type I activin receptor ACVRL1 but not other Alks	2000 (10849432)
BMP-10	–	Secreted (by similarity)	Required for maintaining the proliferative activity of embryonic cardiomyocytes by preventing premature activation of the negative cell cycle regulator CDKN1C/p57KIP and maintaining the required expression levels of cardiogenic factors such as MEF2C and NKX2-5; acts as a ligand for ACVRL1/ALK1, BMPR1A/ALK3, and BMPR1B/ALK6, leading to activation of SMAD1, SMAD5, and SMAD8 transcription factors; inhibits endothelial cell migration and growth	1999 (10072785)

BMP-11	GDF-11	Secreted (probable)	Secreted signal that acts globally to specify positional identity along the anterior/posterior axis during development; plays critical roles in patterning both mesodermal and neural tissues and in establishing the skeletal pattern	1999 (10075854)
BMP-12	GDF-7	Secreted (by similarity)	May play an active role in the motor area of the primate neocortex (by similarity)	1994 (8145850)
BMP-13	GDF-6, CDPM2, MCOP4	Secreted (probable)	Required for normal formation of bones and joints in the limbs, skull, and axial skeleton. Plays a key role in establishing boundaries between skeletal elements during development. (By smilarity)	1994 (8145850)
BMP-14	GDF-5, CDPM1, CDMP1, LAP4, OS5, SYNS2	Secreted	Could be involved in bone and cartilage formation; chondrogenic signaling is mediated by the high-affinity receptor BMPR1B	1994 (8145850)
BMP-15	GDF-9B, ODG2, POF4	Secreted	May be involved in follicular development; oocyte-specific growth/differentiation factor that stimulates folliculogenesis and granulosa cell (GC) growth	1998 (9849956)
BMP-16	Nodal	Secreted (by similarity)	Essential for mesoderm formation and axial patterning during embryonic development by similarity	1999 (US Patent No. 596503)
BMP-17	Lefty1; LeftyB	Secreted	Required for left–right axis determination as a regulator of LEFTY2 and NODAL	2000 (US Patent No. 6027917)

Continued

Table 1 Characteristics of Different BMP Family Members—cont'd

BMP	Names	Subcellular Location	Functions	Year of Description/PMID
BMP-18	Lefty2; LeftyA	Secreted	Required for left–right (L–R) asymmetry determination of organ systems in mammals. May play a role in endometrial bleeding	2000 (US Patent No. 6027917)
GDF-1	DORV, DTGA3	Secreted	May mediate cell differentiation events during embryonic development	1990 (1704486)
GDF-3	KFS3, MCOP7, MCOPCB6	Secreted (probable)	Negatively and positively control differentiation of embryonic stem cells; role in mesoderm and definitive endoderm formation during the pregastrulation stages of development	1993 (8429021)
GDF-8	MSTN, myostatin	Secreted	Acts specifically as a negative regulator of skeletal muscle growth	1997
GDF-9	–	Secreted	Required for ovarian folliculogenesis; promotes primordial follicle development. Stimulates granulosa cell proliferation; promotes cell transition from G0/G1 to S and G2/M phases	1993 (8429021)
GDF-15	MIC-1, MIC1, NAG-1, PDF, PLAB, PTGFB	Secreted (probable)	May be involved in follicular development; oocyte-specific growth/differentiation factor that stimulates folliculogenesis and granulosa cell (GC) growth	1997 (9139826)

Adapted from Carreira, Alves, et al. (2014).

Table 2 Experimentally Solved Structures for the BMP/GDF Family of Cytokines

Protein and PDB Code	Sequence and Mutations (UniProt Accession Numbers)	Ligand	Oligomeric State	PMID
BMP-1				
3EDG	C-terminal 201 residues of P13497	Zn; acetate	Monomer	18824173
3EDH	C-terminal 201 residues of P13497	Zn; acetate; DMSO	Monomer	10074410
BMP-2				
3BMP	C-terminal 114 residues of P12643	MPD	Dimer	0881198
1ES7	C-terminal 116 residues of P12643	BMPR1A	Dimer	15064755
1REW	C-terminal 114 residues of P12643	BMPR1A	Dimer	15064755
1REU	C-terminal 103 residues of P12643	MPD	Dimer	15064755
2GOO	C-terminal 114 residues of P12643	BMPR1A-ECD; ActRII-ECD	Dimer	16672363
2H62	C-terminal 116 residues of P12643	BMPR1A; ActR2B	Dimer	17295905
2H64	C-terminal 116 residues of P12643	BMPR1A; ActR2B	Dimer	17295905
3BK3	C-terminal 114 residues of P12643; F41M, Y91M	Crossveinless-2 (from Zebrafish)	Dimer	18477456
2QJ9	283–397 of P12643	Human BMPR1A v. B1	Dimer	18160401
2QJB	283–397 of P12643	Human BMPR1A v. 1A/1B	Dimer	18160401
2QJA	283–397 of P12643	Human BMPR1A v. B12	Dimer	18160401

Continued

Table 2 Experimentally Solved Structures for the BMP/GDF Family of Cytokines—cont'd

Protein and PDB Code	Sequence and Mutations (UniProt Accession Numbers)	Ligand	Oligomeric State	PMID
BMP-3				
2QCQ	C-terminal 110 residues of P12645		Dimer	17924656
BMP-6				
2R52	375–513 of P22004	IPA	Dimer	18070108
2R53	375–513 of P22004	IPA	Dimer	18070108
2QCW	C-terminal 132 residues of P22004		Dimer	17924656
BMP-7				
1BMP	C-terminal 139 residues of P18075		Dimer	8570652
1M4U	C-terminal 139 residues of P18075	Noggin	Dimer	12478285
1LXI	C-terminal 139 residues of P18075		Dimer	12667445
1LX5	C-terminal 139 residues of P18075	ActRII	Dimer	12667445
BMP-9				
1ZKZ	320–429 of Q9UK05		Dimer	15851468
GDF-5				
2BHK	C-terminal 120 residues of P43026	IPA	Dimer	15752764
1WAQ	387–501 of P43026	MPD	Dimer	15890363
3EVS	387–501 of P43026	BMPR-1B (mouse)	Dimer	19229295

All proteins are human except where indicated. Sequences refer to UniProt accession numbers. ECD, extracellular domain; IPA, isopropyl alcohol; MPD, 4S-2-methylpentane-2,4-diol; v., variant.
Adapted from Rider and Mulloy (2010).

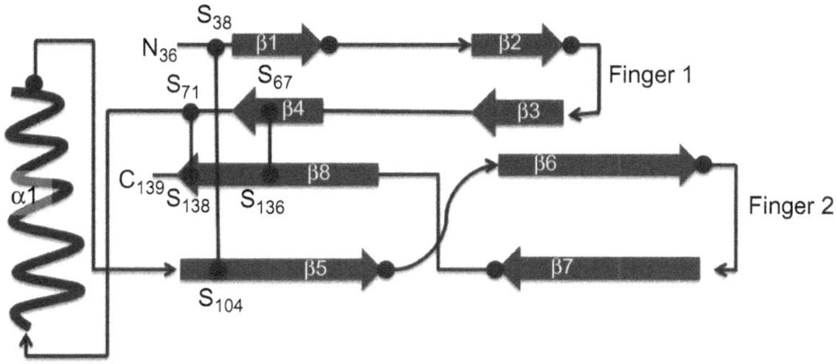

Figure 2 Scheme for BMP-7 structure. Three disulfide bridges stabilize the monomer, initial 67-Cys Cys-136 and Cys-71 Cys-138 form a ring structure, spatially favoring binding of the third disulfide bond, Cys-38 Cys-104. (See the color plate.)

diseases, such as fibrosis, bone and immune disorders, cancer progression, and metastasis. BMP signaling involves specific receptor present on the cell surface. These receptors are transmembrane proteins, which connect the extracellular compartment to the cytoplasmic one, receiving the extracellular signals and transmitting them to specific proteins that are responsible for conducting the signal to the nucleus, where these proteins modulate the activity of transcription factors. In other words, BMPs interact with their receptors (BMPRs) activating them and triggering the signaling cascade. The activated BMPR-II promotes the recruitment and phosphorylation of the BMPR type IA (BMPR-IA or Alk 3) or IB (ALK6 or BMPR-IB). These events provoke the phosphorylation of intracellular messengers (Miyazono, Kamiya, & Morikawa, 2010; Rosenzweig et al., 1995), called R-Smads, which dimerize with Smad4 (also known as common mediator or co-Smad) and are transported to the nucleus, where they regulate the expression of specific genes (Fig. 3).

Type I receptors have no cytoplasmic C-terminal extension, but do possess a 20–30 aa glycine–serine (GS) rich domain between their transmembrane region and kinase domain. The GS region is extremely important for type II–type I receptor interaction. Following ligand binding to the type II receptor homodimer, the type II receptor cross-phosphorylates the type I receptor in the GS region. This phosphorylation activates the type I receptor kinase domain and initiates downstream signaling. A current theory suggests that type I and type II receptors are either constitutively homodimerized or heterotetramerized (i.e., a complex of two homodimers). Heterotetramerization may occur independently of ligand interaction because a constitutively autophosphorylated type II receptor

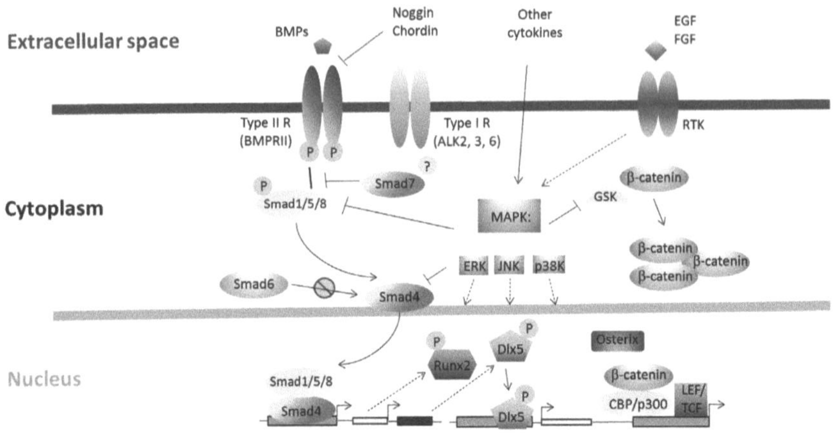

Figure 3 BMP signaling pathways: representation of Smad-dependent and -independent pathways and their mechanisms of regulation. BMPs initiate signaling from the cell surface by binding two different types of receptors (Type I and Type II) that can activate the Smad cytoplasmic proteins. The heterodimeric formation of type I R and type II R may occur before or after BMP binding, which induces signal transduction through Smads. BMPs may also signal through SMAD-independent pathways via mitogen-activated protein kinase (MAPK): Erk (extracellular signal-regulated kinase), JNK (Jun N-terminal kinase), or p38 (p38 MAPK) (Pan et al., 2015). Three classes of Smad proteins include receptor-regulated Smads (R-Smads), such as Smad1, 5, and 8; the common mediator Smad (Co-Smad), Smad4; and inhibitory Smads (I-Smads), Smad6 and 7. Smad6 binds type I BMPR, preventing Smad1, 5, 8 activation (phosphorylation). Phosphorylated Smad1, 5, 8 complexes associate with intracellular Smad4 for translocation into the nucleus, binds to transcription factors, activating target gene transcription. The transcription factors for BMPs may be any of the following: AP-1, bZIP, RUNX, Fox, bHLH, Homeodomain, Sp1, nuclear receptors, or IRF-7. The coactivators which are commonly utilized are the following: CBP/p300, SMIF, MSG1, or ARC105. Smad6 and 7 may suppress signals by preventing association of R-Smad with Co-Smad or R-Smad phosphorylation. BMPs may also act in an autocrine and paracrine manner. Furthermore, specific antagonists (Noggin and Chordin) are also expressed. *Adapted from Carreira, Alves, et al. (2014).* (See the color plate.)

naturally recruits type I receptors. Upon ligand binding, "functional" or signaling heterotetramerization occurs, with the type II receptor phosphorylating serines and threonines in the GS domain of the type I receptor. This phosphorylation activates the type I receptor kinase domain initiating phosphorylation of cytoplasmic Smad proteins and signal transduction.

In turn, Smads are a family of cytoplasmic messengers whose activity is directly dependent upon activation of receptors on the cell surface (Attisano & Lee-Hoeflich, 2001; Cao & Chen, 2005; Sieber, Kopf, Hiepen, & Knaus, 2009). Members of the Smad family of molecules are grouped into three classes: receptor-associated Smads (R-Smads), including

Smad 1, 2, 3, 5, and 8; the common mediator (co-Smad) Smad-4; and inhibitory Smads with antagonistic characteristics (I-Smads): the Smads 6 and 7. When activated by interaction with its ligand, the receptor is able to phosphorylate R-Smads. When phosphorylated, R-Smad dissociates from the receptor to form a heterodimeric complex with a co-Smad (Smad4), allowing its translocation to the nucleus (Whitman, 1998). Once in the nucleus, Smads may enable a wide variety of transcription factors, regulating the activation of specific genes (Attisano & Lee-Hoeflich, 2001; Cao & Chen, 2005; Sieber et al., 2009).

There are extracellular and intracellular regulators that act as antagonist of the BMP signaling pathway, such as noggin and chordin, which may bind to the receptors or sequester the BMP ligand. Several BMP gene members are proven targets of SHH (sonic hedgehog) signaling. BMP-dependent downregulation of Shh is achieved by interfering with the fibroblast growth factor (FGF) and Wnt signaling activities that maintain Shh expression. Evidence is available indicating that BMP downregulation of *Shh* is mediated by a respective downregulation in the Wnt/β-catenin signaling pathway that is normally involved in maintenance of Shh expression. BMPs have also been implicated as potential interactors of the Ihh/PTHrP feedback loop. All of these signaling pathways are involved in steps of mesenchymal, bone, and cartilage cell proliferation and differentiation.

MicroRNAs have emerged as an important field in cell biology, since they play decisive roles during cell reprogramming and differentiation. Grünhagen et al. (2015) identified members of the miR-15 family as being strongly upregulated with age of postnatal bone development *in vivo*. They suggested that miR-497-195 cluster microRNAs act as intracellular antagonists of BMP signaling in bone cells. In addition, it is known that reversible ubiquitylation of key components of the BMP pathway, including type I TGF-β/BMP receptor kinases and SMADs, play a critical role in regulating the outcome of TGF-β signaling (Herhaus & Sapkota, 2014). Herhaus and Sapkota (2014) suggest that the next decade is sure to address the molecular mechanisms by which deubiquitylating enzymes regulate the TGF-β and BMP pathways.

4. BEYOND BONE: MORPHOGENETIC EFFECTS ON CANCER

It has been shown that BMPs are secreted growth factors involved in cancer, since they have been reported to control the differentiation of cancer stem cells (Bleuming et al., 2007; Ehata, Yokoyama, Takahashi, &

Miyazono, 2013; Gao et al., 2012; Karagiannis et al., 2015; Lombardo et al., 2011; Sneddon et al., 2006; Zhang et al., 2012). Cancer stem cells present the ability to self-renew and differentiate, and several studies have suggested TGF-β roles in their maintenance in certain cancer stem cells. In this scenario, Lombardo and colleagues showed BMP involvement during CD133+ colorectal cancer stem cells differentiation. In this context, it has been shown that approximately half of the patients with juvenile polyposis syndrome present a mutation in the *MADH4* (responsible for encoding Smad4) or *BMPR1A* (encoding ALK-2) (Ehata et al., 2013; Howe et al., 1998, 2001, 2004). Altogether, these elegant pieces of work suggest the involvement of BMP signaling in the development of the polyposis syndrome. Another study has reported the importance of Bmpr2 during epithelial hyperplasia in the colon (Beppu et al., 2008).

During tumor progression, endothelial cell proliferation is an important biological event necessary to support tumor cell nutrition. Controversial data have been reported regarding the effects of BMPs signaling in the maintenance of endothelial cells (Koketsu et al., 2015; Laulan & St-Pierre, 2015). It has been shown that TGF-β signaling inhibits endothelial cell growth, while it stimulates angiogenesis *in vivo*. However, increased TGF-β expression is associated with endothelial density in some types of tumors (Ehata et al., 2013; Kiyono et al., 2009; Komuro et al., 2009).

However, there are controversial data regarding the involvement of BMP signaling in angiogenesis. It has been reported to play a pivotal role during metastasis of various types of cancer cells, such as prostate cancer, colon cancer, malignant melanoma, glioblastoma, and adenocarcinomas (Caja, Bellomo, & Moustakas, 2015; Ehata et al., 2013; Hover, Abel, & Owens, 2015; Nolan et al., 2015; Owens et al., 2015; Rajski, Saaf, & Buess, 2015). Katsuno and collaborators demonstrated the importance of both TGF-β and BMP signaling in bone metastasis *in vivo* (Katsuno et al., 2008). Furthermore, Langenfeld and colleagues explored *in vivo* approaches to report the involvement of BMP-2 in A549 lung adenocarcinoma cells.

Based on these statements, BMPs play important roles in development, morphogenesis, and cancer. Rajski et al. (2015) showed that the effects of BMPs and their antagonists, gremlin and noggin, are heterogeneous in different cell types. They also reported that the gene expression pattern induced by BMP-2 in lung fibroblasts significantly correlated with the prognosis of patients with stage I lung adenocarcinomas.

Taking into account, the inhibition of BMP signaling in cancer cells has been proposed, with very interesting results being achieved, mainly reducing bone metastatic foci (Ehata et al., 2013; Feeley et al., 2006).

5. PARACRINE SIGNALING OF BMPs IN MESENCHYMAL STEM CELL METABOLISM: A TISSUE ENGINEERING PERSPECTIVE

Cell signaling molecules exert their actions upon binding to receptors expressed by their target cells. The targeted cells, in turn, establish a negative or positive feedback action, which regulates the release of signaling molecules. The paracrine signaling molecules include four major protein families: (1) FGF, (2) hedgehog, (3) Wnt, and (4) TGF-β. BMPs are secreted to the extracellular compartment, specifically interacting with their transmembrane receptors present in different cell types through autocrine and paracrine mechanisms (reviewed in Carreira, Alves, et al., 2014; Granjeiro, Oliveira, Bustos-Valenzuela, Sogayar, & Taga, 2005). In general terms, the paracrine cell signaling is mediated by signaling molecules that act locally, regulating the behavior of neighboring cells. The main cell targets for BMPs in the paracrine pathway are the endothelial cells, which have been the object of several BMP-mediated angiogenesis studies (Dyer, Pi, & Patterson, 2014). Moreover, BMP is well known to be involved with mesenchymal stem cells (MSCs) differentiation by affecting their metabolism (Vukicevic & Grgurevic, 2009).

BMP involvement in MSC biology has been explored on Regenerative Medicine (Gemini-Piperni et al., 2014; Zambuzzi, Coelho, Alves, & Granjeiro, 2011), since they are multipotent, having the ability to generate a limited number of specialized cells when stimulated by several trophic signals: growth factors involved in biological events of bone formation and repair and of other connective tissues can be grouped as to their biological activities. This includes the proteins of the TGF-β superfamily, the insulin-like growth factors, platelet-derived growth factor, FGF, and epidermal growth factor (Carreira, Lojudice, et al., 2014; Gemini-Piperni et al., 2014).

Among the existing proteins, particularly BMPs currently have an important potential biotechnological and therapeutic approach, mainly related to bone and oral repair. They have a key role in the osteoinductive process of MSCs stimulating the recruitment of MSCs to the bone regrowth site. While the differentiation of MSCs, BMPs interact with their receptors type II by membrane receptors present in the MSCs, that promote the activation of specific proteins that lead the signal to the nucleus, where they are modulated activity of transcription factors. The BMPR-II complex, when activated, promotes the recruitment and phosphorylation of type 1A and 1B BMP receptors. The introduction of the receptors induces phosphorylation of intracellular messengers called R-Smads, which regulate expression of specific genes (Carreira, Alves, et al., 2014).

Thus, it is of great importance the study and scientific research related to biomaterials, whether or not associated with cells. Since the principle of tissue engineering is to produce structures that will replace certain organ or tissue throughout the patient's life or for a certain time, and will provide for better quality of life of the individual to allow the gradual return of tissue or organ functionality. In this sense, BMPs provide important potential biotechnological and therapeutic, mainly related to bone and oral repair indeed.

6. BIOTECHNOLOGY TO APPROACH BMPs BIOLOGY: FROM PURIFICATION TO CLINICAL APPLICATIONS

Several reports show that DBM is able to induce ectopic bone formation upon implantation into animal models. In 1960s, this activity was first related to the organic components of bone (Urist, 1965), and further studies showed it to be mediated by proteins, later named as BMPs (Urist, Iwata, & Ceccotti, 1973; Urist, Mikulski, & Lietze, 1979). However, bone matrix contains many protein components (such as peptide growth factors), suggesting that this observed BMP activity could be a result of these components instead of a new protein factor. With the advent of molecular biology techniques, cloning of the factors responsible for the osteoinductive activity present in bone was made possible. Human BMPs (BMPs 1, 2, and 3) were the first ones to be cloned, sequenced, and expressed in the recombinant form in a heterologous system and all of these purified proteins had their activity confirmed by both *in vitro* and *in vivo* assays (Wozney et al., 1988). Analysis of the amino acid sequence of different BMPs indicated that these proteins present high identity among them and, also, that they belong to the TGF-β superfamily. BMP-2 and BMP-4 sequences are highly related and differ in the amino terminal region (BMP-2 containing a heparin-binding domain). BMP-5, BMP-6, and BMP-7 (also known as OP-1) and BMP-8 (OP-2) form another subgroup of larger proteins, when compared to BMP-2 and BMP-4. A last and more distantly related subgroup is formed by BMP-3 (also known as osteogenin). BMPs have seven cysteine residues, being secreted as dimers. Six of these seven cysteine residues form intrachain disulfide bonds, remaining one interchain bridge, which is important to create the dimer. This dimerization is necessary for bone induction (Matthews, 2005). Glycosylation is also important for BMP activity. Chemical removal of glycans from BMP-2 and BMP-7 resulted in a decrease of the biological activity of these proteins (Sampath, Coughlin, & Whetstone, 1990). BMPs are glycosylated proteins with mature proteins displaying 1–3 potential glycosylation sites (Celeste, Iannazzi, & Taylor, 1990). A glycosylation site is present in the mature form of all BMPs. To date, 20 BMPs have been identified in humans (Hogan, 1996; Sakou, 1998), of which BMP-2, BMP-4, BMP-5, BMP-6, and BMP-7 induced ectopic bone formation in animal models (Matthews, 2005).

The first purification protocols for BMPs used bovine bone as the starting material (Urist et al., 1973). In general, these protocols start with a harsh step treatment, using an acid reagent for removal of the mineral component of bone, followed by protein extraction from the organic bone component with chaotropic agents. The BMP proteins were separated from the pool of proteins extracted by several chromatography steps. Similar protocols were applied using bone of different origins (human, rabbit, rat, mouse, ostrich, emu, guinea pig, and calf) with similar yields, processing kilogram of bone results in microgram quantities of purified BMP proteins (Bessho, Tagawa, & Murata, 1992; Hanamura et al., 1980; Hu et al., 2011; Urist, De Lange, & Finerman, 1983; Urist, Sato, et al., 1983; Urist et al., 1973, 1979; Wozney, 1992). The first BMP identified by this approach was the bovine BMP-3, which was extracted from bone and, after several chromatography steps, a purified protein with osteogenic activity was purified and the trypsin-digested products were sequenced for protein identification (Luyten et al., 1989; Sampath, Muthukumaran, & Reddi, 1987). Using similar protocols, in the late 1980s, the sequence of other BMPs was identified (Wang et al., 1988; Wozney et al., 1988). Before BMP-3, other bovine BMP protein (probably BMP-2) was purified and demonstrated to display osteoinductive activity, but, since its N-terminus was blocked, the protein identification was not possible (Urist, De Lange, et al., 1983; Urist et al., 1984).

Other tissue types, including teeth, as well as bone cell lines, were additionally observed to produce a similar bone-inductive activity. Using similar protocols used for bone extraction, BMPs purified from demineralized dentin matrix of different sources (human, bovine, rabbit, and rat) exhibited bone induction activity (Bang & Urist, 1967; Bessho, Tanaka, Matsumoto, Tagawa, & Murata, 1991; Hideki, Kazuhiko, Moniru, & Hisashi, 1996; Yeoman & Urist, 1967). The Saos-2 cell lineage, a human osteosarcoma, presented osteoinductive ability when bioassayed in nude mice (Anderson et al., 1992). A further report demonstrated that secreted products from the Saos-2 cell line had osteoinductive capacity, and that this material presented a different BMPs content profile (presenting mainly BMP-4 protein), when compared with a nonosteoinductive cell line (U20S) (Anderson et al., 1998). Other human osteosarcoma cell lineage (MG-63) was also used for BMPs purification from conditioned medium, which was submitted to dialysis and size exclusion chromatography. Isolated fractions reacted positively to a monoclonal BMP antibody in Western blot assays and promoted osteogenic activity in an animal model (Hua, Yankui, Anmin, Zhiling, & Jige, 2003).

The initial step of the purification protocols from bone tissue consisted of an organic solvent mixture treatment, such as chloroform–methanol (1:1) to remove lipids and to inhibit endogenous enzymes, followed by an acid treatment with cold hydrochloric acid (HCl 0.5–0.6 M) to demineralize the bone matrix and extract acid-soluble proteins. Some protocols also apply an EDTA (0.5 M) treatment after this step. The insoluble bone matrix gelatin obtained is generally treated with LiCl (8 M) and with $CaCl_2$ (0.5 M) to remove high- and low-molecular-weight protein polysaccharides, respectively. The BMP proteins are extracted from this insoluble matrix with chaotropic agents, such as urea (2–6 M) and guanidine hydrochloride (concentrations above 4 M decrease the osteogenic activity). After this step, a dialysis or ultrafiltration procedure is applied to remove chaotropic agents with a buffer solution containing Tris–HCl or phosphate buffer (50–100 mM and pH $\sim$7.0), NaCl, EDTA, and protease inhibitors, such as benzamidine–HCl, phenylmethanesulfonyl fluoride, and others. Also, some protocols use enzymatic treatment with collagenase and pepsin to expose more BMP proteins to the extraction buffer.

After the extraction step, BMP proteins are separated from other proteins by chromatographic techniques. In the early 1980s, an active bovine BMP protein (probably BMP-2) with a molecular weight (MW) of 18 kDa, in a reduced SDS-PAGE, was purified using hydroxyapatite affinity chromatography, followed by hydrophobic interaction chromatography (Urist et al., 1984). A similar protocol, using only two chromatographic steps (hydroxyapatite affinity and size exclusion) was able to purify a BMP protein with MW around 18 kDa (probably BMP-2) from different bone matrices (bovine, porcine, rabbit, and rat) with osteoinductive activity (Bessho et al., 1992). Other protocols used a reverse phase (C18 silica-based column) step after the affinity chromatography (hydroxyapatite or heparin) to purify bovine osteogenin (Luyten et al., 1989; Sampath et al., 1987).

A more complex protocol was applied to purify bovine BMP-1, BMP-2, and BMP-3 from the same bone matrix extract starting with two sequential ion exchange chromatography (diethylaminoethyl cellulose and carboxymethyl cellulose), followed by affinity chromatography (hydroxyapatite and heparin), size exclusion (Superose), another ion exchange (mono S), and a final step of reverse phase chromatography (C4 silica-based column) (Wang et al., 1988).

As mentioned before, the first recombinant BMPs were obtained in the late 1980s, when human recombinant BMPs (BMP-2 and BMP-3) showed chondrogenic activity (Wozney et al., 1988). Later reports demonstrated that BMP-2, also expressed in CHO cells, could produce a recombinant protein with osteogenic activity. As a secreted protein, BMP-2 was purified from conditioned medium (without serum) and achieved 50-fold enrichment through heparin affinity chromatography and ultrafiltration (Wang et al., 1990; Yamaguchi et al., 1991). To improve the production of recombinant BMP-2 protein in CHO cells, a protocol of methotrexate-mediated gene amplification was applied (Israel, Nove, Kerns, Moutsatsos, & Kaufman, 1992).

Recombinant human BMP-7 (hOP-1) was also produced in CHO cells through gene amplification (methotrexate); this recombinant product was able to induce new bone formation *in vivo*. Purification of this protein from conditioned media (0.5% serum content) was achieved using three chromatographic steps: size exclusion (Sepharose), hydrophobic interaction (phenyl-Sepharose), and reverse phase chromatography (C18 silica-based column) (Sampath et al., 1992). This protein was also produced in HEK293 cells upon antibiotic selection (hygromicin) of transfected cells and the active form of the recombinant product could be partially purified from conditioned medium by a single step of heparin affinity chromatography (yield of 4.5 mg/L) (Bustos-Valenzuela et al., 2010). The N-terminal histidine-tagged BMP-7 was expressed in HEK293 cells and purified from conditioned medium (serum-free) by metal ion chelating chromatography (HiTrap column) and size exclusion chromatography (Sepharose) and concentrated by ultrafiltration, resulting in a yield of 0.5 mg/L (Gregory, Ono, & Charbonneau, 2005).

Clinically approved (FDA) recombinant BMPs, recombinant human BMPs 2 (rhBMP-2) (Infuse®—Medtronic), and rhBMP-7 (OP-1®—Stryker) are produced in CHO DHFR-deficient cell lines via methotrexate-mediated gene amplification. These commercial recombinant proteins are produced from cell clones derived from CHO-DUKX-B11 cell line. The production occurs in bioreactors using defined medium (animal-free origin components) under specific growth conditions. When the cells reach the desired density, the medium is collected and filtered to remove cells and cell debris. Conditioned culture medium containing the recombinant product (hBMP-2 or hBMP-7) is submitted to a heparin-like resin (sulfate resin) followed by a hydrophobic interaction (butyl Sepharose resin) and a cation exchange chromatography (DEAE/Carboxy Sulfon resin). After these chromatographic steps, a viral clearance is carried out (Viresolv®) and a final step of ultrafiltration/diafiltration is performed in order to concentrate the recombinant product prior to formulation (US 8039231 B2 patent).

Recombinant BMPs are effective to treat orthopedic and dental pathological conditions, decreasing surgical morbidity and hospitalization time, and eliminating the need for surgical reinterventions for anatomical corrections improving the patient's quality of life (reviewed in Carreira, Lojudice, et al., 2014; Granjeiro et al., 2005). It is clear now, considering the significant positive results in clinical trials, that recombinant BMPs will be, in the future, the first product of choice for treatment of chronic or complex bone pathologies. Nevertheless, some open questions remain regarding the clinical use of BMPs and how the role of alternative pathways, such as that of MAPKs, and of phosphoproteomics could be of help to better understand these issues (Halcsik et al., 2013). Crescent off-label use of BMP has indicated the need for additional controlled, randomized, double-blind clinical trials for BMPs in order to ensure a better understanding of the side effects and safety of BMPs (Courvoisier, Sailhan, Laffenêtre, & Obert, 2014).

7. APPROACHES FOR DELIVERING RECOMBINANT BMP

Fractures are the most common traumatic injuries to humans, their repair being a major regenerative process involving biological events of embryonic skeletal development. In this scenario, BMPs have been extensively investigated as candidates for the enhancement of skeletal repair (mainly BMP-2 and BMP-7). BMPs have revealed a promising future in the field of Bioengineering and Regenerative Medicine (Einhorn & Gerstenfeld, 2015). Numerous studies support the use of BMPs in nonunions in oral surgery, sinus lift bone grafting, and periodontal regeneration (Table 2; Carreira, Lojudice, et al., 2014; Ronga et al., 2013). Table 3 shows BMPs approved in the United States and European Union. So far, the FDA has only approved two applications of treatments with BMPs for open tibia fractures with intramedullary fixation and long bone nonunion.

Table 3 BMPs Approved in the United States and Europe

Product Name	rhBMP and Expression System	Therapeutic Indication	Date Approved	Company
Osteogenic protein 1	rhosteogenic protein 1 BMP-7 produced in CHO cells	Nonunion of tibia	2001 (EU and USA)	Howmedica (Allendale, NJ, USA) (EU); Stryker (Kalamazoo, MI, USA) (USA)
Infuse	rhbone morphogenic protein-2 produced in CHO cells	Promotes fusion of lower spine vertebrae	2002 (USA)	Medtronic SofamorDanek (Memphis, TN, USA); Olympus
Inductos	Dibotermin alfa, rh bone morphogenic protein-2 produced in CHO cells	Acute tibia fractures	2002 (EU)	Genetics Institute, Wyeth Europa
Infuse bone graft	It contains dibotermin a, an rhBMP-2 produced in CHO cells placed on an absorbable collagen sponge. *Note*: this is the same active ingredient present in the product Infuse	Acute open tibial shaft fracture	2004 (USA)	Wyeth
Opgenra	Eptotermin α, rhBMP-7 produced in CHO cells	Posterolateral lumbar spinal fusion	2009 (EU)	Howmedica (Limerick, Ireland)

Adapted from Carreira, Alves, et al. (2014).

Two products consisting of recombinant BMP-2 and BMP-7 loaded into absorbable collagen sponges are available; however, off-label applications increased complications, including the risk of tumors, have been reported in the literature (Carragee, Hurwitz, & Weiner, 2011; Devine, Dettori, France, Brodt, & McGuire, 2012; Einhorn & Gerstenfeld, 2015). This problem is probably due to an inadequate BMP dosage in the lesion site. Minimally invasive local application methods reduce the risk of ectopic bone formation or high concentrations of circulating BMP or incorrectly placed BMP carriers.

The use of BMPs in nonunions or open fractures has shown promising results, reducing complications, and the healing time. However, high concentrations of BMPs are required at the injured site. The protein needs to be combined with carriers in order to be retained in the site and to warrant controlled release (Lieberman, Daluiski, & Einhorn, 2002). Several studies have shown the use of BMPs added to the allograft to improve the ability for bone-allograft integration and mechanical stability (Cook, Wolfe, Salkeld, & Rueger, 1995; Einhorn & Gerstenfeld, 2015).

Several studies have shown new strategies to improve BMPs' bioactivity, stability, and affinity to the scaffold and/or to the extracellular matrix and cells (Table 4). Other delivering strategies involve the use of viral and nonviral vectors to deliver the BMP gene to the site of injury: vectors could be delivered to *in vivo* sites (*in vivo* gene therapy) or in tissue culture (*ex vivo* gene therapy) (Calori, Donati, Di Bella, & Tagliabue, 2009; Einhorn & Gerstenfeld, 2015; Ulrich-Vinther, 2007). Moreover, it has been suggested to incorporate rhBMPs with their delivery vehicles, such as a type I collagen carrier.

8. BMPs AS A BIOACTIVE MOLECULE FOR COATING BIOMATERIAL'S SURFACES

"Diamond concept" in tissue engineering involves three main elements, namely: a hallmark growth factor to repair bone defects (BMPs), osteoprogenitor cells, and a scaffold to support the cells and maintain the BMPs in place. Therefore, to use BMPs in this repair process, it is necessary to obtain an appropriate scaffold/carrier. The first BMP carrier to be approved by the FDA was an absorbable collagen sponge. The porosity, orientation, and size of fibers and chemical affinity are essential to define the best materials to develop the ideal scaffold for optimum cells adhesion and/or BMPs adsorption. Several materials have been tested as carriers, from synthetic to natural sources, such as hydroxyapatite, bioglass, calcium sulfates, calcium phosphates, β-tricalcium phosphate, calcium carbonate, organic polymers, polylactic acid (PLA), polyglycolic acid (PGA), copolymers (such as polylactic-*co*-polyglycolic acid (PGLA)), methylmethacrylate, hyaluran, fibrina, alginate, silk, agarose and collagen (Carpena, Min, & Lee, 2015; Chen, Liu, Gu, Feng, & Yang, 2015; Dadsetan et al., 2015; Gonçalves et al., 2015; Margolin et al., 1998; Seeherman & Wozney, 2005). The choice of a biomaterial depends on a good compatibility, accessibility, and low immunogenicity. Bioactive membranes are promising options for bone

Table 4 Chemical or Genetic Engineering Modifications for Improving Particular Characteristics of BMPs for Use in Tissue Engineering

Chemical or Genetic Engineering Modifications	Advantages	PMID
Enzymatic cleavage of BMP heparin-binding domain	Reduces nonspecific interactions with extracellular matrix (MEC), leading to increased bioactivity *in vitro*	8065329/11314790
Plasmin enzymatic cleavage of BMP		1419071/10837631
Adding additional heparin-binding domains	Increases retention to MEC or affinity to biomaterial, leading to increased bioactivity *in vitro*	15045531/16170576
Immobilization of synthetic peptides of partial BMP sequences	Improves the stability and specific activity of the BMP and reduces the initial burst release	10737883/14499589/ 15174115/15543633/ 16550532/16506173/ 17323156/17199553
Fusion protein BMP with domain to bind a specific biomaterial	Increases the retention of BMP on carrier, accelerating tissue regeneration; increases stability and activity of the BMP	12168663/15619323
Fusion protein BMP with arginine-glycine-aspartic acid (RGD) for enhanced cell attachment or domain for other specific biological function	Enhances attachment of cells, tissue regeneration, and possibly the biocompatibility of the material	21272672/20537381
Chemical modification of BMP, such as with succinylation, acetylation, or biotinylation, changes the isoelectric point	Enhances the affinity for binding to the biomaterial by changing the peptide solubility, to suit a desired release profile	10837631/0415583/ 10550773/10679688/ 1131479
Formation of heterodimers between different BMPs	The increased bioactivity of BMP heterodimers leads to lower dose	16488673/ 7763240/891903

Adapted from Bessa, Casal, and Reis (2008).

regeneration. Several studies have analyzed BMPs in association with other growth factors or platelet-rich plasma (PRP). As PRP has a proliferative effect, its use, together with BMPs, could improve the role of PRP in the lesion and to accelerate bone repair (Tomoyasu et al., 2007).

9. CONCLUDING REMARKS AND PERSPECTIVES

In the past few years, research on BMPs and TGF-β pathways has drawn a great deal of attention. Since the pioneer work by Urist and colleagues, BMPs have been implicated in a wide variety of roles, including cell proliferation, apoptosis, differentiation, and morphogenesis. Although the existing systematic review and meta-analysis support some therapeutic applications, some adverse effects of their clinical use have been reported, with several important questions remaining to be addressed. As discussed elsewhere, one question is why BMP therapeutics require microgram amounts while endogenous BMPs act within the nanogram level? In bioengineering, BMPs represent promising molecules, acting to accelerate and enhance osteointegration. In this scenario, the development of smart materials, mainly based on biopolymers and bone-like calcium phosphates, appears to provide an attractive alternative for delivering BMPs in a better-controlled fashion. Indeed, based on these controversial data from the literature, our point of view is that further studies are necessary in order to better understand the actual extent of BMPs' potentialities and limitations.

ACKNOWLEDGMENTS

We are grateful to all researches involved in our BMP projects over the years. The financial support of the following Brazilian Federal and State research agencies was essential for generation of this review and development of the work carried out in our laboratory, namely: National Bank for Economical and Social Development (BNDES), National Council for Scientific and Technological Development (CNPq), Coordination for the Improvement of Higher Education Personnel (CAPES), São Paulo Research Foundation (FAPESP), Rio de Janeiro State Foundation Support Research (FAPERJ), Research and Projects Financier (FINEP), Ministry of Science, Technology and Innovation (MCTI), and Ministry of Health of Brazil/Department of Science and Technology (MS-DECIT).

REFERENCES

Anderson, H. C., Hsu, H. H. T., Raval, P., Reynold, P. R., Gurley, D. J., Aguilera, M. X., et al. (1998). The bone-inducing agent in Saos-2 cell extracts and secretions. *Cells and Materials*, *8*, 89–98.

Anderson, H. C., Sugamoto, K., Morris, D. C., Hsu, H. H. T., Anderson, H. C., & Hunt, T. (1992). Bone-inducing agent from cultured human Saos-2 osteosarcoma cells. *Bone and Mineral, 16*, 49–62.

Attisano, L., & Lee-Hoeflich, S. T. (2001). The Smads. *Genome Biology, 2*, reviews3010.1–3010.8.

Bang, G., & Urist, M. R. (1967). Bone induction in excavation chambers in matrix of decalcified dentin. *Archives of Surgery, 94*, 781–789.

Beppu, H., Mwizerwa, O. N., Beppu, Y., Dattwyler, M. P., Lauwers, G. Y., Bloch, K. D., et al. (2008). Stromal inactivation of BMPRII leads to colorectal epithelial overgrowth and polyp formation. *Oncogene, 27*, 1063–1070.

Bessa, P. C., Casal, M., & Reis, R. L. (2008). Bone morphogenetic proteins in tissue engineering: the road from laboratory to clinic, part II (BMP delivery). *Journal of Tissue Engineering and Regenerative Medicine, 2*, 81–96.

Bessho, K., Tagawa, T., & Murata, M. (1992). Comparison of bone matrix-derived bone morphogenetic proteins from various animals. *Journal of Oral and Maxillofacial Surgery, 50*, 496–501.

Bessho, K., Tanaka, N., Matsumoto, J., Tagawa, T., & Murata, M. (1991). Human dentin-matrix-derived bone morphogenetic protein. *Journal of Dental Research, 70*, 171–175.

Bleuming, S. A., He, X. C., Kodach, L. L., Hardwick, J. C., Koopman, F. A., Ten Kate, F. J., et al. (2007). Bone morphogenetic protein signaling suppresses tumorigenesis at gastric epithelial transition zones in mice. *Cancer Research, 67*, 8149–8155.

Bustos-Valenzuela, J. C., Halcsik, E., Bassi, E. J., Demasi, M. A., Granjeiro, J. M., & Sogayar, M. C. (2010). Expression, purification, bioactivity, and partial characterization of a recombinant human bone morphogenetic protein-7 produced in human 293T cells. *Molecular Biotechnology, 46*, 118–126.

Caja, L., Bellomo, C., & Moustakas, A. (2015). Transforming growth factor β and bone morphogenetic protein actions in brain tumors. *FEBS Letters, S0014–5793*(15), 00350–00356.

Calori, G. M., Donati, D., Di Bella, C., & Tagliabue, L. (2009). Bone morphogenetic proteins and tissue engineering: Future directions. *Injury, 40*(Suppl. 3), S67–S76.

Cao, X., & Chen, D. (2005). The BMP signaling and *in vivo* bone formation. *Gene, 357*, 1–8.

Carpena, N. T., Min, Y. K., & Lee, B. T. (2015). Improved *in vitro* biocompatibility of surface-modified hydroxyapatite sponge scaffold with gelatin and BMP-2 in comparison against a commercial bone allograft. *ASAIO Journal, 61*, 78–86.

Carragee, E. J., Hurwitz, E. L., & Weiner, B. K. (2011). A critical review of recombinant human bone morphogenetic protein-2 trials in spinal surgery: Emerging safety concerns and lessons learned. *The Spine Journal, 11*, 471–491.

Carreira, A. C., Alves, G. G., Zambuzzi, W. F., Sogayar, M. C., & Granjeiro, J. M. (2014). Bone morphogenetic proteins: Structure, biological function and therapeutic applications. *Archives of Biochemistry and Biophysics, 561*, 64–73.

Carreira, A. C., Lojudice, F. H., Halcsik, E., Navarro, R. D., Sogayar, M. C., & Granjeiro, J. M. (2014). Bone morphogenetic proteins: Facts, challenges, and future perspectives. *Journal of Dental Research, 93*, 335–345.

Celeste, A. J., Iannazzi, J. A., & Taylor, R. C. (1990). Identification of transforming growth factor b family members present in bone-inductive protein purified from bovine bone. *Proceedings of the National Academy of Sciences of the United States of America, 87*, 9843–9847.

Chen, L., Liu, H. L., Gu, Y., Feng, Y., & Yang, H. L. (2015). Lumbar interbody fusion with porous biphasic calcium phosphate enhanced by recombinant bone morphogenetic protein-2/silk fibroin sustained-released microsphere: An experimental study on sheep model. *Journal of Materials Science. Materials in Medicine, 26*, 126.

Cook, S. D., Wolfe, M. W., Salkeld, S. L., & Rueger, D. C. (1995). Effect of recombinant human osteogenic protein-1 on healing of segmental defects in non-human primates. *The Journal of Bone and Joint Surgery. American Volume, 77*, 734–750.

Courvoisier, A., Sailhan, F., Laffenêtre, O., & Obert, L. (2014). French study group of BMP in orthopedic surgery. Bone morphogenetic protein and orthopaedic surgery: Can we legitimate its off-label use? *International Orthopaedics, 38*, 2601–2605.

Dadsetan, M., Guda, T., Runge, M. B., Mijares, D., LeGeros, R. Z., LeGeros, J. P., et al. (2015). Effect of calcium phosphate coating and rhBMP-2 on bone regeneration in rabbit calvaria using poly(propylene fumarate) scaffolds. *Acta Biomaterialia, 18*, 9–20.

Dawson, E., Bae, H. W., Burkus, J. K., Stambough, J. L., & Glassman, S. D. (2009). Recombinant human bone morphogenetic protein-2 on an absorbable collagen sponge with an osteoconductive bulking agent in posterolateral arthrodesis with instrumentation. A prospective randomized trial. *The Journal of Bone and Joint Surgery. American Volume, 91*, 1604–1613.

Devine, J. G., Dettori, J. R., France, J. C., Brodt, E., & McGuire, R. A. (2012). The use of rhBMP in spine surgery: Is there a cancer risk? *Evidence-Based Spine-Care Journal, 3*, 35–41.

Dyer, L. A., Pi, X., & Patterson, C. (2014). The role of BMPs in endothelial cell function and dysfunction. *Trends in Endocrinology and Metabolism, 25*, 472–480.

Ehata, S., Yokoyama, Y., Takahashi, K., & Miyazono, K. (2013). Bi-directional roles of bone morphogenetic proteins in cancer: Another molecules Jekyll and Hyde? *Pathology International, 63*, 287–296.

Einhorn, T. A., & Gerstenfeld, L. C. (2015). Fracture healing: Mechanisms and interventions. *Nature Reviews. Rheumatology, 11*, 45–54.

Feeley, B. T., Krenek, L., Liu, N., Hsu, W. K., Gamradt, S. C., Schwarz, E. M., et al. (2006). Overexpression of noggin inhibits BMP-mediated growth of osteolytic prostate cancer lesions. *Bone, 38*, 154–166.

Gao, H., Chakraborty, G., Lee-Lim, A. P., Mo, Q., Decker, M., Vonica, A., et al. (2012). The BMP inhibitor Coco reactivates breast cancer cells at lung metastatic sites. *Cell, 150*, 764–779.

Gemini-Piperni, S., Takamori, E. R., Sartoretto, S. C., Paiva, K. B., Granjeiro, J. M., de Oliveira, R. C., et al. (2014). Cellular behavior as a dynamic field for exploring bone bioengineering: A closer look at cell-biomaterial interface. *Archives of Biochemistry and Biophysics, 561*, 88–98.

Giannoudis, P. V., Kanakaris, N. K., Dimitriou, R., Gill, I., Kolimarala, V., & Montgomery, R. J. (2009). The synergistic effect of autograft and BMP-7 in the treatment of atrophic nonunions. *Clinical Orthopaedics and Related Research, 467*, 3239–3248.

Gonçalves, F., Bentini, R., Burrows, M. C., Carreira, A. C. O., Kossungue, P., Sogayar, M. C., et al. (2015). Hybrid membranes of PLLA/collagen for bone tissue engineering: A comparative study of scaffold production techniques for optimal mechanical properties and osteoinduction ability. *Materials, 8*, 408–423.

Granjeiro, J. M., Oliveira, R. C., Bustos-Valenzuela, J. C., Sogayar, M. C., & Taga, R. (2005). Bone morphogenetic proteins: From structure to clinical use. *Brazilian Journal of Medical and Biological Research, 38*, 1463–1473.

Gregory, K. E., Ono, R. N., & Charbonneau, N. L. (2005). The pro-domain of BMP-7 targets the BMP-7 complex to the extracellular matrix. *The Journal of Biological Chemistry, 280*, 27970–27980.

Griffith, D. L., Keck, P. C., Sampath, T. K., Rueger, D. C., & Carlson, W. D. (1996). Three-dimensional structure of recombinant human osteogenic protein 1: Structural paradigm for the transforming growth factor beta superfamily. *Proceedings of the National Academy of Sciences of the United States of America, 93*, 878–883.

Grünhagen, J., Bhushan, R., Degenkolbe, E., Jäger, M., Knaus, P., Mundlos, S., et al. (2015). MiR-497∼195 cluster microRNAs regulate osteoblast differentiation by targeting BMP signaling. *Journal of Bone and Mineral Research, 30*, 796–808.

Halcsik, E., Forni, M. F., Fujita, A., Verano-Braga, T., Jensen, O. N., & Sogayar, M. C. (2013). New insights in osteogenic differentiation revealed by mass spectrometric assessment of phosphorylated substrates in murine skin mesenchymal cells. *BMC Cell Biology*, *14*, 47.

Hanamura, H., Higuchi, Y., Nakagawa, M., Iwata, H., Nogami, H., & Urist, M. R. (1980). Solubilized bone morphogenetic protein (BMP) from mouse osteosarcoma and rat demineralized bone matrix. *Clinical Orthopaedics and Related Research*, *148*, 281–290.

Herhaus, L., & Sapkota, G. P. (2014). The emerging roles of deubiquitylating enzymes (DUBs) in the TGFβ and BMP pathways. *Cellular Signalling*, *26*, 2186–2192.

Hideki, M., Kazuhiko, M., Moniru, U., & Hisashi, I. (1996). A study of the bone morphogenetic protein derived from bovine demineralized dentin matrix. *Nagoya Journal of Medical Science*, *59*, 37–47.

Hogan, B. L. (1996). Bone morphogenetic proteins: Multifunctional regulators of vertebrate development. *Genes & Development*, *10*, 1580–1594.

Hover, L. D., Abel, T. W., & Owens, P. (2015). Genomic analysis of the BMP family in glioblastomas. *Translational Oncogenomics*, 7, 1–9.

Howe, J. R., Bair, J. L., Sayed, M. G., Anderson, M. E., Mitros, F. A., Petersen, G. M., et al. (2001). Germline mutations of the gene encoding bone morphogenetic protein receptor 1A in juvenile polyposis. *Nature Genetics*, *28*, 184–187.

Howe, J. R., Roth, S., Ringold, J. C., Summers, R. W., Jarvinen, H. J., Sistone, P., et al. (1998). Mutations in the SMAD4/DPC4 gene in juvenile polyposis. *Science*, *280*, 1086–1088.

Howe, J. R., Sayed, M. G., Ahmed, A. F., Ringold, J., Larsen-Haidle, J., Merg, A., et al. (2004). The prevalence of MADH4 and BMPR1A mutations in juvenile polyposis and absence of BMPR2, BMPR1B, and ACVR1 mutations. *Journal of Medical Genetics*, *41*, 484–491.

Hu, Z., Peel, S. A., Lindholm, T. C., Sandor, G. K., Clokie, C. M., & Su, Y. (2011). Osteoinductivity of partially purified bovine, ostrich and emu bone morphogenetic proteins *in vitro*. *Journal of Biomedical Materials Research*, *98*, 473–477.

Hua, W. U., Yankui, L., Anmin, C., Zhiling, Q. U., & Jige, C. (2003). Extraction, purification and identification of bone morphogenetic protein in conditioned medium of osteosarcoma cell (MG-63). *The Chinese-German Journal of Clinical Oncology*, *2*, 234–236.

Israel, D. I., Nove, J., Kerns, K. M., Moutsatsos, I. K., & Kaufman, R. J. (1992). Expression and characterization of bone morphogenetic protein-2 in Chinese hamster ovary cells. *Growth Factors*, 7, 139–150.

Jung, R. E., Windisch, S. I., Eggenschwiler, A. M., Thoma, D. S., Weber, F. E., & Hämmerle, C. H. (2009). A randomized-controlled clinical trial evaluating clinical and radiological outcomes after 3 and 5 years of dental implants placed in bone regenerated by means of GBR techniques with or without the addition of BMP-2. *Clinical Oral Implants Research*, *20*, 660–666.

Karagiannis, G. S., Musrap, N., Saraon, P., Treacy, A., Schaeffer, D. F., Kirsch, R., et al. (2015). Bone morphogenetic protein antagonist gremlin-1 regulates colon cancer progression. *Biological Chemistry*, *396*, 163–183.

Katsuno, Y., Hanyu, A., Kanda, H., Ishikawa, Y., Akiyama, F., Iwase, T., Ogata, E., Ehata, S., Miyazono, K., & Imamura, T. (2008). Bone morphogenetic protein signaling enhances invasion and bone metastasis of breast cancer cells through Smad pathway. *Oncogene*, *27*, 6322–6333.

Kessler, E., Takahara, K., Biniaminov, L., Brusel, M., & Greenspan, D. S. (1996). Bone morphogenetic protein-1: The type I procollagen C-proteinase. *Science*, *271*, 360–362.

Kiyono, K., Suzuki, H. I., Morishita, Y., Komuro, A., Iwata, C., Yashiro, M., et al. (2009). c-Ski overexpression promotes tumor growth and angiogenesis through inhibition of

transforming growth factor-beta signaling in diffuse-type gastric carcinoma. *Cancer Science*, *100*, 1809–1816.
Koketsu, K., Yoshida, D., Kim, K., Ishii, Y., Tahara, S., Teramoto, A., et al. (2015). Gremlin, a bone morphogenetic protein antagonist, is a crucial angiogenic factor in pituitary adenoma. *International Journal of Endocrinology*, *2015*, 834137. http://dx.doi.org/10.1155/2015/834137. Epub 2015 Mar 5.
Komuro, A., Yashiro, M., Iwata, C., Morishita, Y., Johansson, E., Matsumoto, Y., et al. (2009). Diffuse-type gastric carcinoma: Progression, angiogenesis, and transforming growth factor b signaling. *Journal of the National Cancer Institute*, *101*, 592–604.
Laulan, N. B., & St-Pierre, Y. (2015). Bone morphogenetic protein 4 (BMP-4) and epidermal growth factor (EGF) inhibit metalloproteinase-9 (MMP-9) expression in cancer cells. *Oncoscience*, *2*, 309–316.
Lieberman, J. R., Daluiski, A., & Einhorn, T. A. (2002). The role of growth factors in the repair of bone. Biology and clinical applications. *The Journal of Bone and Joint Surgery. American Volume*, *84*(A(6)), 1032–1044.
Lombardo, Y., Scopelliti, A., Cammareri, P., Todaro, M., Iovino, F., Ricci-Vitiani, L., et al. (2011). Bone morphogenetic protein 4 induces differentiation of colorectal cancer stem cells and increases their response to chemotherapy in mice. *Gastroenterology*, *140*, 297–309.
Luyten, F. P., Cunningham, N. S., Ma, S., Muthukumaran, N., Hammonds, R. G., Nevins, W. B., et al. (1989). Purification and partial amino acid sequence of osteogenin, a protein initiating bone differentiation. *The Journal of Biological Chemistry*, *264*, 13377–13380.
Margolin, M. D., Cogan, A. G., Taylor, M., Buck, D., McAllister, T. N., Toth, C., et al. (1998). Maxillary sinus augmentation in the non-human primate: A comparative radiographic and histologic study between recombinant human osteogenic protein-1 and natural bone mineral. *Journal of Periodontology*, *69*, 911–919.
Matthews, S. J. E. (2005). Biological activity of bone morphogenetic proteins (BMP's). *Injury*, *36*, S34–S37.
Mi, L. Z., Brown, C. T., Gao, Y., Tian, Y., Le, V. Q., Walz, T., et al. (2015). Structure of bone morphogenetic protein 9 procomplex. *Proceedings of the National Academy of Sciences of the United States of America*, *112*, 3710–3715.
Miyazono, K., Kamiya, Y., & Morikawa, M. (2010). Bone morphogenetic protein receptors and signal transduction. *Journal of Biochemistry*, *147*, 35–51.
Nolan, K., Kattamuri, C., Luedeke, D. M., Angerman, E. B., Rankin, S. A., Stevens, M. L., et al. (2015). Structure of neuroblastoma suppressor of tumorigenicity 1 (NBL1): Insights for the functional variability across bone morphogenetic protein (BMP) antagonists. *The Journal of Biological Chemistry*, *290*, 4759–4771.
Owens, P., Pickup, M. W., Novitskiy, S. V., Giltnane, J. M., Gorska, A. E., Hopkins, C. R., et al. (2015). Inhibition of BMP signaling suppresses metastasis in mammary cancer. *Oncogene*, *34*, 2437–2449.
Pan, H., Li, X., Wang, J., Zhang, K., Yang, H., Li, Z., et al. (2015). LIM mineralization protein-1 enhances bone morphogenetic protein-2 mediated osteogenesis through activation of ERK1/2 MAPK pathway and upregulation of Runx2 transactivity. *Journal of Bone and Mineral Research*. http://dx.doi.org/10.1002/jbmr.2481. [Epub ahead of print].
Rajski, M., Saaf, A., & Buess, M. (2015). BMP2 response pattern in human lung fibroblasts predicts outcome in lung adenocarcinomas. *BMC Medical Genomics*, *8*, 16.
Rider, C. C., & Mulloy, B. (2010). Bone morphogenetic protein and growth differentiation factor cytokine families and their protein antagonists. *The Biochemical Journal*, *429*, 1–12.

Ronga, M., Fagetti, A., Canton, G., Paiusco, E., Surace, M. F., & Cherubino, P. (2013). Clinical applications of growth factors in bone injuries: Experience with BMPs. *Injury*, *44*(Suppl. 1), S34–S39.

Rosenzweig, B. L., Omammura, T., Okadome, T., Cox, G. N., Yamashita, H., ten Dijke, P., et al. (1995). Cloning and characterization of a human type II receptor for bone morphogenetic proteins. *Proceedings of the National Academy of Sciences of the United States of America*, *92*, 7632–7636.

Sakou, T. (1998). Bone morphogenetic proteins: From basic studies to clinical approaches. *Bone*, *22*, 591–603.

Sampath, T. K., Coughlin, J. E., & Whetstone, R. M. (1990). Ovine osteogenic protein is composed of dimers of OP-1 and BMP-2A, two members of the transforming growth factor-b superfamily. *The Journal of Biological Chemistry*, *265*, 13198–13205.

Sampath, T. K., Maliakal, J. C., Hauschka, P. V., Jones, W. K., Sasak, H., Tucker, R. F., et al. (1992). Recombinant human osteogenic protein-1 (hOP-1) induces new bone formation *in vivo* with a specific activity comparable with natural bovine osteogenic protein and stimulates osteoblast proliferation and differentiation *in vitro*. *The Journal of Biological Chemistry*, *267*, 20352–20362.

Sampath, T. K., Muthukumaran, N., & Reddi, A. H. (1987). Isolation of osteogenin, an extracellular matrix-associated, bone-inductive protein, by heparin affinity chromatography. *Proceedings of the National Academy of Sciences of the United States of America*, *84*, 7109–7113.

Seeherman, H., & Wozney, J. M. (2005). Delivery of bone morphogenetic proteins for orthopedic tissue regeneration. *Cytokine & Growth Factor Reviews*, *16*, 329–345.

Sieber, C., Kopf, J., Hiepen, C., & Knaus, P. (2009). Recent advances in BMP receptor signaling. *Cytokine & Growth Factor Reviews*, *20*, 343–355.

Sneddon, J. B., Zhen, H. H., Montgomery, K., van de Rijn, M., Tward, A. D., West, R., et al. (2006). Bone morphogenetic protein antagonist gremlin 1 is widely expressed by cancer-associated stromal cells and can promote tumor cell proliferation. *Proceedings of the National Academy of Sciences of the United States of America*, *103*, 14842–14847.

Spinella-Jaegle, S., Roman-Roman, S., Faucheu, C., Dunn, F., Kawai, S., Galléa, S., et al. (2001). Opposite effects of bone morphogenetic protein-2 and transforming growth factor-beta 1 on osteoblast differentiation. *Bone*, *29*, 323–330.

Tomoyasu, A., Higashio, K., Kanomata, K., Goto, M., Kodaira, K., Serizawa, H., et al. (2007). Platelet-rich plasma stimulates osteoblastic differentiation in the presence of BMPs. *Biochemical and Biophysical Research Communications*, *14*, 62–67.

Ulrich-Vinther, M. (2007). Gene therapy methods in bone and joint disorders. Evaluation of the adeno-associated virus vector in experimental models of articular cartilage disorders, periprosthetic osteolysis and bone healing. *Acta Orthopaedica. Supplementum*, *78*, 1–64.

Urist, M. R. (1965). Bone formation by autoinduction. *Science*, *150*, 893–899.

Urist, M. R., De Lange, R. J., & Finerman, G. A. (1983). Bone cell differentiation and growth factors. *Science*, *220*, 680–686.

Urist, M. R., Huo, Y. K., Brownell, A. G., Hohl, W. M., Buyske, J., Lietze, A., et al. (1984). Purification of bovine bone morphogenetic protein by hydroxyapatite chromatography. *Proceedings of the National Academy of Sciences of the United States of America*, *81*, 371–375.

Urist, M. R., Iwata, H., & Ceccotti, P. L. (1973). Bone morphogenesis in implants of insoluble bone gelatin. *Proceedings of the National Academy of Sciences of the United States of America*, *70*, 3511–3515.

Urist, M. R., Mikulski, A., & Lietze, A. (1979). Solubilized and insolubilized bone morphogenetic protein. *Proceedings of the National Academy of Sciences of the United States of America*, *76*, 1828–1832.

Urist, M. R., Sato, K., Brownell, A. G., Malinin, T., Lietze, A., Huo, Y. K., et al. (1983). Human bone morphogenetic protein (hBMP). *Proceedings of the Society for Experimental Biology and Medicine, 173*, 194–198.

Vukicevic, S., & Grgurevic, L. (2009). BMP-6 and mesenchymal stem cell differentiation. *Cytokine Growth Factor Reviews, 20*, 441–448.

Wang, E. A., Rosen, V., Cordes, P., Hewick, R. M., Kriz, M. J., Luxenberg, D. P., et al. (1988). Purification and characterization of other distinct bone inducing factors. *Proceedings of the National Academy of Sciences of the United States of America, 85*, 9484–9488.

Wang, E. A., Rosen, V., D'Alessandro, J. S., Bauduy, M., Cordes, P., Harada, T., et al. (1990). Recombinant human bone morphogenetic protein induces bone formation. *Proceedings of the National Academy of Sciences of the United States of America, 87*, 2220–2224.

Whitman, M. (1998). Smads and early developmental signaling by the TGFβ superfamily. *Genes & Development, 12*, 2445–2462.

Wozney, J. M. (1992). The bone morphogenetic protein family and osteogenesis. *Molecular Reproduction and Development, 32*, 160–167.

Wozney, J. M., Rosen, V., Celeste, A. J., Miltsock, L. M., Whitters, M. J., Kriz, R. W., et al. (1988). Novel regulators of bone formation: Molecular clones and activities. *Science, 242*, 1528–1534.

Yamaguchi, A., Katagari, T., Ikeda, T., Wozney, J. M., Rosen, V., Wang, E. A., et al. (1991). Recombinant human bone morphogenetic protein-2 stimulates osteoblastic maturation and inhibits myogenic differentiation *in vitro*. *The Journal of Cell Biology, 113*, 681–687.

Yeoman, J. D., & Urist, M. R. (1967). Bone induction by decalcified dentin implanted into oral osseous and muscle tissue. *Archives of Oral Biology, 12*, 999–1008.

Zambuzzi, W. F., Coelho, P. G., Alves, G. G., & Granjeiro, J. M. (2011). Intracellular signal transduction as a factor in the development of "smart" biomaterials for bone tissue engineering. *Biotechnology and Bioengineering, 108*, 1246–1250.

Zhang, L., Sun, H., Zhao, F., Lu, P., Ge, C., Li, H., et al. (2012). BMP4 administration induces differentiation of CD133+ hepatic cancer stem cells, blocking their contributions to hepatocellular carcinoma. *Cancer Research, 72*, 4276–4285.

INDEX

Note: Page numbers followed by "*f*" indicate figures and "*t*" indicate tables.

U

V

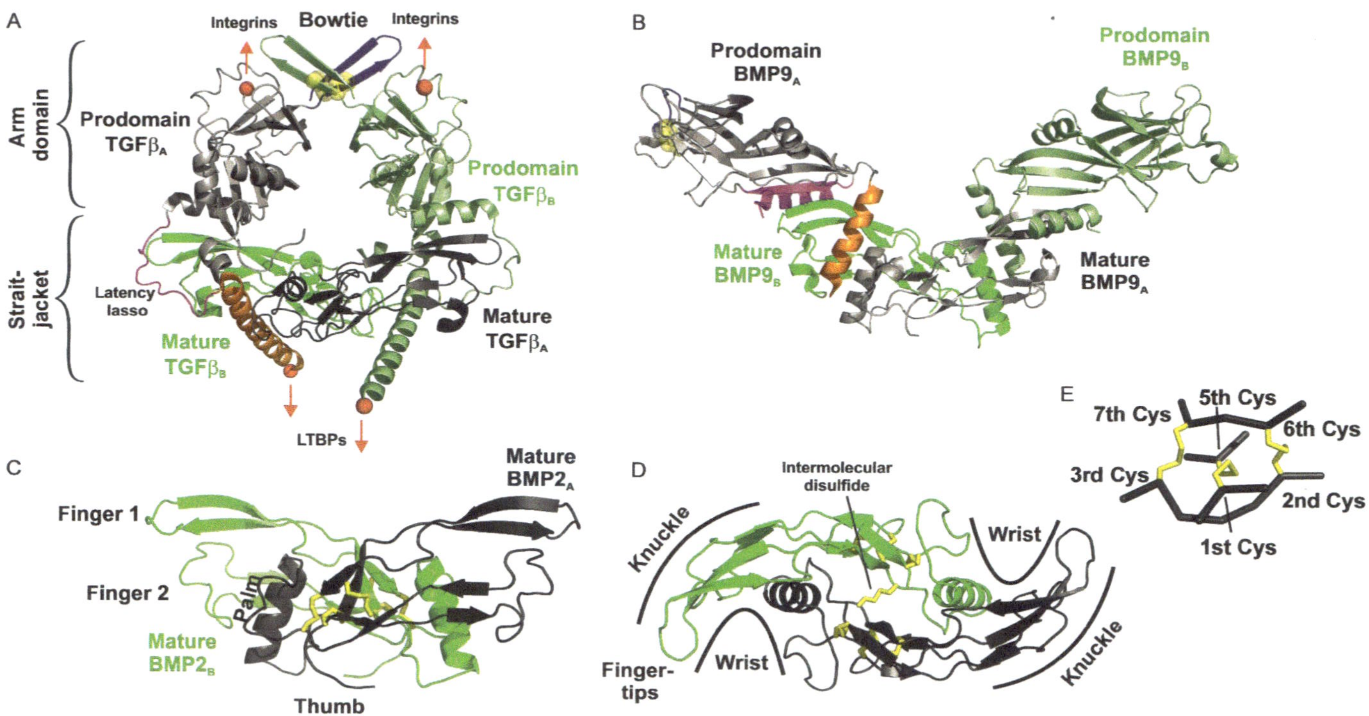
A
Integrins
Bowtie
Integrins
Arm domain
Prodomain TGFβA
Prodomain TGFβB
Strait-jacket
Latency lasso
Mature TGFβB
Mature TGFβA
LTBPs
B
Prodomain BMP9A
Prodomain BMP9B
Mature BMP9B
Mature BMP9A
C
Finger 1
Mature BMP2A
Finger 2
Palm
Mature BMP2B
Thumb
D
Intermolecular disulfide
Knuckle
Wrist
Wrist
Finger-tips
Knuckle
E
7th Cys
5th Cys
6th Cys
3rd Cys
2nd Cys
1st Cys

Thomas D. Mueller, Figure 2 (A) Structure of the proprotein complex of TGFβ1. The mature region is marked in green and dark gray. The prodomain can be separated into the arm domain and the straitjacket. Whereas the latter wraps around the mature region thereby blocking binding to the TGFβ type I and type II receptors, the former domain is involved in dimerization of two prodomains. The prodomain dimer is then stabilized by two disulfide bonds formed in the bowtie segment. Activation of the TGFβ mature region is likely to occur via tensile forces using the RGD motifs in the arm domain to bind to integrins and two cysteine residues at the N-terminus to connect to LTBP proteins. (B) Structure of the proprotein complex of BMP9 showing that the architecture vastly differs from TGFβ1. Although type I and type II receptor binding is abrogated by the prodomains, the elements conferring this blockage are different. In contrast to the prodomains of TGFβ1 the prodomains of BMP9 do not dimerize. (C) The butterfly-shaped architecture of the mature region of BMP2 resembles two hands assembled palm-to-palm, the β-sheets represent the fingers, the α-helix and the dimer interface form the palm. (D) The convex side of the fingers present the type II receptor binding sites (termed knuckle) of BMP/GDFs and Activins, the TGFβs bind their type II receptors via the fingertips. The type I receptor binding in BMPs/GDFs and Activins occurs in the wrist, which is formed by the concave side of the finger and the palm of the other monomer. (E) The cystine-knot is built from three disulfides of which two disulfide bonds between the second and fifth and the third and seventh form a ring, which is then penetrated by the third disulfide bond between the first and the fourth cysteine residue.

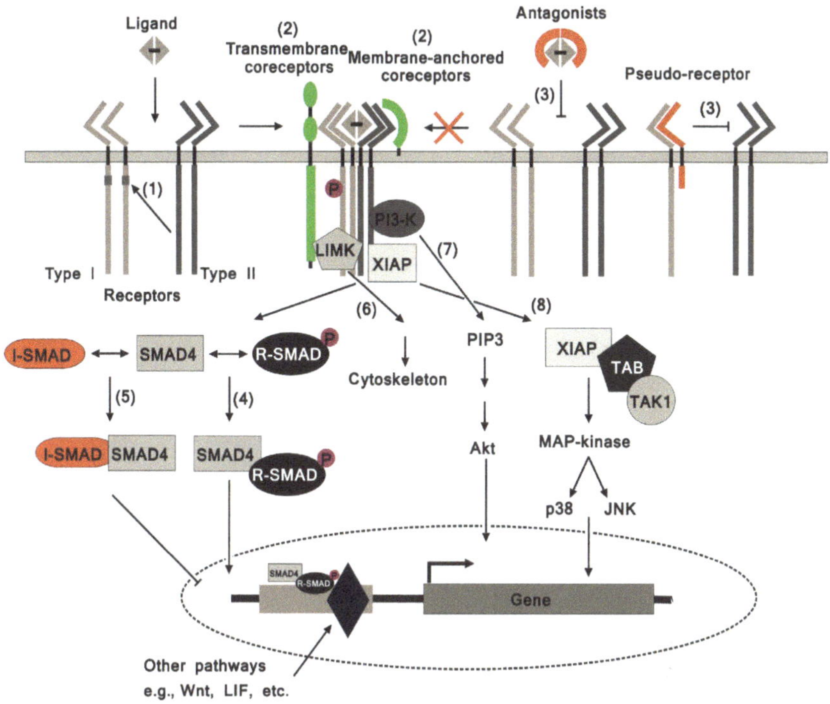

Thomas D. Mueller, Figure 3 Schematic representation of the TGFβ signaling cascade. Activation occurs by binding of a TGFβ ligand to TGFβ type I and type II receptors leading to transphosphorylation of the type I receptor by the type II receptor kinase (1). Activation can be modulated by coreceptors (2), or inhibited by modulator proteins/secreted antagonists and negatively regulating pseudo-receptors (3). Activation can be transferred to R-SMAD proteins, which then hetero-oligomerize with SMAD4 (4), translocate into the nucleus and regulate gene transcription. The SMAD pathway can be inhibited by inhibitory SMADs (5) that interfere with R-SMAD/Co-SMAD complex formation. Also SMAD-independent signaling cascades can be triggered by TGFβ ligands, such as the LIMK kinase (6), the phosphatidylinositol-3-kinase (7), or the MAP kinase pathway (8).

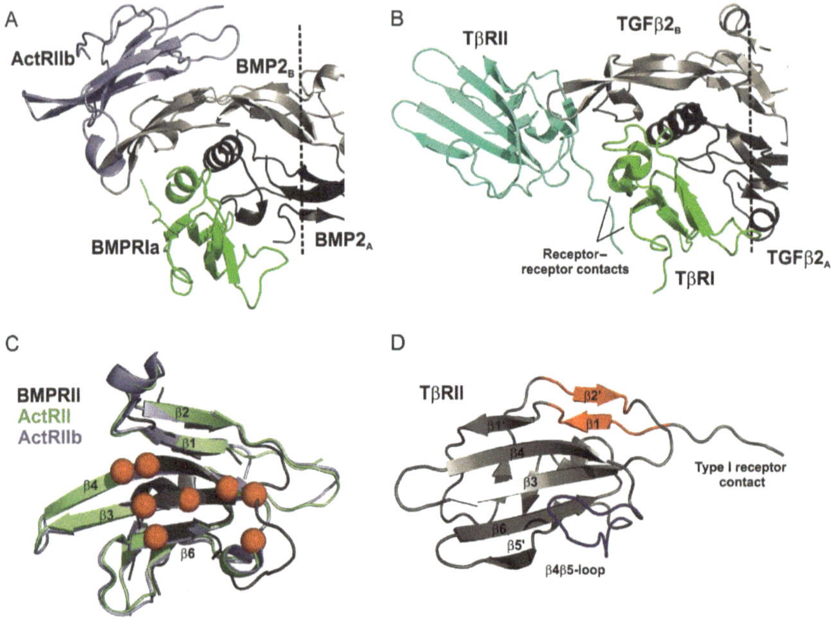

Thomas D. Mueller, Figure 4 Type II receptor specificity between BMPs/GDFs and TGFβs is implemented via different receptor binding sites. (A) In BMPs/GDFs and Activins/Inhibins, the type II receptor (marked in light blue) binds to the ligand's knuckle epitope and seems not to share any contacts with the type I receptor (marked in green). (B) In contrast in TGFβs, the type II receptor binds to the fingertips, the type I receptor moves toward the type II receptor engaging in receptor–receptor contacts likely explaining cooperativity and high receptor selectivity. (C) The core structure as represented by the two- and three-stranded β-sheet involved in the ligand contact (indicated by red spheres) is almost identical for the three type II receptors BMPRII, ActRII, and ActRII. Only the loops connecting the strands differ in length and conformation allowing implementing ligand specificity where needed. (D) Binding of the type II receptor TβRII to BMPs/GDFs and Activins is impeded by the different length and conformation of the β4β5-loop (marked in blue), which would protrude into the BMP/GDF ligand interface. Binding to TGFβs is achieved by interaction of the β-strands β1 and β2 (marked in red) with the TGFβ fingertips.

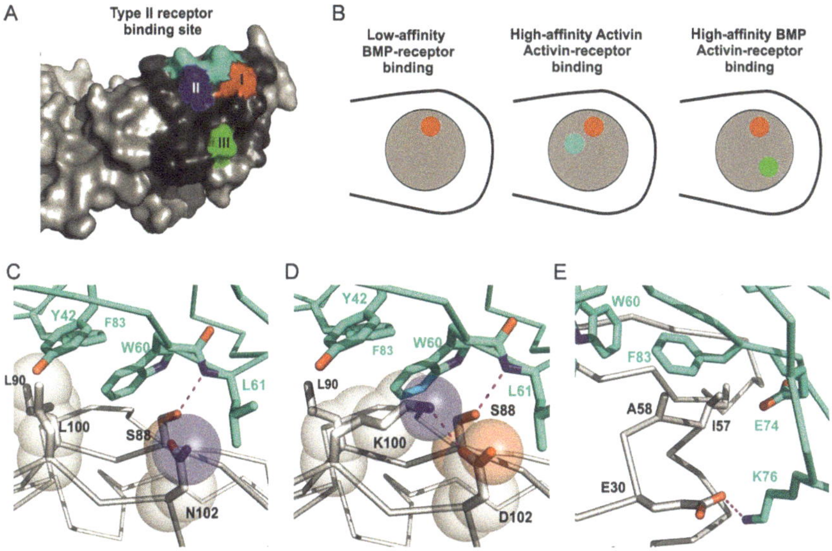

Thomas D. Mueller, Figure 5 (A) The type II receptor interface in BMPs is modular. Different (combinations) of hot spots of binding (indicated with I, II, and III) allow modulating binding affinity and receptor specificity. In the low-affinity binding of BMP2 to its type receptors, only a single hot spot is utilized (B), mutating two residues that affect shielding (marked in cyan) activates a second potential hot spot of binding and ensures high-affinity binding of BMP2 to ActRIIb (B, middle panel). In BMP3 and BMP7, another hot spot of binding (indicated in green) is used to discriminate between the type II receptors ActRIIb and ActRII (B, right panel). (C) The hydrogen bond between Ser88 of BMP2 and the backbone amide of L61 in ActRIIb does not contribute significantly to the overall binding in wild-type BMP2 due to inefficient shielding of the polar bond from the access of solvent. (D) Mutating Leu100 and N102 to the equivalent amino acids as found in ActivinA activates this hot spot. (E) The charge–charge interaction between various BMPs (e.g., BMP3, BMP7) and ActRII and ActRIIb allows discriminating between the two highly similar Activin type II receptors.

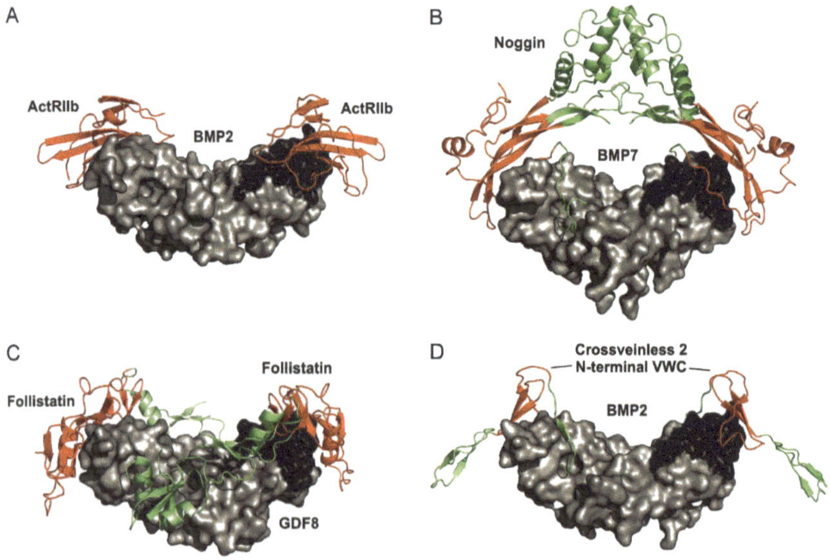

Thomas D. Mueller, Figure 6 The concept of a modular hot spot of binding allows highly different structures to bind to the same interface area. The type II receptor-binding site in the knuckle epitope somewhat presents an interaction hub, with the almost identical epitope in the ligand being recognized by very different binding partner and despite the fact that the architecture and orientation of the interacting element differs significantly between ActRIIb (A), Noggin (B), Follistatin (C), and the Von Willebrand type C domain of the Chordin-modulator family member Crossveinless 2 (D).

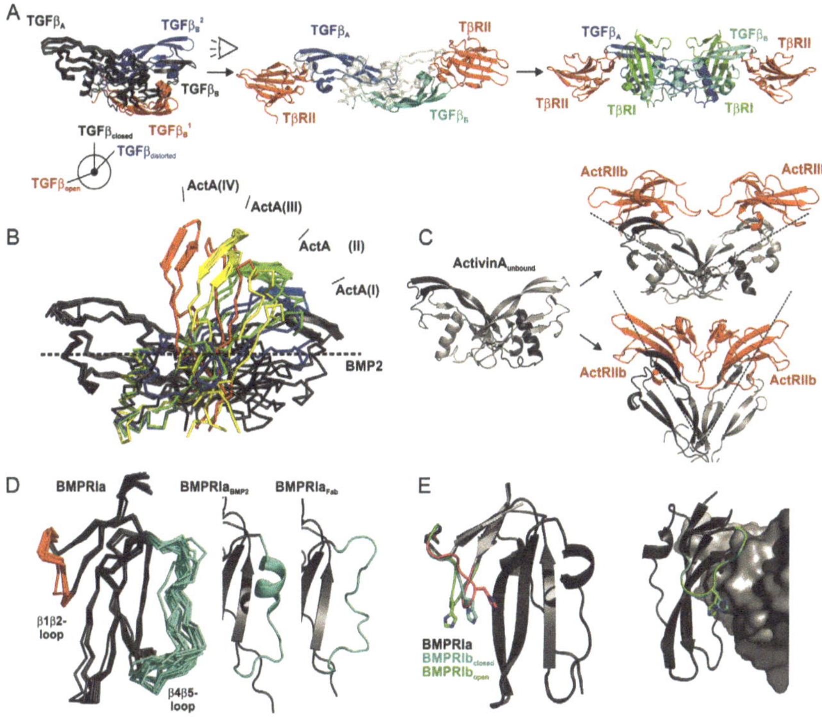

Thomas D. Mueller, Figure 7 Structural variability can be utilized to generate specificity and promiscuity. (A) Data suggest that the butterfly-shaped architecture of the *sensu stricto* TGFβs is flexible and exists in an open and a closed conformation depending on its binding partner or environment. Structure analysis of TGFβ2 bound to its TβRII revealed an open conformation for TGFβ2 (middle panel), whereas in the ternary complex of TGFβ3 bound to TβRII and TβRI the ligand adopts the canonical closed conformation. The inherent flexibility likely contributes to type I receptor specificity and also possibly ensures a defined order of receptor binding. (B) ActivinA also exhibits a highly flexible architecture with the two monomers adopting different angles to each. Considering the ligand architecture as a butterfly, the wings are fully spread in BMPs (BMP2 shown in gray), but adopt other conformations with the wings being in a more closed form (blue: ActivinA when bound to Follistatin, PDB 2B0U; green: ActivinA in its unbound conformation, PDB 2ARV; yellow: ActivinA when bound to Follistatin Fs12, PDB 2ARP; red: ActivinA when bound to ActRIIb, PDB 1NYS). (C) Similar to TGFβ2 also for ActivinA different ligand conformations were found when bound to its type II receptor ActRIIb suggesting that also for ActivinA the type II/type I receptor-binding order might be due to the type I receptor interface being formed only when the ligand is in complex with its type II receptor on the cell surface. (D) In the BMP type I receptors, a dynamic and flexible loop (left panel, BMPRIa in its unbound conformation), enable adaptation of the type I receptor to its respective binding partner via adopting different structures (α-helical upon binding to BMPs or extended when bound to a neutralizing antibody fragment). Here, flexibility is used to generate promiscuity. (E) The β1β2-loop in the BMP type I receptors BMPRIa and BMPRIb also exist in different conformations. In BMPRIa, the loop folds into a locked conformer, which attenuates binding to GDF5 due to strong van der Waals contacts. In contrast in BMPRIb, the loop adopts two more open conformations, which both fold away from the GDF5 surface thereby leaving sufficient space for bulky residues in GDF5 and thus allowing high-affinity binding of BMPRIb to GDF5.

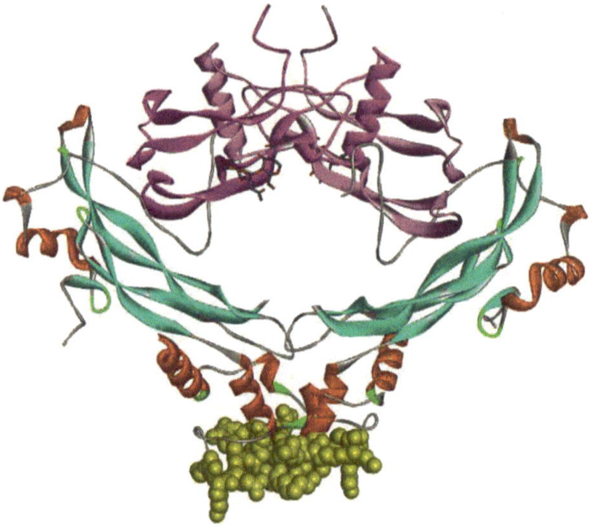

B. Mulloy and C.C. Rider, Figure 1 The crystal structure of the BMP antagonist Noggin complexed with BMP-7 (PDB: 1M4U). The BMP-7 dimer is represented as a magenta-colored ribbon at the top of the diagram, and the Noggin dimer as a ribbon colored by secondary structure: red, α-helices; aquamarine, β-strands; and, green, turns. The heparin-binding site, in amino acids 133–144, which encompasses a cluster of eight basic arginine and lysine residues (at the bottom of the diagram), has been represented in CPK space-filling format and colored yellow.

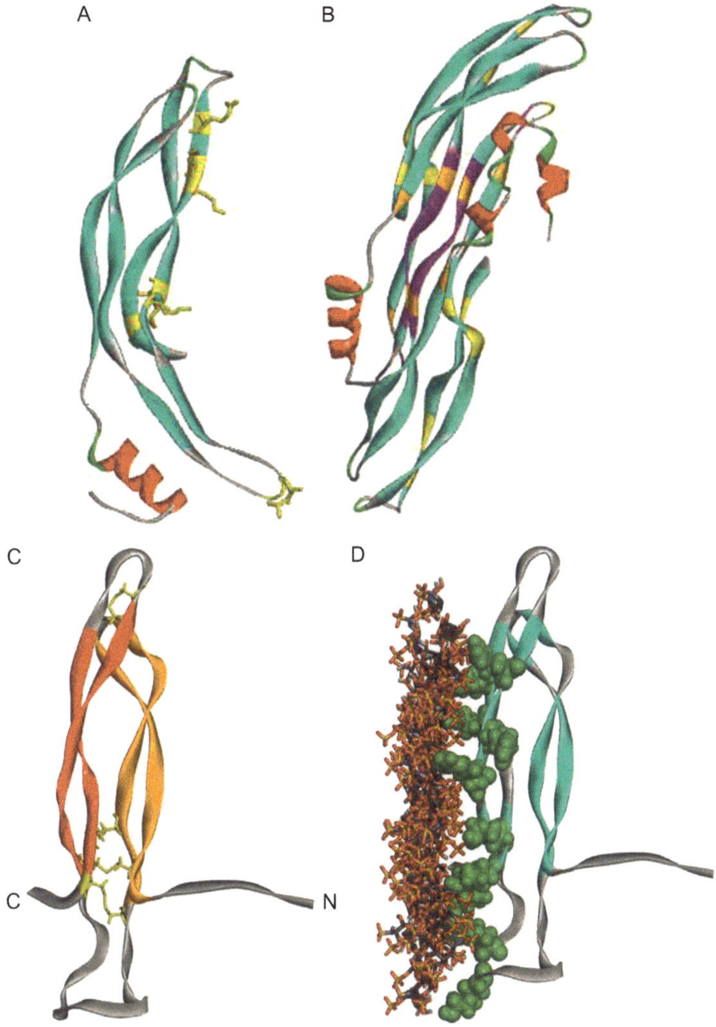

B. Mulloy and C.C. Rider, Figure 2 Ribbon representations of DAN family members, colored red for α-helices, aquamarine for β-strands, and green for turns. (A) Monomer of PRDC with possible heparin-binding residues in yellow stick representation. (B) Dimer of PRDC, taken from the crystal structure PDB: 4JPH; with heparin-binding site colored yellow, dimer interface purple, and BMP-binding site orange. (C) Ribbon representation of sclerostin structure, one of the ensemble of NMR structures determined by Veverka et al. (2009). N- and C-termini are marked and colors distinguish the β-strand fingers 1 (on the right: orange) and 3 (on the left: red); the disorganized loop 2 is colored white. Cystine bridges are shown in yellow. (D) The same sclerostin polypeptide structure (β-strands colored aquamarine) docked with a heparin oligosaccharide (on the left: stick representation); the results of multiple docking calculations are superimposed. The sidechains of contact arginines and lysines are shown in green space-fill.

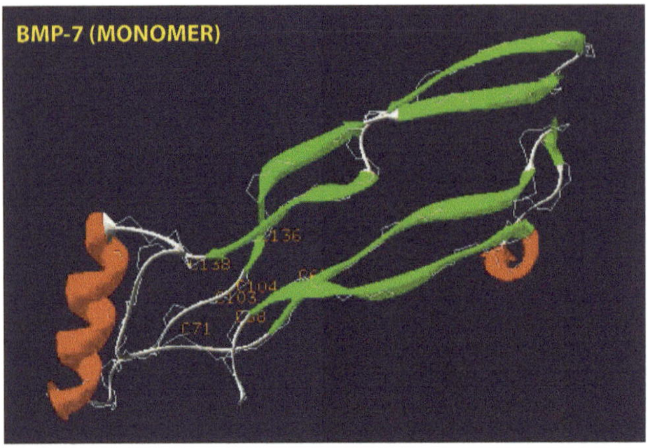

Scott R. Manson *et al.*, Figure 2 *The cysteine-knot fold tertiary structure of BMP-7.* The BMP-7 monomer folds into the cysteine-knot fold tertiary structure that is characteristic of TGF-β/BMP family members. As indicated, this structure is stabilized by three intrapeptide disulfide bonds that connect Cys_{38}–Cys_{104}, Cys_{67}–Cys_{136}, and Cys_{71}–Cys_{138}. An additional interchain disulfide bond links monomers by joining the Cys_{103} residues on each chain. *Structural information was obtained from entry* 1BMP *(Griffith, Keck, Sampath, Rueger, & Carlson, 1996) in the Protein Data Bank at the Research Collaboratory for Structural Bioinformatics and then a ribbon diagram was created using DeepView PDB-Viewer.*

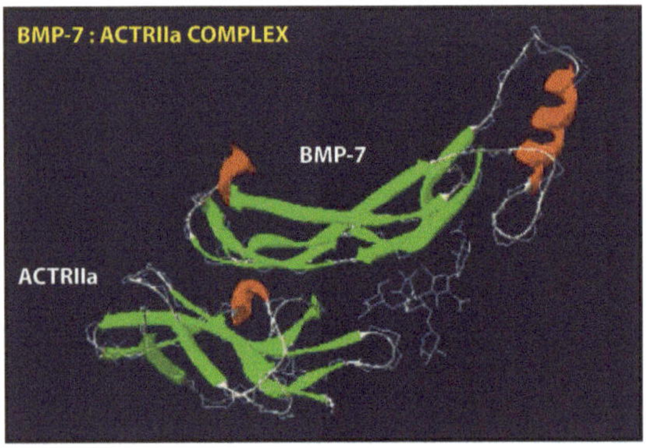

Scott R. Manson *et al.*, Figure 4 *Binding of BMP-7 to the Type II BMP receptor ActRIIa.* The crystal structure of the complex formed between BMP-7 and the extracellular domain of the Type II receptor ActRIIa demonstrates that specificity is mediated by interactions between the side chains on the ligand and its receptor. *Structural information was obtained from entry* 1LX5 *(Greenwald et al., 2003) in the Protein Data Bank at the Research Collaboratory for Structural Bioinformatics and then a ribbon diagram was created using DeepView PDB-Viewer.*

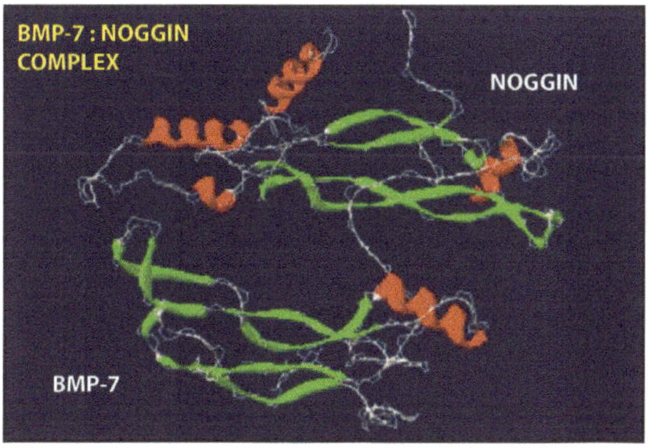

Scott R. Manson *et al.*, Figure 6 *Binding of the BMP Antagonist Noggin to BMP-7.* Noggin folds into a cysteine-knot fold structure that resembles the structure of TGF-β/BMP family members. A Noggin homodimer then binds to a BMP-7 homodimer to form a ring-like structure where the finger/clip regions extend from Noggin and bind to BMP-7 to block key residues on BMP-7 from interacting with its receptors. *Structural information was obtained from entry* 1M4U *(Groppe et al., 2002) in the Protein Data Bank at the Research Collaboratory for Structural Bioinformatics and then a ribbon diagram was created using DeepView PDB-Viewer.*

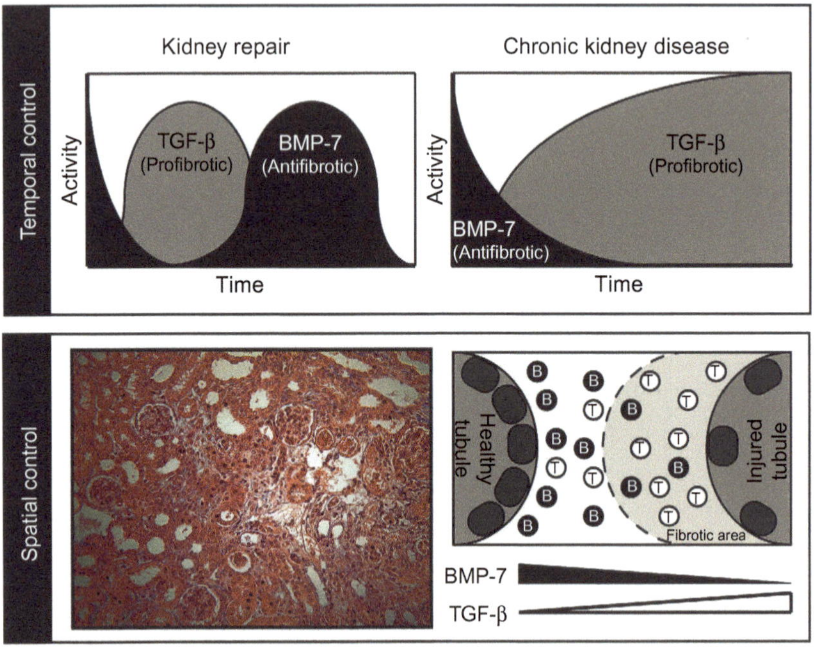

Scott R. Manson *et al.*, Figure 8 *A model for the regulation of TGF-β-dependent, profibrotic responses by spatiotemporal gradients of BMP-7.* During kidney repair, the physiologic functions of TGF-β are opposed by BMP-7 to provide mechanisms for the spatiotemporal regulation of wound-healing responses. The restoration of epithelial differentiation is paralleled by an increase in BMP-7 expression that leads to the cessation of the TGF-β response. Gradients of BMP-7 may also restrict wound-healing responses to sites of injury, as illustrated by the sporadic scarring of the kidney following obstruction-induced renal injury. During the progression of CKD, persistent activation of TGF-β and suppression of BMP-7 leads to the erosion of the innate repair mechanisms of the kidneys, renal fibrosis, and ESRD.

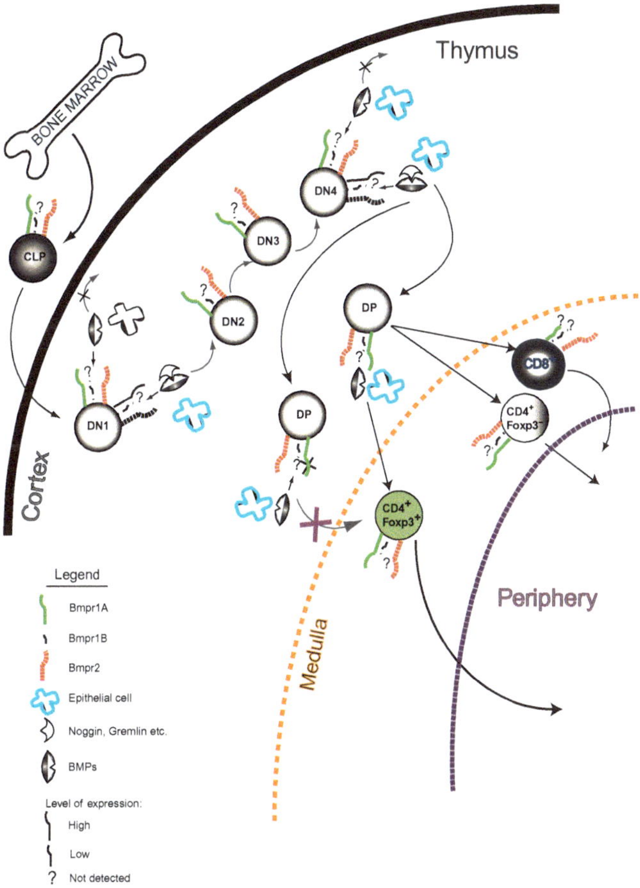

Michal Kuczma and Piotr Kraj, Figure 1 Bone morphogenetic proteins (BMPs) affect thymopoiesis at multiple stages. Bone marrow-derived common lymphoid progenitor (CLP) cell colonizes thymus and undergoes series of changes to become mature T cell. Upon binding of BMPs (mainly BMP2 and 4) produced by multiple cells in the thymus (i.e., epithelial cells) at double negative (DN) stages 1 and 4, thymocyte development in the cortex is arrested. Natural inhibitors of BMP signaling (i.e., Noggin, Gremlin) bind to BMPs sequestering them from their type I (mainly BMPR1A, expression of BMPR1B was not reported in the thymus) and type II (BMPR2) receptors. Regulatory T cells ($CD4^+Foxp3^+$) develop only upon signal mediated by BMPs in the context of BMPR1A, and their generation is largely affected when DP (double positive) thymocytes lack its expression. $CD8^+$ T cells express very low level of BMPR1A, and their development and function is minimally affected by its deletion.

Michal Kuczma and Piotr Kraj, Figure 2 Effect of BMP signaling in the tumor microenvironment. BMPs act locally to enhance self-renewal/stemness of cancer stem cells (CSCs) by induction of STAT3 expression. This in turn enhances the metastatic potential of tumor cells. BMPs also induce Foxp3 expression in conventional CD4 cells ($CD4^{+}Foxp3^{-}$) to become regulatory T cells (induced T_{reg}, iT_{reg}). T_{reg} (both iT_{reg} and thymic-derived natural T_{reg} (nT_{reg})) and MDSC suppress antitumor response targeting granzyme/perforin producing $CD8^{+}$ T cells, NK cells, and $CD4^{+}$ cells of Th1 lineage by cell-dependent contact and expression of suppressive cytokines or metabolites (TGF-β, IDO, arginase, etc.). MDSC, myeloid-derived suppressor cell; NK, natural killer; B, B cell; MΦ, macrophage; DC, dendritic cell.

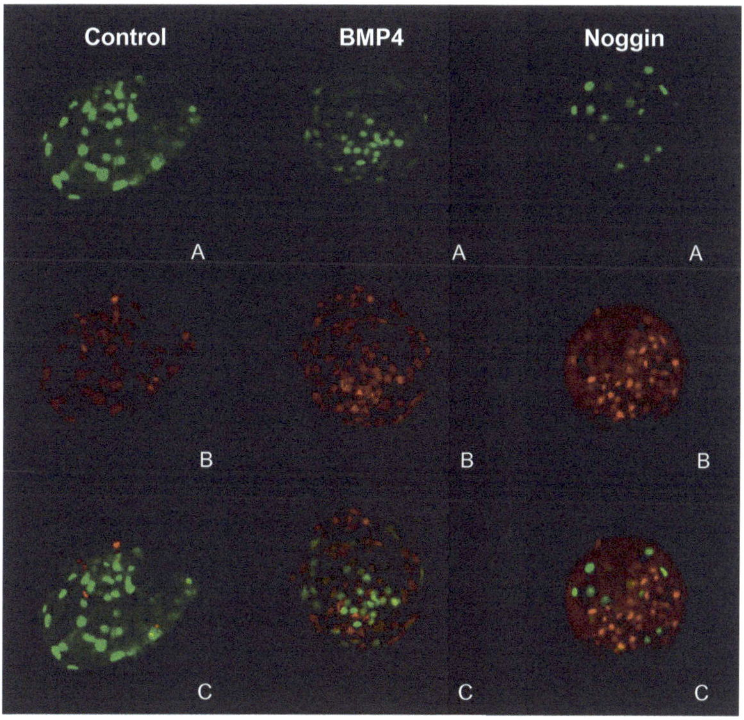

Isabel La Rosa, Figure 2 Oct-4 expression in bovine blastocysts. Oct-4 immunostaining of blastocysts cultured with 100 ng/ml of either BMP4 or Noggin; Controls were cultured without supplementation. Confocal microscopy, augmentation: 20× and zoom: 2. (A) Oct-4-positive cells (green), (B) total nuclei (red) and (C) Oct-4-positive cells and total nuclei (merged).

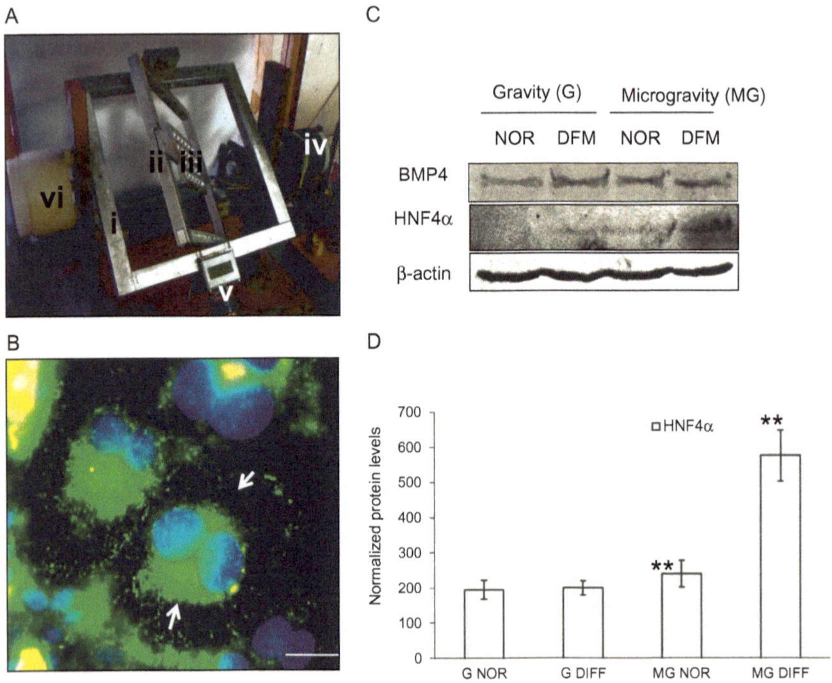

Jamila H. Siamwala *et al.*, Figure 1 (A) Clinostat for inducing microgravity. Outer frame (i) rotates unidirectionally at 20 rpm and an inner frame (ii) with a sample holder (iii) which rotates bidirectionally at 10 rpm. Motor (iv) rotates the outer frame and the inner frame (v), and its rotational direction and velocity are controlled by the controller (vi) connected to a computer with a clinostat control program. (B) HNF4α expression in BMOL cells. Differentiated BMOL cells subjected to microgravity are stained with HNF4α-fluorescein isothiocyanate (FITC) primary antibody (green) and counterstained with 4',6-diamidino-2-phenylindole (DAPI), a nuclear stain (blue). White arrows indicate binding of antibodies to HNFα in differentiated BMOL cells. (C) Western blot analysis of BMP4, HNF4α, and β-actin performed on BMOL cells. The samples are probed with BMP4, HNF4α, and β-actin and the band intensities are estimated using densitometry quantification function of ImageJ. (D) The differences in protein expression are expressed as mean ± SE, $^{**}p < 0.001$ by one-way ANOVA.

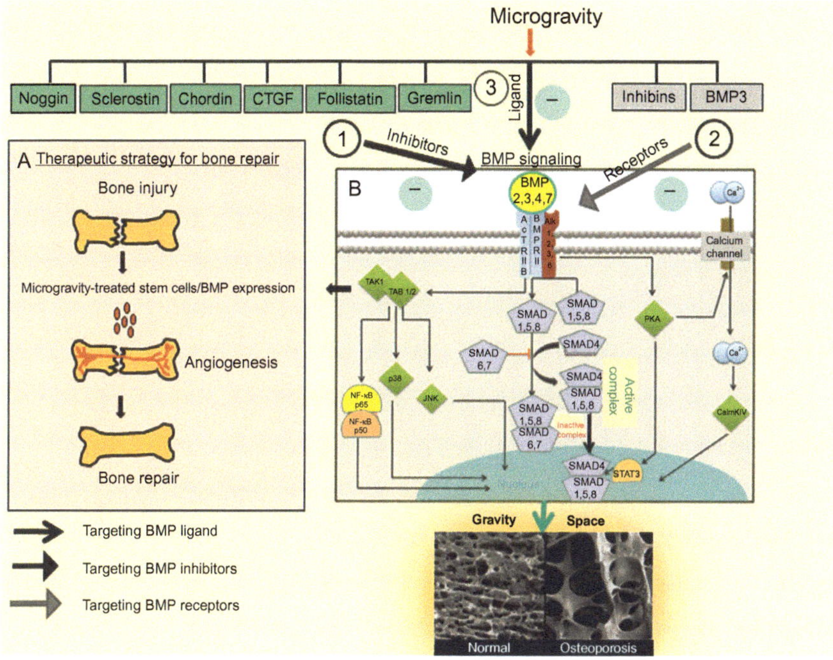

Jamila H. Siamwala *et al.*, Figure 2 The schematic diagram represents (A) possible therapeutic approach to facilitate bone repair in space or Earth through BMP signaling and angiogenesis (left). (B) Probable BMP signaling events leading to bone weakening in space. The main strategies highlighted in the box (right) are (1) suppression of natural inhibitors of BMPs, (2) suppression of natural inhibitors of BMP receptors, and (3) reduction of BMP expression and secretion. *The BMP pathway is reproduced with permission from Affymetrix, Inc., Copyright (2012), USA and NASA, USA.*

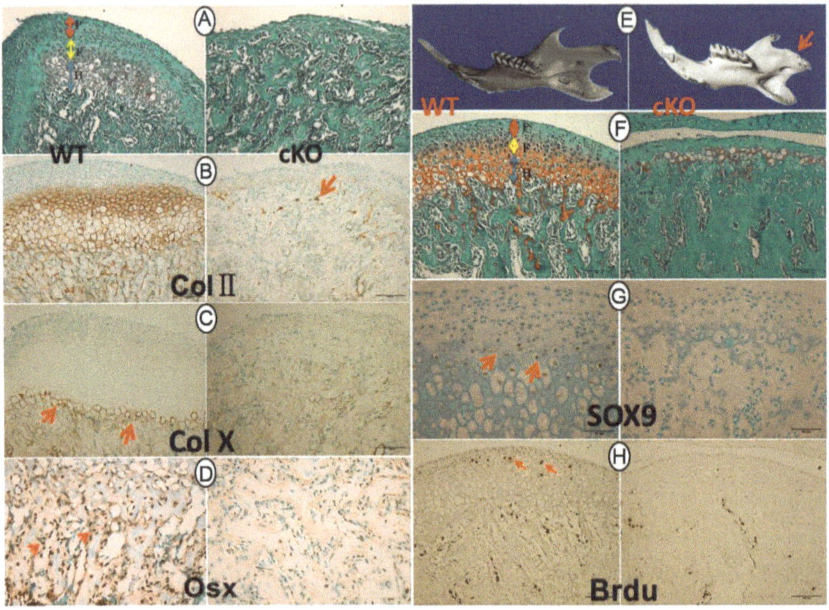

Junjun Jing *et al.*, Figure 1 Phenotype of mandibular condyle in *Bmpr1a*$^{fl/fl}$; *Aggrecan-CreERT2* mice. (A) No indication of proteoglycans in the knockout MCC (2 weeks of age). (B) Col II immunoreactivity. (C) col X immunoreactivity. (D) Osterix immunoreactivity (small red arrows). (E) Micro-CT images demonstrating the shorter (red arrow) condyle in *Bmpr1a*-deficient mice (4 weeks of age). (F) Safranin O staining. (G) SOX9 immunoreactivity. (H) BrdU immunoreactivity (red arrows). P, polymorphic layer; F, flattened layer; H, hypertrophic layer. *Reproduced with permissions from Jing J, Hinton RJ, Mishina Y, Liu Y, Zhou X, Feng JQ. Critical role of Bmpr1a in mandibular condyle growth, Connective Tissue Research, 2014 Aug;55 Suppl 1:73-8.*

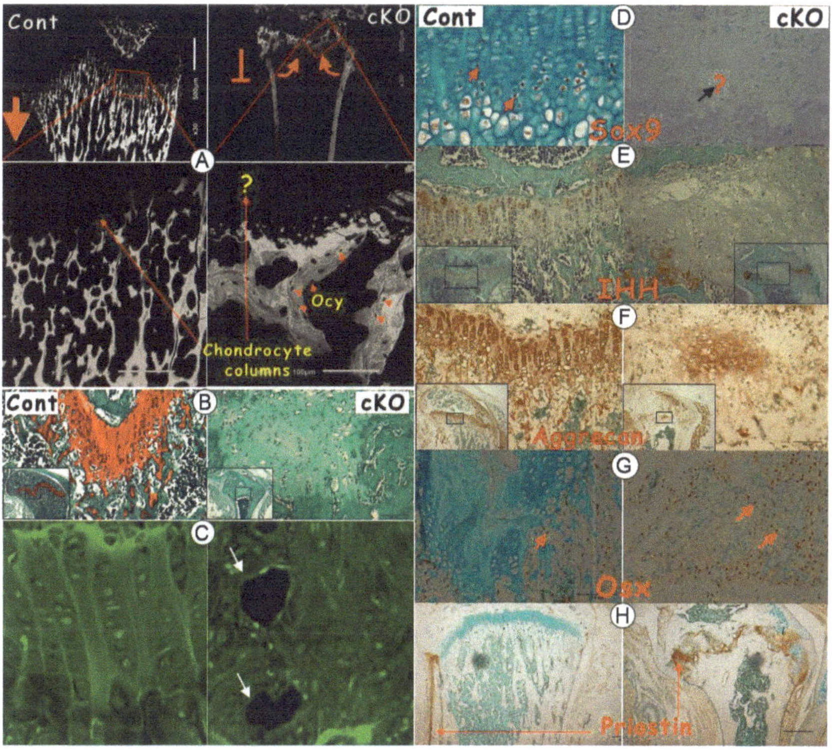

Junjun Jing *et al.*, Figure 2 Long bone phenotype in *Bmpr1a*$^{fl/fl}$; *Aggrecan-CreERT2* mice. Representative backscattered SEM (A) images showed a lack of chondrocyte columns and trabeculae in the cKO femur. Instead, there were numerous spherical mineral bodies adjacent to a few large clusters of mature bones, in which the osteocytes (Ocy) were well formed; the Safranin O stained images revealed lack of proteoglycan in the mutant growth plate (B) and the FITC confocal images revealed blood vessels and fibrous-like tissues in the cKO growth plate (C); IHC-stained images in the cKO growth plate indicated either a lack of signals of SOX9 (D) or IHH (E) separately, or faint signals of Aggrecan (F), or ectopic expressions of OSX (G), and Periostin (H). Cont, control; cKO, conditional knockout. *Reproduced with permissions from Jing J, Ren Y, Zong Z, Chuanju Liu C, Nobuhiro Kamiya N, Mishina Y, Liu Y, Zhou X, and Feng JQ. BMP Receptor 1A determines the cell fate of the postnatal growth plate Int. J. Biol. Sci. 2013; 9(9):895-906.*

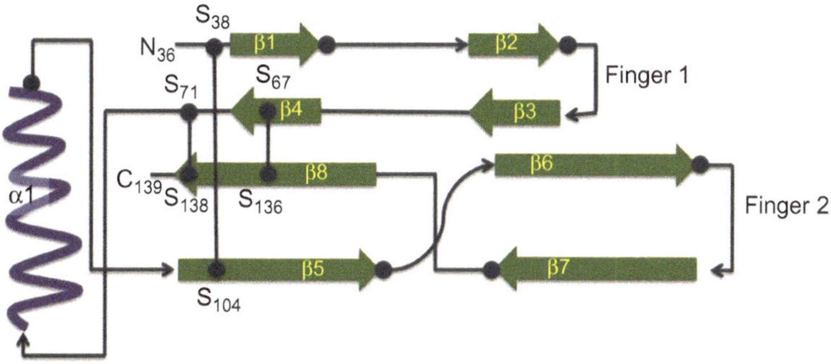

Ana Claudia Oliveira Carreira *et al.*, Figure 2 Scheme for BMP-7 structure. Three disulfide bridges stabilize the monomer, initial 67-Cys Cys-136 and Cys-71 Cys-138 form a ring structure, spatially favoring binding of the third disulfide bond, Cys-38 Cys-104.

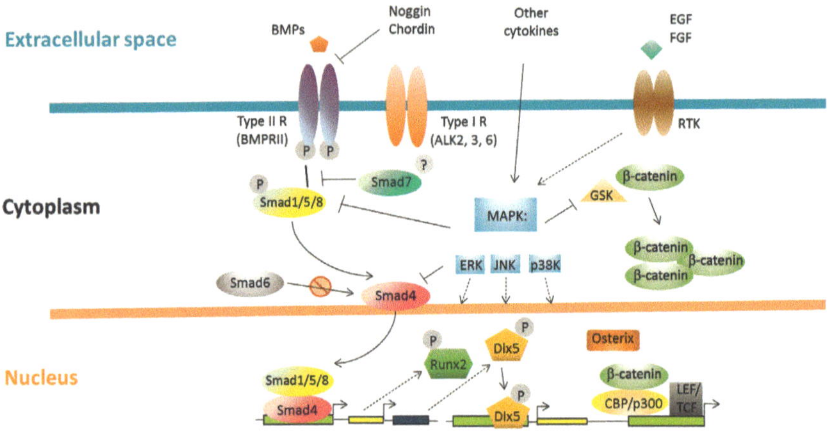

Ana Claudia Oliveira Carreira *et al.*, Figure 3 BMP signaling pathways: representation of Smad-dependent and -independent pathways and their mechanisms of regulation. BMPs initiate signaling from the cell surface by binding two different types of receptors (Type I and Type II) that can activate the Smad cytoplasmic proteins. The heterodimeric formation of type I R and type II R may occur before or after BMP binding, which induces signal transduction through Smads. BMPs may also signal through SMAD-independent pathways via mitogen-activated protein kinase (MAPK): Erk (extracellular signal-regulated kinase), JNK (Jun N-terminal kinase), or p38 (p38 MAPK) (Pan et al., 2015). Three classes of Smad proteins include receptor-regulated Smads (R-Smads), such as Smad1, 5, and 8; the common mediator Smad (Co-Smad), Smad4; and inhibitory Smads (I-Smads), Smad6 and 7. Smad6 binds type I BMPR, preventing Smad1, 5, 8 activation (phosphorylation). Phosphorylated Smad1, 5, 8 complexes associate with intracellular Smad4 for translocation into the nucleus, binds to transcription factors, activating target gene transcription. The transcription factors for BMPs may be any of the following: AP-1, bZIP, RUNX, Fox, bHLH, Homeodomain, Sp1, nuclear receptors, or IRF-7. The coactivators which are commonly utilized are the following: CBP/p300, SMIF, MSG1, or ARC105. Smad6 and 7 may suppress signals by preventing association of R-Smad with Co-Smad or R-Smad phosphorylation. BMPs may also act in an autocrine and paracrine manner. Furthermore, specific antagonists (Noggin and Chordin) are also expressed. *Adapted from Carreira, Alves, et al. (2014).*

MIX
Papier | Fördert gute Waldnutzung
FSC® C083411

Kolibri 360 GmbH
Ferdinand-Jühlke-Straße 7
99095 Erfurt, Deutschland
produktsicherheit@kolibri360.de